POWER LAWS IN ASTROPHYSICS

Research applications of complex systems and nonlinear physics are rapidly expanding across various scientific disciplines. A common theme among them is the concept of "self-organized criticality systems," which this volume presents in detail for observed astrophysical phenomena, such as solar flares, coronal mass ejections, solar energetic particles, solar wind, stellar flares, magnetospheric events, planetary systems, and galactic and black-hole systems. The author explores fundamental questions: Why do power laws, the hallmarks of self-organized criticality, exist? What power law index is predicted for each astrophysical phenomenon? Which size distributions have universality? What can waiting time distributions tell us about random processes? This is the first monograph that tests comprehensively astrophysical observations of self-organized criticality systems for students, postdocs, and researchers. A highlight is a paradigm shift from microscopic concepts, such as the traditional cellular automaton algorithms, to macroscopic concepts formulated in terms of physical scaling laws.

MARKUS ASCHWANDEN studied observations and theoretical models of self-organized criticality systems over the last 40 years as a researcher at Lockheed Martin Solar and Astrophysics Laboratory, analyzing space data from many NASA missions. He has organized self-organized criticality-dedicated workshops at ISSI Bern and Leiden University. He served as Editor of *Space Science Reviews* and is a member of the American Geophysical Union, the American Astronomical Society's Solar Physics Division, and the Committee of European Solar Radio Astronomers.

POWER LAWS IN ASTROPHYSICS

Self-Organized Criticality Systems

MARKUS ASCHWANDEN
Lockheed Martin

CAMBRIDGE
UNIVERSITY PRESS

CAMBRIDGE
UNIVERSITY PRESS

Shaftesbury Road, Cambridge CB2 8EA, United Kingdom

One Liberty Plaza, 20th Floor, New York, NY 10006, USA

477 Williamstown Road, Port Melbourne, VIC 3207, Australia

314–321, 3rd Floor, Plot 3, Splendor Forum, Jasola District Centre,
New Delhi – 110025, India

103 Penang Road, #05–06/07, Visioncrest Commercial, Singapore 238467

Cambridge University Press is part of Cambridge University Press & Assessment,
a department of the University of Cambridge.

We share the University's mission to contribute to society through the pursuit of
education, learning and research at the highest international levels of excellence.

www.cambridge.org
Information on this title: www.cambridge.org/9781009562935

DOI: 10.1017/9781009562942

When citing this work, please include a reference to the DOI 10.1017/9781009562942

First published 2025

A catalogue record for this publication is available from the British Library

A Cataloging-in-Publication data record for this book is available from the Library of Congress

ISBN 978-1-009-56293-5 Hardback

Dedicated to Carol J. Kersten

Contents

Foreword

Science, as "a way of knowing," has been extremely successful in providing a quantitative explanatory framework to describe, understand, and sometimes predict a host of natural phenomena. Yet many still defy understanding, for a variety of reasons.

In some cases, we do not even know if our physical theories are adequate and complete enough to describe a given phenomenon. In astrophysics, the so-called dark matter and dark energy are currently good cases in point, the majority opinion being that some sort of "new physics" is required. Likewise, understanding the singularities inside black holes requires a merging of quantum physics and general relativity that has not yet been achieved after a century, and not for lack of trying. In other words, we don't even know which equations we should be solving!

By contrast, sometimes we know very well the underlying physical theories, but the phenomena continue to defy physical modeling and understanding. Falling into this category are a broad class of systems in which the spatiotemporal evolution unfolds over a very wide range of spatial and/or temporal scales, so that direct numerical simulations remain impractical despite continuing increase of computing power.

Earthquakes offer a classic example. Mechanical stress builds up over decades or centuries along seismic fault lines, and an earthquake occurs when one or more meter-sized rocky contact points between adjacent plates break off, leading to slippage. The devastating 1906 San Francisco earthquake was powered by the slippage of the San Andrea fault over a contact surface estimated to have spanned over 400 km in length, and 12 km in depth. A multidecadal global numerical simulation of this event would thus require resolving in time and space a vast collection of meter-size irregularly shaped structures distributed along a contact surface of some 5,000 km^2, undergoing slow plastic deformation up to rupture, the latter occurring on timescales of seconds; impossible in practice, even though the governing physical equations are well known.

In such cases, two options are in fact available to pursue modeling. Option 1 is to truncate the scales; eliminating in this way the largest spatial scales leads to local models requiring carefully formulated boundary conditions, and restricts the size of structures and dynamical events that can be captured by the model; while evacuating the smallest scales demands the astute design of so-called subgrid models so as to properly mimic their dynamical effects at the smallest scale actually resolved by the model. Sometimes this can work quite well, for example, in simulating the decay of homogeneous, isotropic fluid turbulence. In other situations, it does not, such as when energy injected into the system can cascade toward both smallest (direct cascade) and largest (inverse cascade) scales. In such a situation, scale truncation may well severely distort the global dynamics, making the simulation behave very differently from the phenomenon being modeled.

The second option is to retain the full range of spatial and temporal scales characterizing the phenomenon but replace physical laws by very simple dynamical evolutionary rules, so as to yield a computationally tractable simulation. This must be emphasized: We are not talking about just simplifying (geometrically and/or dynamically) the model or the physical laws describing its behavior; this is common and accepted practice and admittedly a bit of an art. Instead, the physical backdrop of the phenomenon and associated physical laws governing its behavior are flat out replaced by a much simpler underlying structure and ad hoc set of evolutionary rules. Models of this variety exist in a variety of guises and go under many names, such as cellular automata, lattice simulations, sandpile models, or avalanche models, to list but a few.

Energy release events such as earthquakes and solar flares rely on very different physical mechanisms of energy loading and release, yet share the same power law form in their event size statistics; in the case of earthquakes, this is known as the Gutenberg–Richter law. Such power law distributions are indicative of *scale invariance*, for which *criticality* is a well-understood generator. Criticality, in turn, is known to materialize in systems undergoing a *phase transition*, which typically requires very fine tuning of one or more control parameters. This can be readily achieved in the laboratory under controlled conditions, but in natural systems, such outside control is evidently absent.

Self-organized criticality (SOC) took statistical physics by storm in the late 1980's. The name refers to systems in which the critical point is an attractor of the underlying dynamics, so that scale-invariant behavior does not require external fine tuning of any control parameter. This offers a very robust explanatory framework for the occurrence of scale-invariant behavior in natural systems.

Computational models implementing SOC dynamics are often almost ridiculously simple: A SOC model of earthquakes energy release reproducing the Gutenberg–Richter law can be written in less than 100 lines of computer code.

To argue that such a simple simulation is be taken seriously as a physical model of such a complex phenomenon is likely to send many a geophysicists rolling on the floor, gasping for air in a choking fit of hysterical laughter; and to further inform the poor soul that a few simple changes to a few lines of this code could turn it into a credible model of energy release in solar flares would likely deliver the *coup de grâce*.

This is a most natural and expected reaction. There is a gaping abyss separating these simplified models from the real underlying physics subsumed in the model's simple evolutionary rules. Going beyond global statistical behavior, this greatly restricts the scope of reliable physical inferences and/or predictions that are quantitatively testable against observational data.

For the past 25 years, Markus Aschwanden has been at the forefront of efforts to bridge this gap. In this book, after introducing the reader to the required physical concepts of criticality, scale invariance, fractal geometry, power laws, and their associated mathematical representations, he literally takes us on a grand tour of the universe. Starting from the Earth magnetosphere and working his way up to galaxies, he investigates the applicability of SOC dynamics to a host of astrophysical systems, along with the observational evidence backing his claim that SOC is indeed the most robust an credible generator of scale invariance and power law behavior across most of astrophysics.

Time to get going; raise that back seat, fold up that tray table, and fasten your seat belts!

Paul Charbonneau,
Université de Montréal, Canada

Preface

This book is about nonlinear physics and complexity physics. It started with the notion of fractal geometry (see textbooks of Mandelbrot 1977; Feder 1988; Schroeder 1991; Baryshev and Teerikorpi 2002) and has been developed into the concept of self-organized criticality (SOC) (Katz 1986; Bak et al. 1987, 1988; Bak 1996), which mostly deals with the statistics of nonlinear events, also called avalanches or catastrophes. On the observational side, it was found that power law size distributions (or occurrence frequency distributions) represent reliable hallmarks of SOC avalanche processes, while on the theoretical side, cellular automaton models appear to mimic SOC processes adequately on a microscopic level (Pruessner 2012; Charbonneau 2017; Jensen 2023).

The reader may ask why we need another book on SOC phenomena or SOC models? The main motivation of this endeavor is the aim to obtain a deeper physical understanding of SOC models, which requires theoretical models that are sufficiently detailed to produce quantitative theoretical predictions. At this point we observe many phenomena with power law-like size distributions in nature (in astrophysics, magnetospheric physics, geophysics, biophysics, sociophysics, etc.), but we do not understand why power laws exist, why they have specific values for the power law slopes, which parameters have universality, and what is the role of waiting time distributions. At the end of the book, we hope to have settled these questions in a quantitative manner, in terms of the so-called standard fractal-diffusive self-organized criticality model (FD-SOC). The previously published textbook *"Self-Organized Criticality in Astrophysics: The Statistics of Nonlinear Processes in the Universe"* (Aschwanden 2011a) contains many complementary aspects of SOC models but does not contain any treatment of the FD-SOC model, which has been published later on (Aschwanden 2012a, 2014; Aschwanden et al. 2016).

This book contains three major parts: Chapters 1–3 contain introductions into the fundamentals of SOC (Part I), while Chapters 4–15 describe applications of

SOC models to astrophysical phenomena (Part II), and Chapter 16 summarizes the Conclusions (Part III). Chapter 1 deals with SOC definitions and introductions into SOC avalanche models. Chapter 2 summarizes the most important characteristics of power law size distributions. Chapter 3 includes a summary of waiting time distribution functions and their theoretical treatment. Once the reader is familiar with these three chapters, the theoretical predictions and observational interpretations in the following chapters should be straightforward to understand. The body of the book is the application of SOC models to observations, such as solar flare hard X-rays (Chapter 4); solar flare soft X-rays (Chapter 5); solar *extreme ultra-violet (EUV)* nanoflares (Chapter 6); solar photospheric events (Chapter 7); solar radio bursts (Chapter 8); coronal mass ejections (CMEs) (Chapter 9); solar energetic particle events (SEP) (Chapter 10); solar wind fluctuations (Chapter 11); magnetospheric phenomena (Chapter 12); planetary systems, lunar craters, meteorites, Martian fluvial systems, the Saturnian ring system, extrasolar planets (Chapter 13); stellar systems, stellar flares, Kepler observations, pulsar glitches, soft gamma-ray repeaters, magnetars, black holes around accretion disks (Chapter 14); and galactic systems, cosmology, cosmic rays, and extragalactic fast radio bursts (Chapter 15). The modeling with the FD-SOC model provides quantitative theoretical predictions of the statistical size distribution functions of SOC avalanche processes, which exhibit universality across the entire universe.

The author is most indebted to invaluable discussions, comments, and reviews from colleagues (listed in alphabetical order): Sandra Chapman, Paul Charbonneau, Norma Crosby, Michaila Dimitropoulou, Manolis Georgoulis, Manuel Güdel, Stefan Hergarten, Henrik Jeldtoft Jensen, James McAteer, Alexander Milovanov, Shin Mineshige, Laura Morales, Naoto Nishizuka, Gunnar Pruessner, Raul Sanchez, Jeff Scargle, Surja Sharma, Antoine Strugarek, Vadim Uritsky, Virginia Trimble, and Nicolas Watkins. Extensive usage of scientific literature was enabled by the NASA Astrophysics Data System (ADS), operated by the Smithsonian Astrophysical Observatory (SAO), as well as by numerous *Wikipedia* and *Google* searches. Special thanks goes also to my family, my children Pascal Dominique and Alexander Julian, and particularly to my wife Carol J. Kersten for the enthusiastic support of this project.

Abbreviations

1-D, 2-D, 3-D	one-, two-, three-dimensional
2dFGRS	2dF Galaxy Redshift Survey
ACE	Advanced Composition Explorer
ADS	Astrophysics Data System
AE	Auroral Electron jet index
AGILE	Astro-Rivelatore Gamma a Immagini Leggero (spacecraft)
AGN	Active Galactic Nuclei
AIA	Atmospheric Imaging Assembly (on SDO)
AMATERAS	Atacama Large Millimeter/SubMillimeter Array (Chile)
ARTEMIS	Acceleration, Reconnection, Turbulence, and Electrodynamics of Moon's interaction with the Sun (multichannel solar radio spectrograph, Athens)
ASKAP	Australian Square Kilometer Array Pathfinder (radio telescope)
AU	Astronomical Unit
BATSE	Burst and Transient Source Experiment (on CGRO)
BBSO	Big Bear Solar Observatory (ground-based)
BCS	Bragg Crystal Spectrometer (on Yohkoh)
BTW	Bak–Tang–Wiesenfeld (1987) model
CA	Cellular Automaton model
CACTUS	Computer Aided CME tracking software (method)
CCC	Cross-Correlation Coefficient
CDAW	Coordinated Data Analysis Workshop
CGRO	Compton Gamma Ray Observatory (spacecraft)
CHIME	Canadian Hydrogen Intensity Mapping Experiment (radio telescope)
CME	Coronal Mass Ejection
COBE	Cosmic Background Explorer (spacecraft)
CORIMP	Coronal Imaging Processing (method)
CSHKP	Carmichael–Sturrock–Hirayama–Kopp–Pneuman (flare model)
CV	Cataclysmic Variable stars
DCIM	DeCIMetric bursts
DCIM-P	Decimetric Pulsations
EBTEL	Enthalpy-Based Thermal Evolution of Loops model
EIT	Extreme-ultraviolet Imaging Telescope (on SoHO)
EM	Emission Measure
EUV	Extreme UltraViolet

EUVE	Extreme UltraViolet Explorer (spacecraft)
EUVI	Extreme-UltraViolet Imager (on SECCHI/STEREO)
EXOSAT	European X-Ray Observatory Satellite (spacecraft)
FD-SOC	Fractal-Diffusive Self-Organized Criticality model
FM	Finite system-size effect model
FWHM	Full Width Half Maximum
FRB	Fast Radio Burst (galactic)
FSOC	Forced Self-Organized Criticality model
FSRQ	Flat Spectrum Radio Quasars
FUV	Far UltraViolet imager (on IMAGE spacecraft)
GBT	Green Bank Telescope
GEOTAIL	magnetospheric satellite
GOES	Geostationary Orbiting Earth Satellite (spacecraft)
GHz	Giga Hertz
GPS	Global Positioning Systems
GRANAT	International Astrophysical Observatory (Russian spacecraft)
GRB	Gamma-Ray Burst spectrometer (on ULYSSES spacecraft)
GSFC	Goddard Space Flight Center (NASA)
Hα	hydrogen line (6562.8 Å)
HILDCAA	High-Intensity, Long-Duration, Continuous AE Activity
HMI	Helioseismic and Magnetic Imager (on SDO)
HMXB	High-Mass X-ray Binaries
HSP	High-Speed Photometer (on HST spacecraft)
HST	Hubble Space Telescope (spacecraft)
HXRBS	Hard X-Ray Burst Spectrometer (on SMM)
HXT	Hard X-ray Telescope (on Yohkoh)
ICE	International Cometary Explorer (ISEE-3 spacecraft)
IMAGE	Imager for Magnetopause-to-Aurora Global Exploration (spacecraft)
INTEGRAL	INTErnational Gamma-Ray Astrophysical Laboratory (ESA spacecraft)
IMF	Interplanetary Magnetic Field
IMP	Interplanetary Monitoring Platform (spacecraft)
IPS	InterPlanetary Scintillation (data)
IRIS	Interface Region Imaging Spectrograph (spacecraft)
ISEE-3	International Sun/Earth Explorer 3 (ICE spacecraft)
IT3	Interplanetary Type III burst
IT	Intermittent Turbulence
LA	Logistic Avalanche model
LASCO	Large Angle Spectrometric COronagraph (on SOHO)
LMC	Large Magellanic Cloud (a galaxy)
LMXB	Low-Mass X-ray Binary star
MASER	Microwave Amplification by Stimulated Emission of Radiation
MDI	Michelson Doppler Imager (on SoHO)
MFI	Magnetic Field Investigations (on WIND)
MHD	Magneto-HydroDynamics
MK	Mega Kelvin
MLE	Maximum Likelihood estimator
MVT	Minimum Variation Timescale
MW	MicroWaves
MW-S	MicroWave Spike bursts
NASA	National Aeronautics and Space Administration

NGC	New General Catalogue (of nebulae and star clusters)
NGDC	National Geophysical Data Center (USA)
NIXT	Normal Incidence X-Ray Telescope (rocket instrument)
NOAA	National Oceanic and Atmospheric Administration (USA)
NuSTAR	Nuclear Spectroscopic Telescope Array (spacecraft)
OGO	Orbiting Geophysical Observatory
OSO	Orbiting Solar Observatory (satellites)
OVV	Optical Violent Variable (quasar)
PHEBUS	Payload for High Energy BUrst Spectroscopy (on GRANAT)
PM	Pareto distribution model
POLAR	Polar satellite
PSP	Parker Solar Probe
PSR	PulSaR
Q15	Quarter 15 (Kepler spacecraft)
QPO	Quasi-Periodic Oscillations (in stellar data)
RHESSI	Reuven Ramaty High Energy Solar Spectroscopic Imager (spacecraft)
RTV	Rosner–Tucker–Vaiana (1978) model
RV	Rosner–Vaiana (1978) model
RXTE	Rossi X-Ray Timing Explorer (spacecraft)
SAA	South Atlantic Anomaly
SAO	Smithsonian Astrophysical Observatory
SDO	Solar Dynamics Observatory
SDSS	Sloan Digital Sky Survey (ground-based telescope)
SECCHI	Sun Earth Connection Coronal and Heliospheric Investigation (on STEREO)
SEEDS	Solar Eruption Event Detection System (method)
SEP	Solar Energetic Particle events
SFXT	Supergiant Fast X-ray Transient
SGD	Solar Geophysical Data
SGR	Soft Gamma-ray Repeaters
SJI	Slit Jaw Image (on IRIS)
SMM	Solar Maximum Mission (spacecraft)
SNR	SuperNova Remnant
SO	Solar Orbiter
SOC	Self-Organized Criticality
SOT	Solar Optical Telescope (on Hinode)
SOHO	SOlar and Heliospheric Observatory (spacecraft)
SP	Spectro-Polarimeter (on SOT/Hinode)
SPE	Solar Proton Enhancements
SSC	Sudden Storm Commencement (magnetospheric events)
SST	Swedish 1-m Solar Telescope
STEREO	Solar TErrestrial RElations Observatory (spacecraft)
SUMER	Solar Ultraviolet Measurements of Emitted Radiation (on SOHO)
SuperDARN	Super Dual Auroral Radar Network
SWAVES	STEREO/WAVES instrument (on STEREO spacecraft)
SWE	Solar Wind Experiment (on WIND)
Swift	spacecraft to observe gamma-ray bursts (NASA)
SXT	Soft X-ray Telescope (on Yohkoh)
TESIS	Solar Telescope/Imaging Spectrometer (on CORONAS/PHOTON)
TESS	Transiting Exoplanet Survey Satellite (spacecraft)

TGF	Terrestrial Gamma-ray Flashes
TRACE	Transition Region and Coronal Explorer (spacecraft)
TV	TeleVision
UBV	Ultraviolet Blue Visual spectroscopy
UCB	University of California, Berkeley
Ulysses	interplanetary spacecraft
ULF	Ultra-Low Frequency
UV	ultraviolet
UVI	UltraViolet Imager (onboard POLAR spacecraft)
WATCH	Wide Angle Telescope for Cosmic Hard X-Rays (on GRANAT)
WIND	interplanetary spacecraft
WMAP	Wilkinson Microwave Anisotropy Probe (spacecraft)
WTD	Waiting Time Distribution
XEST	XMM Extended Survey of the Taurus Molecular Cloud
XMM	X-ray Multi-Mirror Misson (spacecraft Newton)
XRS	X-Ray Sensor (instrument on GOES)
XUV	eXtreme UltraViolet

Part I

Fundamentals of SOC

1
Fundamentals

In this introductory chapter, we review the fundamentals of astrophysical processes that are governed by self-organized criticality (SOC), briefly called "SOC avalanches." We will discover that most complex systems with many interacting components display some nonlinear behavior that is governed by SOC. This book is focused on numerical, analytical, and theoretical modeling of SOC in avalanching astrophysical systems. The major novelty of the described SOC framework is a paradigm shift from *microscopic* cellular automaton simulations to *macroscopic* modeling of physical scaling laws.

The concept of SOC has been pioneered by Per Bak and was first published in the seminal paper by Bak, Tang, and Wiesenfeld (1987), which has been cited already over 4,000 times. A slightly earlier application of SOC is due to Katz (1986). The brief abstract of Bak et al. (1987) succinctly summarizes the quintessence of SOC: "We show that certain extended dissipative dynamical systems naturally evolve into a critical state, with no characteristic time or length scales. The temporal 'fingerprint' of the SOC self-organized critical state is the presence of 1/f noise; its spatial signature is the emergence of scale-invariant (fractal) structure" (Bak et al. 1987). The authors demonstrate the principle of SOC with a simple cellular automaton model, in analogy to the sandpile paradigm (Figure 1.1), which produces avalanches of arbitrary sizes that can be statistically sampled and exhibit a size distribution close to a power law function. The power law shape of size distributions became the hallmark and principal diagnostic of SOC phenomena. However, as we will see in Sections 2.7–2.14, power law distributions are neither necessary nor sufficient conditions for SOC processes. Popular accounts of SOC phenomena can be found in the lucidly written book *How Nature Works* by Bak (1996) and in the article "Self-Organized Criticality" by Bak and Chen (1991). Comprehensive treatments of SOC phenomena can be found in Table 1.2 at the end of this chapter.

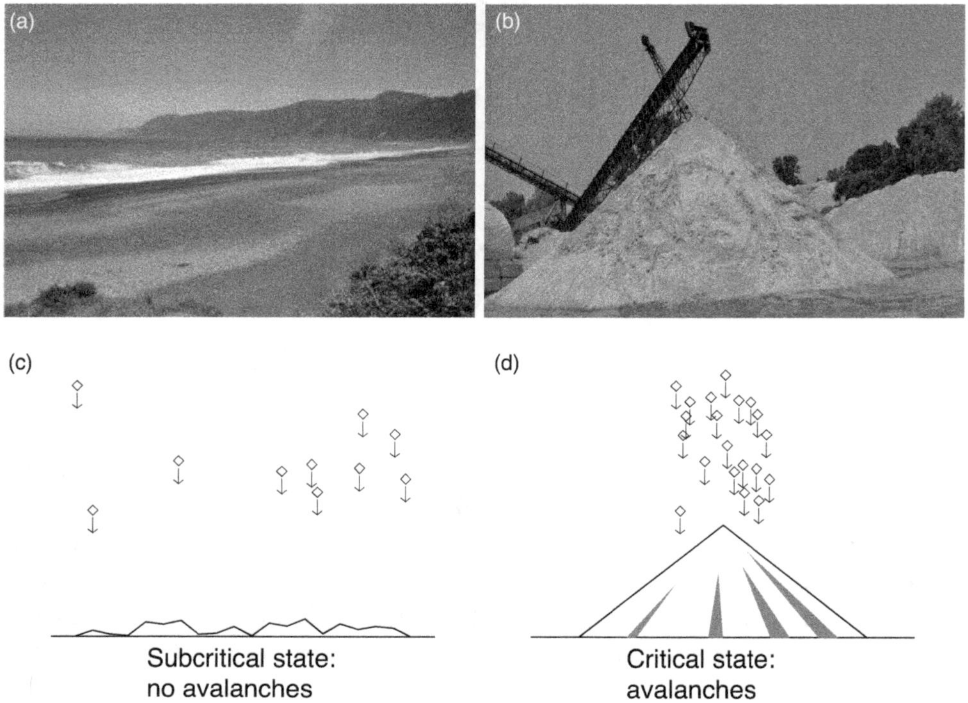

Figure 1.1 A static equilibrium produces no avalanche events (c), such as the flat sand beach in northern California (a), while sand, randomly dripping onto a sandpile, produces a state of SOC, where avalanches occur (d), such as with the conveyer belt of the Indian River Enterprises (b) (Aschwanden 2011a).

1.1 SOC Definitions

The original definition of the term self-organized criticality was inspired by a numerical lattice simulation of a dynamical system with spatially complex patterns, mimicking avalanches of a sandpile, which became the Bak–Tang–Wiesenfeld (BTW) model (Bak et al. 1987), and demonstrated that:

> Dynamical systems with extended spatial degrees of freedom naturally evolve into self-organized critical structures of states which are barely stable. Flicker noise, or 1/f noise, can be identified with the dynamics of the critical state. This picture also yields insight into the origin of fractal objects. (Bak et al. 1987)

In this first seminal paper, the authors had already fractal structures such as cosmic strings, mountain landscapes, and coastal lines as potential applications in mind and concluded: *We believe that the new concept of SOC can be taken much further and might be* **the** *underlying concept of dissipative systems with extended degrees of freedom* (Bak et al. 1987). In this spirit, the application of the SOC concept has been broadened substantially over the last 38 years.

If we read a recent definition of SOC, we find:

Self-organized criticality is regarded as scale invariance without external tuning of a control parameter, but with all the features of the critical point of an ordinary phase transition, in particular long range (algebraic) spatio-temporal correlations (Pruessner 2012).

Another definition of SOC can be gleaned from *Wikipedia*:

In physics, self-organized criticality is a property of (classes of) dynamical systems which have a critical point as an attractor. The macroscopic behavior thus displays the spatial and/or temporal scale-invariance characteristic of the critical point of a phase transition, but without the need to tune control parameters to precise values.

In the same vein, it is stated in the original paper of the SOC creators: *The criticality in our theory is fundamentally different from the critical point at phase transitions in equilibrium statistical mechanics, which can be reached by tuning of a parameter, for instance the temperature* (Bak et al. 1987). The aspect of self-tuning in SOC systems is the most crucial difference to (second-order) phase transitions, where fine-tuning is necessary and is not automatically arranged by nature. The implications and theoretical details of this peculiar feature are discussed in Watkins et al. (2016). However, whenever there is a threshold for instabilities, the threshold value itself could be called a "critical point" that decides whether an instability, also called a nonlinear energy dissipation event, or avalanche, happens or not. Over the past 38 years, a lot of applications of the SOC concept have been made to slowly driven systems with a critical threshold, especially in solar and astrophysics, which is studied in this book. We therefore like to use a more pragmatic and empirical definition of SOC systems, as depicted in Figure 1.2:

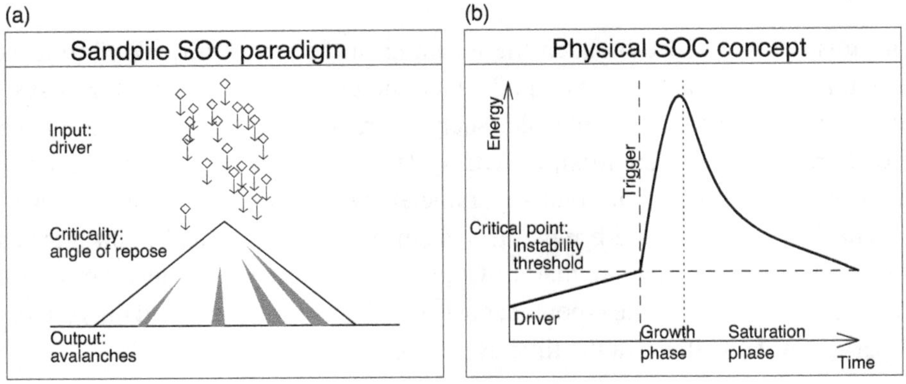

Figure 1.2 (a) The original sandpile SOC paradigm, consisting of the (input) driver, the SOC mechanism (self-tuning angle of repose), and the (output) avalanches. (b) In a physical SOC concept, the driver is a slow and continuous energy input rate, the criticality mechanism is replaced by a critical point in form of a global instability threshold, where an avalanche is triggered, usually consisting of a nonlinear growth phase and a subsequent saturation phase (Aschwanden et al. 2016).

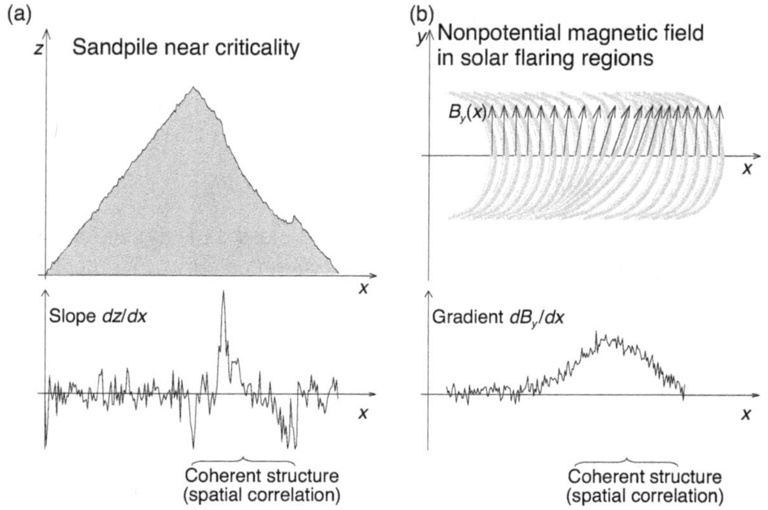

Figure 1.3 (a) A sandpile in a SOC state in the vicinity of criticality is shown in terms of a vertical component $z(x)$, with the slope (or repose angle) dz/dx (bottom), exhibiting short-range fluctuations due to noise, and long-range correlations due to local deviations from the mean critical slope. (b) The analogy of a solar flaring region is visualized in terms of a loop arcade straddling along a neutral line in x-direction, consisting of loops with various shear angles that are proportional to the gradient of the field direction B_x/B_y, showing some local (nonpotential) deviations from the potential magnetic field (bottom) (Aschwanden et al. 2016).

SOC is a critical state of a nonlinear energy dissipation system that is slow and continuously driven toward a critical value of a system-wide instability threshold, producing scale-free, fractal-diffusive, and intermittent avalanches with power law-like size distributions. (Aschwanden 2014)

With this definition, we broaden the meaning of the term "criticality" to a more general meaning of a "critical point," which includes almost any nonlinear system with a (global) instability threshold, such as gravity (Figure 1.3a) or a magnetic field (Figure 1.3b). In addition, a SOC system has to be self-organizing or self-tuning without an external control parameter, which is accomplished by a *slow and continuous driver* and brings the system back to the critical point after each avalanche. Thus, we can say that a SOC system has energy balance between the slowly driven input and the (spontaneously) avalanching output, and thus energy is conserved in the system (in the time average).

1.2 SOC Drivers

The driver is the input part of a SOC system. Without a driver, avalanching would die out and the system becomes subcritical and static (Figure 1.1a). On the other side, the driver must be slowly and continuous, so that the critical state is restored in the asymptotic limit, while a strong driver would lead the system into a catastrophic collapse and may destroy the system. In the classical BTW model, sand grains are dripped under the action of gravity at a slow rate, at random locations of the sandpile, which refill and restore dents from previous avalanches toward the critical angle of repose (Figure 1.3a).

In astrophysical systems (Table 1.1), the driver or energy input of a SOC system may be gravity (in galaxy formation, star formation, black holes, planet formation, and asteroid formation), gravitational disturbances (in Saturn ring), or creation and stressing of magnetic flux (Figure 1.3b) (in solar flares, stellar flares, neutron stars, and pulsar glitches). The driver must bring the system back to the critical point after each major avalanche, which means that the system is locally pushed toward the instability threshold again, so that further avalanching can occur. In the slowly driven limit, the time duration of an avalanche is shorter than the (waiting) time intervals between two subsequent events, which warrants a separation of time scales. In some natural systems, the driver may temporarily or permanently stop, such as the solar dynamo during the Maunder minimum, which stopped solar flaring. Likewise, the final stage of the sweep-up of debris, left over from the formation of the solar system 4.0 billion years ago, stopped lunar cratering.

An overview of terrestrial SOC phenomena (top half of Table 1.1), astrophysical SOC phenomena (bottom half of Table 1.1), drivers that provide a steady energy input (second column in Table 1.1), instability thresholds at the critical state (third column in Table 1.1), and energy dissipation in form of intermittent avalanches (right column in Table 1.1), are juxtaposed in Table 1.1. In summary, SOC drivers discussed here include gravity fields, magnetic fields, mechanical stress forces, biological growth, chemical reactions, as well as randomly dripping of sand grains.

1.3 Instability and Criticality

We broaden the meaning of "criticality" in the original BTW model to a system-wide "instability threshold," which does not need to be tuned by external parameters, since an "instability threshold" is established by common physical conditions throughout a system. For instance, an earthquake is triggered at a critical stressing break point that may have a similar threshold in different tectonic plates around the globe, due to similar geophysical conditions (i.e., the gravity force at the same distance from Earth center, similar continental drift rates, rock constitutions, and crust fracturing conditions). In analogy, a magnetic instability leading to magnetic reconnection is caused by similar physical threshold conditions in solar active regions (such as the kink instability, the torus instability, or the tearing mode instability), and thus solar or stellar flares occur whenever such global instability thresholds are exceeded locally. Such instabilities occur naturally because the driver brings the

Table 1.1 *Hypothetical physical mechanisms operating in self-organized criticality systems.*

Phenomenon	Energy input (steady driver)	Instability threshold (criticality)	Energy output (intermittent avalanches)
SOC-related systems:			
Sandpile	Gravity (dripping sand)	Angle of repose	Sand avalanches
Superconductor	Magnetic field change	Phase transition	Vortex avalanches
Ising model	Temperature increase	Phase transition	Atomic spin-flip
Tea kettle	Temperature increase	Boiling point	Vapor bubbles
Earthquakes	Tectonic stressing	Dynamical friction	Rupture area
Forest fire	Tree growth	Fire ignition point	Burned area
BTW cellular automaton	Random nodes	Critical threshold	Next-neighbor redistribution
Astrophysics:			
Lunar craters	Meteorite production	Lunar collision	Lunar impact craters
Asteroid belt	Planetesimals	Critical mass density	Asteroid
Saturn ring	Gravitational disturbances	Collision rate	Saturn ring particles
Magnetospheric substorm	Solar wind	Magnetic reconnection	Auroral bursts
Radiation belt	Solar wind	Magnetic trapping	Electron bursts
Solar flares	Magnetic stressing	Magnetic reconnection	Nonthermal particles
Stellar flares	Magnetic stressing	Magnetic reconnection	Nonthermal particles
Pulsar glitches	Neutron star spin-up	Vortex unpinning	Neutron starquakes
Soft gamma-ray repeaters	Magnetic stressing	Star crust fracture	Neutron starquakes
Black-hole objects	Gravity	Accretion and inflow	X-ray bremsstrahlung pulses
Blazars	Quasar jets	Jet direction jitter	Optical radiation pulses
Cosmic rays	Galactic magnetic fields	Acceleration threshold	High-energy particles

system continuously back to the instability threshold. In sandpiles, the dripping of additional sand grains rises the angle of repose wherever it is subcritical. In earthquakes, the continental drift is continuously driven by forces that are rooted below the Earth crust. In solar flares, differential rotation, emergence of magnetic flux, and braiding of magnetic fields by random motion in the subphotospheric magneto-convection layer continuously build up nonpotential free magnetic energy that can be released in subsequent avalanches. The analogy of unstable coherent structures in a near-critical state in sandpiles and solar flares is visualized in Figure 1.3.

A selection of plasma instabilities that have been theoretically modeled and identified in astrophysical phenomena is shown in Figure 1.4. These instabilities are controlled by gravitational, hydrodynamic, and magneto-hydrodynamic (MHD) forces. Although the geometric and spatiotemporal structures of these (plasma) instabilities are quite different from case to case, they share common dynamic characteristics, such as an exponentially growing phase (after exceeding a system-wide threshold), followed by a saturation phase, which we will use to calculate scaling laws between the underlying physical parameters. In our avalanche scenario, the duration of an avalanche is commensurable with the exponential-growth time, and the peak value of a time profile is an approximate measure of the dissipated energy in an avalanche event (Figure 1.2).

1.4 Avalanching

Avalanches are defined as nonlinear energy dissipation events, which occur in our generalized SOC definition whenever and wherever a local instability threshold is exceeded. Avalanches are the output part of a SOC system, which balance the energy input rate in the time average for conservative SOC systems. Avalanches are detectable events, which can be obtained in astrophysical observations with large parameter statistics, such as obtained from their observed length scales (L), time scales or durations (T), fluxes (F), and fluences or energies (E). The occurrence frequency distributions of these observables tend to be power law-like functions, a hallmark of SOC systems, but deviations from power law functions can be explained by measurement bias effects (such as incomplete sampling, finite system-size effects, and truncations of distributions) or could reflect multiple physical processes. Unnecessary to say that these observables and their size distributions and underlying scaling laws provide important tests of SOC models.

The time evolution of avalanches contains essential information on the underlying spatiotemporal transport process (i.e., diffusion, fractal diffusion, percolation, turbulence, etc.). A generic time evolution is an initially nonlinear (i.e., exponential) growth phase (Figure 1.2b), followed by a quenching or saturation phase (as expressed in the popular saying "No tree grows in the sky!"). In solar flares,

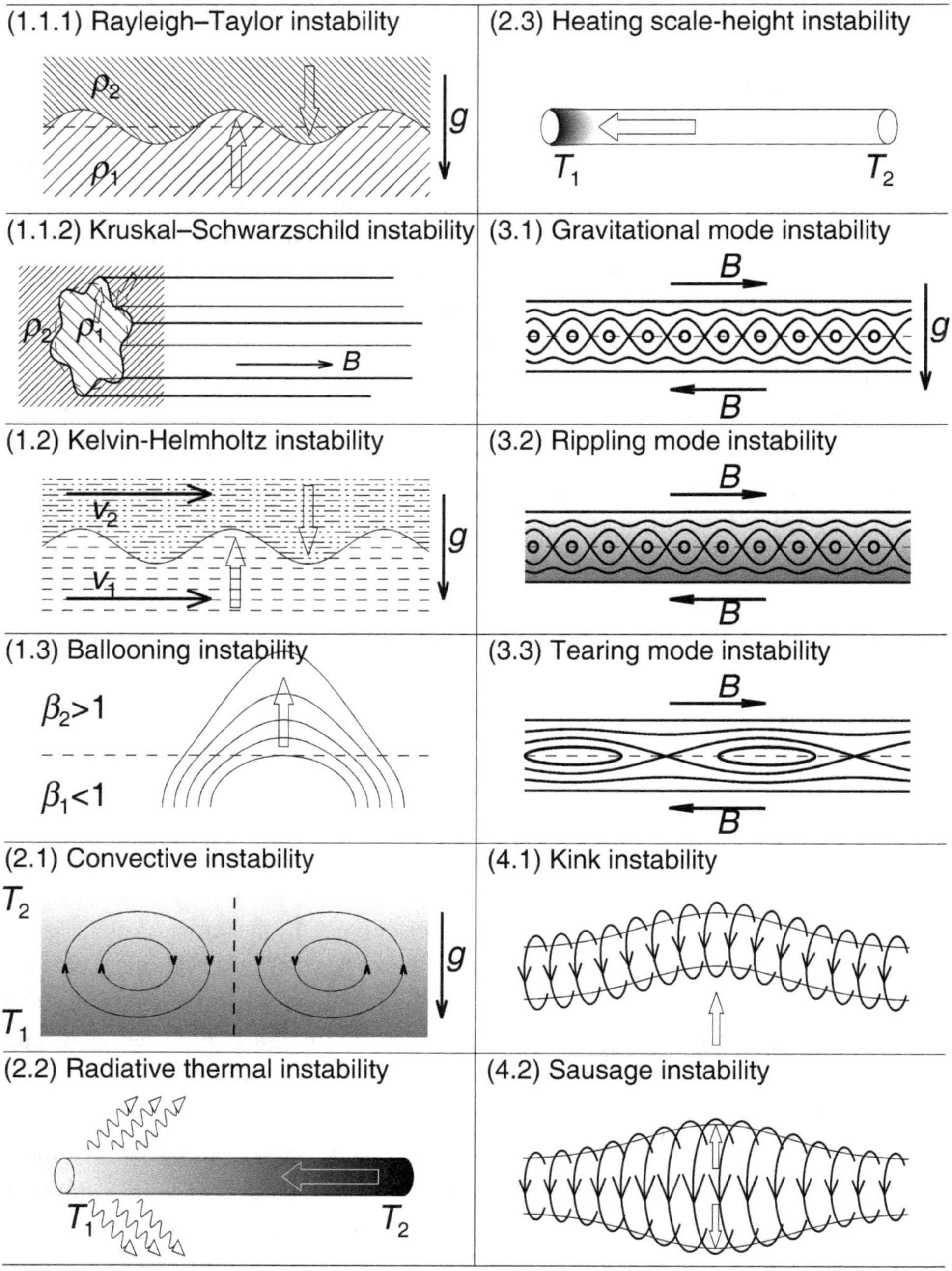

Figure 1.4 Hydrodynamic and MHD instabilities that can occur in plasma physics are illustrated. Different densities (ρ_1, ρ_2) are rendered with hatched linestyle, different velocities (v_1, v_2) with dashed linestyle, temperature gradients (T_1, T_2) with grayscales, longitudinal magnetic field lines (B_0) with thin solid lines, azimuthal magnetic field components (B_φ) with thick solid lines, and radiation with wiggly lines. The directions of the disturbances that lead to an instability are indicated with thick white arrows (Aschwanden 2004).

the initial growth phase is called "impulsive phase," and the subsequent saturation phase is called "postflare phase." In earthquakes, the terms "precursors" and "after-shocks" are common.

1.5 Microscopic Structure

Self-organized criticality systems are a means to study complexity, systems with extended degrees of freedom. Ultimately, a real-world object consists of as many degrees of freedom as the Avogadro number of atoms, that is, $N_A = 6.0 \times 10^{23}$. Such large numbers prevent us from modeling complex nonlinear systems in a deterministic way. In order to deal with SOC systems, we have to resort to numerical simulations with far fewer degrees of freedom, and we have to approximate the complexity of microscopic structures by macroscopic parameters and statistical probability distributions. For example, the complex microscopic structure of the solar chromosphere (Figure 1.5a,) can be rendered with a binary lattice on a much coarser scale (Figure 1.5b). The question is, whether the basic physics that governs the dynamics of a real-world system can also be adequately represented by numerical lattice simulations. In the example shown in Figure 1.5, one binary node of a lattice corresponds to a cube with 1,000-km length scale on the solar surface,

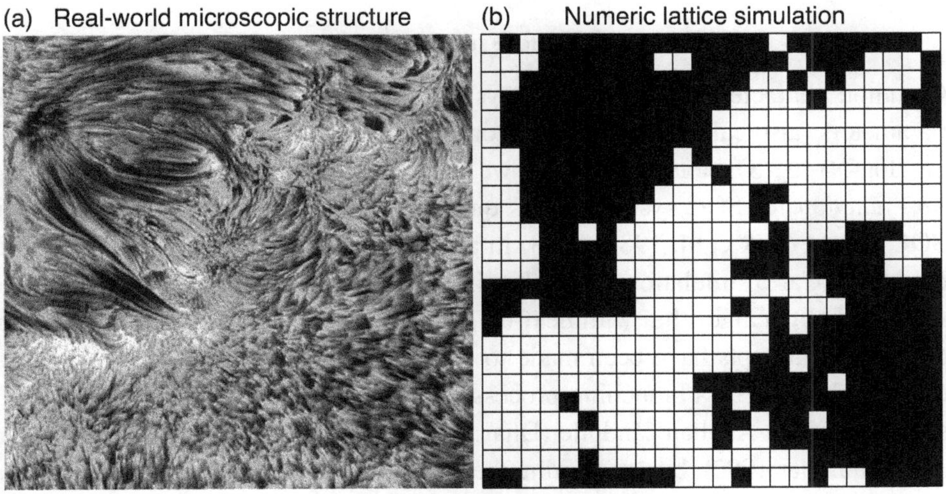

Figure 1.5 (a) A high-resolution image (480 × 480 pixel) of chromospheric spiculae in solar active region 10380, observed on 2003, June 16, with the Swedish 1-m Solar Telescope (SST) on La Palma, Spain, using a tunable filter, tuned to the blue-shifted line wing of the Hα 6536 Å line (Courtesy of Bart De Pontieu). (b) A digitized binary (two-level) version of the left solar image is shown, using a lattice grid with a size of 24 × 24 nodes (or cells). The left image shows the microscopic structure of real-world data, while the right image shows the rendering of numerical lattice simulations used in SOC models (Aschwanden et al. 2016).

where the complex plasma dynamics driven by MHD processes exceeds the information content of a binary lattice node by far, so that it appears to be hopeless to mimic the dynamics of a SOC system with numerical cellular automaton simulations. Interestingly however, numerical lattice simulations do reproduce the emergent complex behavior in physical systems to some extent, regardless of the vast discrepancy of spatial scales and information content. For instance, the statistical size distribution of solar flares can be reproduced with cellular automatons for various physical parameters (spatial, temporal scales, flux, and energy), as demonstrated by Lu and Hamilton (1991). Therefore, SOC models have the powerful ability to give us insight into system dynamics in complex systems, regardless of the intricate details of real-world microscopic fine structure. On the other side, the mathematical world of numerical lattice simulations created a whole new cosmos of complex spatial patterns (i.e., Wolfram 2002) and cellular automaton toy models (Pruessner 2012), which appear to have nothing in common with real-world microscopic fine structure, except that they provide practical means to simulate the same dynamic behavior of complex nonlinear systems. Consequently, in this text on solar and astrophysical SOC applications, the emphasis is not on mathematical and numerical SOC models (except when they were specifically designed for astrophysical applications), although they dominate the extant SOC literature.

1.6 Scale-Free Probability

Common characterizations of SOC systems are statistical distributions of SOC parameters (also called "size distributions," "occurrence frequency distributions," or "log(N)-log(S) plots"). How do we derive a statistical *probability distribution function* for SOC systems? This question has been answered in the original SOC papers (Bak et al. 1987, 1988) in an empirical way, by performing numerical Monte Carlo simulations of avalanches in Cartesian lattice grids, according to the well-known algorithm with next-neighbor interactions (BTW model). Several theoretical attempts have been made to derive statistical probabilities, by considering avalanches as a branching process (Christensen and Olami 1993), by exact solutions of the Abelian sandpile (Dhar and Ramaswamy 1989; Dhar 1999; Dhar and Majumdar 1990), by considering the BTW cellular automaton as a discretized diffusion process using the Langevin equations (Wiesenfeld et al. 1990; Zhang 1989), or by renormalization group theory (Pietronero et al. 1994; Vespignani et al. 1995; Loreto et al. 1995). Most of these analytical theories represent special solutions to a particular set of mathematical redistribution rules but predict different power law exponents for the probability distribution functions obtained with each method, and thus lack the generality to interpret the ubiquitous and omnipresent SOC phenomena observed in nature.

A simple approach to estimate the size distributions of SOC avalanche sizes has recently been proposed by making a simple statistical probability argument, called the *scale-free probability conjecture* (Aschwanden 2012a, 2014, 2015), which predicts the functional form of power laws for most observable SOC parameters, and predicts specific values for their power law slopes (or exponents). The derivation goes as follows. If we consider the derivation of a normal or Gaussian distribution function, we can toss a number of dice and enumerate all possible statistical outcomes, ending up with a binomial distribution function, which converges to a Gaussian distribution function for a large number of dice, and thus characterizes a maximum likelihood distribution. Similarly, we can enumerate all statistically possible sizes L of avalanches in a system bound by a finite size L_{max}, which is simply a number density that is reciprocal to the volume $V = L^d$ of avalanches with size L, that is,

$$N(L)dL \propto L^{-d}dL \qquad \text{for } L \leq L_{max}, \tag{1.1}$$

where d is the Euclidean dimension of the SOC system, having values of $d = 1$, 2, or 3. Figure 1.6 illustrates how the number of 1-D, 2-D, and 3-D avalanches (approximated by spheres) increases reciprocally with the size of the avalanches. This distribution function is based on the principle of statistical maximum likelihood, which follows from braking up a finite system volume into smaller pieces. This distribution function is also related to *packing rules* (e.g., sphere packing) in geometric aggregation problems. A similar approach using geometric scaling laws was also applied to earthquakes (Main and Burton 1984). Of course, for slowly driven SOC systems, only one avalanche happens at a time, and thus the whole SOC system is not fully "packed" with avalanches occurring at once, but the statistical likelihood probability for an avalanche of a given size is nevertheless proportional to the packing density, for a statistically representative subset of all possible avalanche sizes (in a system with $L \leq L_{max}$). This basic scale-free probability conjecture (Eq. 1.1) straightforwardly predicts the size distribution of length scales of SOC avalanches, namely, $N(L) \propto L^{-3}$ in 3-D space, and can be used to derive the size distributions of other geometric parameters.

1.7 Fractal Geometry

"Fractals in nature originate from self-organized critical dynamical processes" (Bak and Chen 1989). The fractal geometry has been postulated for SOC processes by the first proponents of SOC. However, the geometry of fractals has been explored about a decade earlier than the SOC concept (Mandelbrot 1977, 1983, 1985). An extensive discussion of measuring the fractal geometry in SOC systems associated with solar and planetary data is given in Aschwanden (2011a, chapter 8) and McAteer (2013).

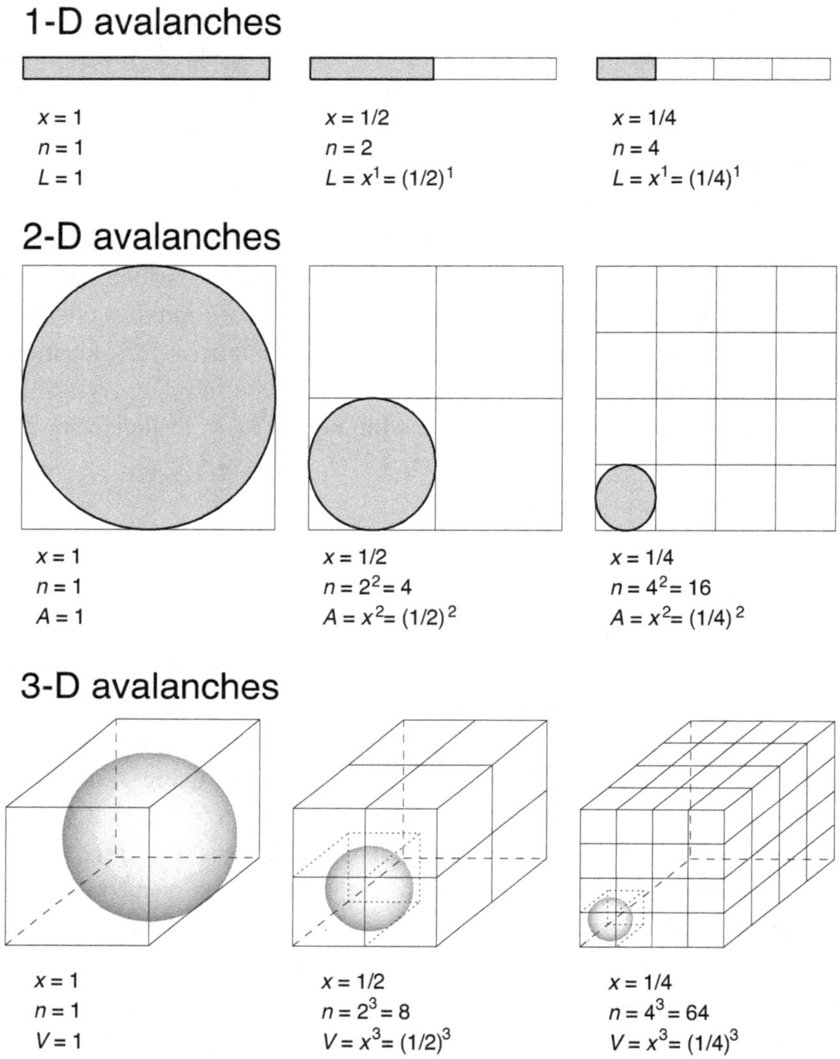

Figure 1.6 Schematic diagram of the Euclidean volume scaling of spherical avalanches, visualized in the three Euclidean space dimensions $d = 1, 2, 3$. The Euclidean length scale x of subcubes decreases by a factor 2 in each step ($x_i = 2^{-i}$, $i = 0, 1, 2$), while the number of subcubes increases by $n_i = (2^i)^d$, defining a probability of $N(x_i) \propto x_i^{-d}$ for each avalanche size with size x_i (Aschwanden 2012a).

The simplest fractal is the Hausdorff dimension D_d, which is a monofractal and depends on the Euclidean space dimension $d = 1, 2, 3$. The Hausdorff dimension D_V for the 3-D Euclidean space ($d = 3$) is

$$D_V = \frac{\log V_f}{\log (L)}, \tag{1.2}$$

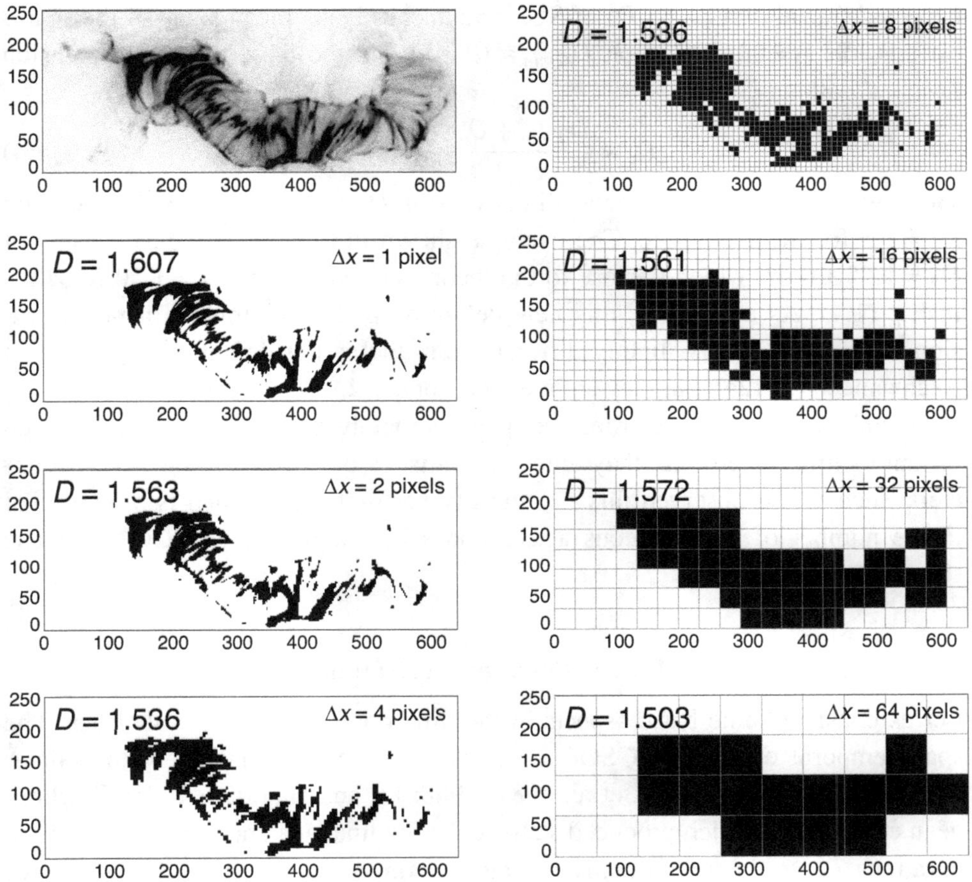

Figure 1.7 Measurement of the fractal area of a solar flare, observed by TRACE 171 Å on 2000, Jul 14, 10:59:32 UT. The Hausdorff dimension is evaluated with a box-counting algorithm for pixels above a threshold of 20% of the peak flux value, yielding a mean of $D_A = 1.55 \pm 0.03$ for the seven different spatial scales ($\Delta x = 1, 2, 4, \ldots, 64$ pixels) shown here (Aschwanden and Aschwanden 2008a, 2008b).

and analogously for the 2-D Euclidean space ($d = 2$),

$$D_A = \frac{\log A_f}{\log (L)}, \tag{1.3}$$

with A_f and V_f being the fractal area and volume of a SOC avalanche. These fractal dimensions can be determined by a box-counting method, where the area fractal D_A can readily be obtained from images of the real world (e.g., for a solar flare as shown in Figure 1.7), while the fractal volume D_V is generally not available (except in numerical simulations), unless one infers the corresponding 3-D information from stereoscopic triangulation or tomography. A good approximation for the expected fractal dimension D_d of SOC avalanches is the mean value of the smallest possible fractal dimension $D_{d,min} \approx (d-1)$ and the largest fractal dimension $D_{d,max} = d$, that is,

$$D_d \approx \frac{D_{d,min} + D_{d,max}}{2} = d - \frac{1}{2}. \tag{1.4}$$

Thus, we expect a mean fractal dimension of $D_V = 2.5$ for the 3-D space and $D_A = 1.5$ for the 2-D space. The example shown in Figure 1.7 yields a value of $D_A = 1.55 \pm 0.03$, which is close to the theoretical prediction $D_A = 1.50$ for $d = 2$ in Eq. (1.4). As a footnote, this new definition of the minimum fractal dimension $D_{d,min} = (d-1)$ is slightly different from the earlier definition $D_{d,min} = 1$ in Aschwanden (2012a), but they are identical for $d = 2$.

Fractals are measurable from the spatial structure of an avalanche at a given instant of time. Therefore, they enter the statistics of time-evolving SOC parameters, such as the observed flux or intensity per time unit, which is proportional to the number of instantaneous active nodes in a lattice-based SOC avalanche simulation.

1.8 Spatiotemporal Evolution

Let us consider some basic aspects in the time domain of SOC avalanches. The spatiotemporal evolution of SOC avalanches has been simulated with cellular automaton simulations (Bak et al. 1987, 1988; Lu and Hamilton 1991; Charbonneau et al. 2001), which produced statistics of the final avalanche sizes L and time durations T, but there is virtually no statistics on the spatiotemporal evolution of the instantaneous avalanche size or radius $r(t)$ as a function of time t, which could characterize the macroscopic transport process. Statistics on this spatiotemporal evolution is important to establish spatiotemporal correlations and scaling laws between L and T, which defines the macroscopic transport process.

Ignoring the complexity of the microscopic transport, which is quantified by an iterative redistribution rule in cellular automaton simulations, we can measure the radius $r(t) = \sqrt{A(t)/\pi}$ of a circular 2-D area $A(t)$ as a function of time t, which corresponds to the solid (Euclidean) area that is equivalent to the time-integrated fractal avalanche area. This has been performed for BTW cellular automaton simulations (Aschwanden 2012a), as well as for solar flare data (Aschwanden 2012b; Aschwanden and Shimizu 2013; Aschwanden et al. 2013b), and was found to fit a diffusion-type relationship,

$$r(t) = \kappa (t - t_0)^{\beta/2}, \tag{1.5}$$

where t_0 is the onset time of the instability, κ is the diffusion coefficient, and β is the diffusive spreading exponent: A value of $\beta \lesssim 1$ corresponds to subdiffusion, $\beta = 1$

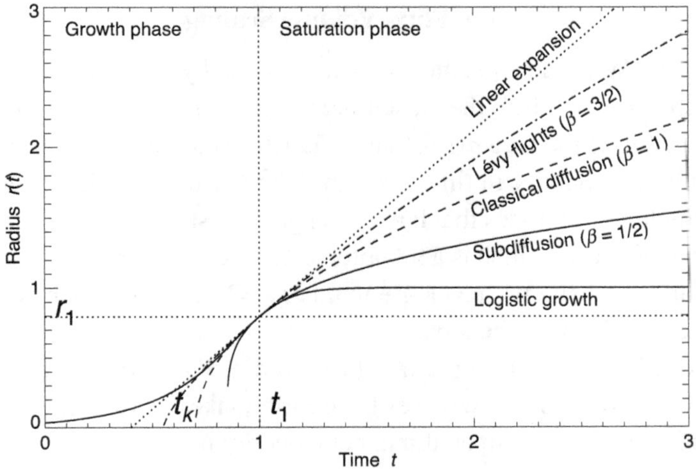

Figure 1.8 Comparison of spatiotemporal evolution models: Logistic growth with parameters $t_1 = 1.0, r_\infty = 1.0, \tau_G = 0.1$, subdiffusion ($\beta = 1/2$), classical diffusion ($\beta = 1$), Lévy flights or hyper-diffusion ($\beta = 3/2$), and linear expansion ($r \propto t$) (Aschwanden 2012a).

to classical diffusion, $\beta \gtrsim 1$ to hyper-diffusion or Lévy flight, and $\beta = 2$ to linear expansion (Figure 1.8). From this macroscopic evolution, we expect a statistical scaling law of the form,

$$L \propto T^{\beta/2}, \tag{1.6}$$

for the final sizes L and durations T of SOC avalanches. Substituting this scaling law $L(T)$ (Eq. 1.6) into the probability distribution function $N(L)$ of length scales L (Eq. 1.1), we establish a power law distribution function for time scales,

$$N(T)dT = N(L[T])\frac{dL}{dT}dT = T^{-[1+(d-1)\beta/2]}\, dT = T^{-\alpha_T}\, dT, \tag{1.7}$$

with the power law slope of

$$\alpha_T = 1 + \frac{(d-1)\beta}{2}, \tag{1.8}$$

which has a value of $\alpha_T = 2.0$ for 2-D Euclidean space ($d = 2$) and a value of $\alpha_T = 1.5$ for 2-D Euclidean space ($d = 2$). This power law slope for avalanche time scales is a prediction of universal validity, since it is only based on the scale-free probability conjecture (Eq. 1.1), $N(L) \propto L^{-d}$, and the statistical property of random walk (or classical diffusion) in the transport process.

The transport process in avalanching systems is found to be diffusion-dominated in systems with extremely strong perturbations (Bregman 2007).

1.9 Flux–Volume Scaling

The original BTW model specifies avalanche sizes by the total number of active nodes, which corresponds to the cluster area of an avalanche in a 2-D lattice. If we want to characterize the area $a(t)$ of an avalanche as a function of time, which is a highly fluctuating quantity in time, we can define also a time-integrated final area $a(< t)$ that includes all nodes that have been gone unstable at least once during the course of an avalanche, which is a monotonically increasing quantity and quantifies the size of an avalanche with a single number $A = a(t = T)$, which we simply call the time-integrated avalanche area.

In real-world data, we observe a signal from a SOC avalanche in form of an intensity flux $f(t)$ (e.g., seismic waves from earthquakes, hard X-ray flux from solar flares, or the amount of dissipated magnetic energy in a solar flare. Let us assume that this intensity flux is proportional to the volume of active nodes in the BTW model, which corresponds to the instantaneous fractal volume $V_f(t)$ (Eq. 1.2) in a macroscopic SOC model (Aschwanden 2012a, 2014, 2015),

$$f(t) \propto V_f(t) \propto r(t)^{D_d} . \tag{1.9}$$

The flux time profile $f(t)$ is expected to fluctuate substantially in real-world data as well as in lattice simulations, because the approximation of the instantaneous volume of a SOC avalanche implies a highly variable fractal dimension $D_d(t)$, which can vary in the range of $D_{d,min} \approx (d - 1)$ and $D_{d,max} = d$, with a mean value $D_d = (d - 1/2)$ (Eq. 1.4). Occasionally, the instantaneous fractal dimension may reach its maximum value, that is, $D_d(t) \lesssim d$, which defines an expected upper limit $f_{max}(t)$,

$$f_{max}(t) \propto V(t) \propto r(t)^d . \tag{1.10}$$

This is an important quantity that corresponds to the peak flux $f_{max}(t)$ of an avalanche, which is often measured in astrophysical observations.

1.10 Coherent and Incoherent Radiation

Self-organized criticality models can be diagnosed and tested by means of statistical distributions, for example, by the omnipresent power law or power law-like size distributions, and by the underlying scaling laws that relate the power law slopes of different observables to each other (see also McAteer et al. 2016 for a description of methods). The original paradigm of a SOC model, the BTW cellular automaton simulations (Bak et al. 1987, 1988), produce power law distributions of two variables, the size S, and the time duration T. The size S is simply defined by the time-integrated area A of active nodes (pixels) in 2-D lattice simulations, or by the time-integrated fractal volume V_f of active nodes (voxels) in 3-D lattice simulations.

In astrophysical observations, however, the volume of an avalanche cannot be measured directly, but rather a flux intensity F_λ in some wavelength regime λ is observed, which is not necessarily proportional to the fractal volume V_f, depending on the emission mechanism that is dominant at wavelength λ. Therefore, for astrophysical observations in particular, we have to introduce a relationship between the observed flux F_λ and the emitting volume V_f that is fractal for a SOC avalanche process. For sake of simplicity we characterize this relationship with a power exponent γ (Aschwanden 2012b),

$$F_\lambda \propto (V_f)^\gamma . \tag{1.11}$$

This definition allows us to distinguish two categories of physical processes: *incoherent processes* that have a linear relationship between the emitting flux and volume ($\gamma = 1$) and *coherent processes* that have a nonlinear relationship ($\gamma > 1$),

$$F_\lambda \propto (V_f)^\gamma , \qquad \begin{cases} \gamma > 1 & \text{(coherent process)} \\ \gamma = 1 & \text{(incoherent process)} \end{cases} \tag{1.12}$$

Incoherent processes are, for instance, free–free emission in optically thin plasma, bremsstrahlung, or gyrosynchrotron emission. Free–free emission is a common emission mechanism in soft X-rays and EUV, where the total flux scales with the emission measure *EM* and is integrated over the entire (fractal) source volume V_f. Coherent processes on the other hand, can occur by wave–particle interactions in collisionless plasmas, such as loss-cone instabilities, electron-beam instabilities, or microwave amplification by stimulated emission of radiation (MASER) emission. The flux level of coherent waves amplifies exponentially or with a nonlinear power to the spatial scale of the source and thus with a nonlinear power to the source volume.

1.11 Separation of Time Scales

Most of the original numerical simulations of SOC systems were performed in the slowly driven limit, which warrants a strict separation of time scales. In lattice-type cellular automaton simulations, the separation of time scales is enforced by dropping only one single sand grain at a time, or disturbing only one single lattice node at a time. If nothing happens, the algorithm proceeds with the next input of a disturbance. In the alternative case, when a disturbance triggers an avalanche, the incremental input function is stopped until the avalanche process ends, after which another input step is continued. This asymptotic limit of strict time-scale separation between the waiting time scale and durations of subsequent avalanches is also called a "slowly-driven SOC system." This ideal, but unnatural condition is, however, not necessarily always enforced in nature. Especially for SOC systems with time-variable drivers, the trigger rate can get so high that multiple avalanches are triggered near-simultaneously and small avalanches occur at various places while a previously triggered large avalanche is still evolving. If we encounter such a "multi-avalanching system," or multi-avalanching behavior during some busy periods of time, we may call it a "fast-driven" or "strongly driven" system. We can adopt the terminology of a slow or fast driver as a synonym for the existence or nonexistence of time-scale separation, which can be expressed by the ratio of the avalanche duration T to the waiting time τ,

$$\begin{array}{rl} \text{Slow driver:} & \mapsto \tau \gtrsim T \\ \text{Fast driver:} & \mapsto \tau \lesssim T \end{array} . \tag{1.13}$$

For a fast driver, the question arises how this affects the observed (power law-like) size distributions that we calculate in the slowly driven limit. The answer depends very much on the event detection method. Ideally one would use imaging information so that the spatial locations of two temporally overlapping events can be separately determined and the time profiles of the two events can be properly disentangled. In practice, especially in the case of astrophysical observations, spatial sources of co-temporaneous events cannot be resolved and a time-series analysis is the only available method. In that case, superimposed time profiles of different events can still be separated if they have a characteristic shape, for instance, a rapid rise and an exponential decay, using a deconvolution method. If no proper deconvolution method is applied, which is unfortunately the case in almost all published studies with event statistics applied to SOC models, there will be a systematic bias of underestimating the time duration of long events, especially when the rule is applied that a previous event has to end before the next event is detected. This leads predictably to steeper power law slopes in the time-scale distribution $N(T)$. We will see later on that an increase in the event rate (for instance, the flaring rate during the maximum of the solar cycle) will lead to substantially steeper power law slopes of the time-scale duration (for solar flare events detected in soft X-rays), from a value of $\alpha_T \approx 2$ in the slowly driven regime (during solar cycle minimum) to a large value of $\alpha_T \lesssim 5$ in the fast-driven regime (during solar cycle maximum). Interestingly, the size distribution of fluxes was not affected in the slowly driven regime. In another study, it was demonstrated that low-resolution observations of a time profile causes an exponential cutoff at large values of the time-scale distribution, which also leads to steeper power law slopes (Isliker and Benz 2001). Thus, we generally expect steeper power laws or exponential distributions of time scales in the limit of strong driving with clustered events that violate the separation of time scales, although there are also reports with flatter power law slopes during episodes of higher event rates (e.g., Bai 1993; Georgoulis and Vlahos 1996, 1998).

1.12 Universality of SOC

About 38 years ago, the concept of SOC emerged (Bak et al. 1987), initially envisioned to explain the ubiquitous 1/f-power spectra, which can be characterized by a power law spectrum $P(\nu) \propto \nu^{-1}$. The term *1/f power spectra* or *flicker noise* should actually be understood in broader terms, including power spectra with pink noise ($P(\nu) \propto \nu^{-1}$), red noise ($P(\nu) \propto \nu^{-2}$), and black noise ($P(\nu) \propto \nu^{-3}$), essentially everything except white noise ($P(\nu) \propto \nu^{0}$). While white noise represents traditional random processes with uncorrelated fluctuations, 1/f power spectra are a synonym for time series with nonrandom structures that exhibit long-range correlations. These nonrandom time structures represent the avalanches in Bak's paradigm of sandpiles. Consequently, Bak's seminal paper in 1987 triggered a host of numerical simulations of sandpile avalanches, which all exhibit power law-like size distributions of avalanche sizes and durations. These numerical simulations were, most commonly, cellular automatons in the language of complexity theory, which are able to produce complex spatiotemporal patterns by iterative application of a simple mathematical redistribution rule. The numerical algorithms of cellular automatons are extremely simple, basically a one-liner that defines the lattice redistribution rule, with an iterative loop around it, but can produce the most complex dynamical patterns, similar to the beautiful geometric patterns created by Mandelbrot's fractal algorithms (Mandelbrot 1977, 1983, 1985). An introduction and exhaustive description of cellular automaton models that simulate SOC systems is given in Pruessner (2012), and a review of cellular automaton models applied to solar physics is given in Charbonneau et al. (2001).

Four years after introduction, Bak's SOC concept was applied to solar flares, which were known to exhibit similar power law size distributions for hard X-ray peak fluxes, total fluxes, and durations as the cellular automaton simulations produced for avalanche sizes and durations (Lu and Hamilton 1991). This discovery enabled a host of new applications of the SOC concept to astrophysical phenomena, such as solar and stellar flare statistics, magnetospheric substorms, X-ray pulses from accretion disks, pulsar glitches, and so forth. A compilation of SOC applications to astrophysical phenomena is given in a recent textbook (Aschwanden 2011a), as well as in recent review articles (Aschwanden 2013b; Aschwanden et al. 2016; Crosby 2011). The successful spreading of the SOC concept in astrophysics mirrored the explosive trend in other scientific domains, such as the application of SOC in magnetospheric physics (auroras, substorms; Sharma et al. 2016), in geophysics (earthquakes, mountain and rock slides, snow avalanches, forest fires; Hergarten 2002), in biophysics (evolution and extinctions, neuron firing, spread of diseases), in laboratory physics (Barkhausen effect, magnetic domain patterns, Ising model, tokamak plasmas; Jensen 1998), financial physics (stock market crashes; Sornette 2004), and social sciences (urban growth, traffic, global networks, internet) or sociophysics (Galam 2012). This wide range of applications elevated the SOC concept to a truly interdisciplinary research area, which inspired Bak's vision to explain "how nature works" (Bak 1996). What is common to all these systems is the statistics of nonlinear processes, which often ends up in power law-like size distributions. Other aspects that are in common among the diverse applications are complexity, contingency, and criticality (Bak and Paczuski 1995), which play a grand role in complexity theory and systems theory. Feedback plays a very important role in the phenomena of SOC, as it ensures a steady state, where the system is marginally stable against a disturbance (Kadanoff 1989).

What became clear over the last 38 years of SOC applications is the duality of (i) an universal statistical aspect and (ii) a special physical system aspect. The universal aspect is a statistical argument that can be formulated in terms of the scale-free probability conjecture (Aschwanden 2012a), which explains the power law function and the values of the power law slopes of most occurrence frequency distributions of spatiotemporal parameters in avalanching systems. This statistical argument for the probability distributions of nonlinear systems is as common as the statistical argument for binomial or Gaussian distributions in linear random systems. In this sense, solar flares, earthquakes, and stock market systems have a statistical commonality (e.g., de Arcangelis et al. 2006). On the other hand, each SOC system may be governed by different instabilities that are unique to each observed SOC phenomenon, such as plasma magnetic reconnection physics in solar flares, mechanical stressing of tectonic plates in earthquakes, or the networking of brokers in stock market crashes. So, one should always be aware of this duality of model components when creating a new SOC model. There is no need to reinvent the universal statistical aspects or power law probability distributions each time, while the modeling of physical systems may be improved with more accurate measurements and model parameterizations in every new SOC application.

There is another duality in the application of SOC: the numerical world of lattice simulation toy models and the real world of quantitative observations governed by macroscopic physical laws. The world of lattice simulations has its own beauty in producing complexity with mathematical simplicity, but it cannot capture the physics of a SOC system. It can be easily designed, controlled, modified, and visualized. It allows us to perform Monte Carlo simulations of SOC models and may give us insights about the universal statistical aspects of SOC. Real-world phenomena, in contrast, need to be observed and measured with large statistics and reliable parameters that have been cleaned from systematic bias effects, incomplete sampling, and unresolved spatial and temporal scales, which is often hard to achieve. However, computer power has increased drastically over the last 38 years, exponentially according to Gordon Moore's law, so that enormous databases with up to $\approx 10^9$ events have been gathered per dataset from some SOC phenomena, such as from solar small-scale phenomena for instance (McIntosh and Gurman 2005).

1.13 SOC Textbooks and Reviews

At the time of writing we count at least 18 textbooks related to the topic of SOC, which are listed in Table 1.2 in chronological order. The titles of these textbooks reflect the early interest in fractals, before SOC concepts came "en vogue."

Besides the text book references, the topic of SOC is also well documented with over 37 reviews (at the time of writing). However, there is no review that

Table 1.2 *Textbooks on the topic of SOC.*

Book title	References
The Fractal Geometry of Nature	Mandelbrot (1977)
Fractals	Feder (1988)
Fractals, Chaos, Power Laws: Minutes from an Infinite Paradise	Schroeder (1991)
How Nature Works: The Science of Self-Organized Criticality	Bak (1996)
Self-Organized Criticality: Emergent Complex Behavior in Physical and Biological Systems	Jensen (1998)
Cellular Automatons	Ilachinski (2001)
Self-Organized Criticality in Earth Systems	Hergarten (2002)
Discovery of Cosmic Fractals	Baryshev and Teerikorpi (2002)
Critical Phenomena in Natural Sciences: Chaos, Fractals, Self-Organization and Disorder: Concepts and Tools	Sornette (2004)
The Nonlinear Universe: Chaos, Emergence, Life	Scott (2007)
Lectures in Supercomputational Neuroscience: Dynamics in Complex Brain Networks	BeimGraben et al. (2008)
Theory of Zipf's Law and Beyond	Saichev et al. (2009)
Self-Organized Criticality in Astrophysics. The Statistics of Nonlinear Processes in the Universe	Aschwanden (2011a)
Self-Organised Criticality: Theory, Models and Characterisation	Pruessner (2012)
Sociophysics: A Physicist's Modeling of Pseudo-Political Phenomena	Galam (2012)
Self-Organized Criticality Systems	Aschwanden, ed. (2013b)
Natural Complexity: A Modeling Handbook	Charbonneau (2017)
Complexity Physics	Jensen (2023)

covers all interdisciplinary fields and aspects of SOC sciences. Nevertheless, the partial reviews cover SOC in a nutshell (Nagler et al. 1999; p. 516 in Aschwanden 2019); natural complexity (Bak and Paczuski 1995; Watkins and Freeman 2008; Sanchez and Newman 2018; Newman 2011; Toledo-Roy et al. 2019); the issue of how nature works (Raup 1997; Bhattacharya 1997; Bak and Weissman 1997; Finnigan 2001; Buchanan 2015; Watkins et al. 2016); geophysics (Turcotte 1999; Sachs et al. 2013; Hergarten 2013a); forest fires (Clar et al. 1996; Hergarten 2013b); networks (Zou et al. 2013); computer simulations (Pruessner 2013); percolation models (Milovanov 2013); fractals (McAteer 2013; McAteer et al. 2016); ecology (Kelty-Stephen et al. 2013); branching processes (Corral and Font-Clos 2013); turbulence (Vlahos and Isliker 2016); solar flares (Charbonneau et al. 2001; Charbonneau 2013; Crosby 2013); space and laboratory plasma (Sharma et al. 2016); and astrophysics (Aschwanden 2013b, 2013c; Aschwanden et al. 2016; Turolla et al. 2015).

1.14 Summary

Self-organized criticality (SOC) is a theoretical concept that describes the statistics of nonlinear processes. It is a fundamental principle that is common to many nonlinear dissipative systems in the universe. Due to its universality and ubiquity, SOC is a law of nature, for which we derive the theoretical framework and specific macroscopic physical models in this book. The SOC concept was introduced by Bak, Tang, and Wiesenfeld in 1987 and has been applied to laboratory experiments of sandpiles, to human activities such as population growth, language, economy, traffic jams, or wars, to biophysics, geophysics (earthquakes, landslides, forest fires), magnetospheric physics, solar physics (flares), stellar physics (flares, cataclysmic variables, accretion disks, black hole object, pulsar glitches, gamma ray bursts), and to galactic physics and cosmology. From an observational point of view, the hallmark of SOC behavior is the power law shape of occurrence frequency distributions of spatial, temporal, and energy scales, implying scale-free nonlinear processes. Power laws are neither a necessary nor a sufficient condition for SOC behavior, because intermittent turbulence produces power law-like size distributions also. A novel trend that is ongoing in current SOC research is a paradigm shift from “microscopic” scales (in numerical cellular automaton simulations) toward “macroscopic” modeling based on physical scaling laws.

2

Power Law Size Distributions

Power law slopes of the size distributions of avalanche parameters are the hallmarks of self-organized criticality (SOC) systems. Although the parameterization of SOC avalanches has mostly been derived from the electromagnetic emission observed in astrophysical phenomena (e.g., solar or stellar flares), the definition of avalanching systems with SOC behavior has much wider applications (e.g., earthquakes, forest fires, or pandemics) and can be defined universally. We consider a SOC avalanche as a spatiotemporal energy dissipation event with nonlinear growth during the initial rise phase, followed by gradual decay during the saturation phase (Figure 1.2).

2.1 Differential Size Distribution

The data input for an occurrence frequency distribution is usually a list or a catalog of events, characterized by some size parameter x_i for $i = 1, .., n$ events, regardless whether the list was generated by visual inspection or by an automated computer algorithm. How do we construct a log–log histogram from an event catalog? There are essentially two ways, either a logarithmically binned histogram if large statistics is available, or a rank-order plot if the size of the statistical sample is rather small.

If we have large statistics (at least $n \gtrsim 10^2, \ldots, 10^3$), we can first establish a logarithmic binning axis bound between the minimum and maximum values, $x_{min} \le x_i \le x_{max}$, with n_j bins,

$$x_j^{bin} = x_{min} \left(\frac{x_{max}}{x_{min}} \right)^{(j-1)/(n_j-1)} , \qquad j = 1, \ldots, n_j \tag{2.1}$$

which is equidistant on a logarithmic scale but has variable intervals on a linear scale,

$$\Delta x_j^{bin} = x_{j+1}^{bin} - x_j^{bin} = x_j^{bin} \left[\left(\frac{x_{max}}{x_{min}} \right)^{1/(n_j-1)} - 1 \right] . \tag{2.2}$$

In a next step, we can count the number of events N_j^{bin} that fall in each bin with interval $x_j^{bin} \leq x_i \leq x_{j+1}^{bin}$. We have to be aware that this number N_j^{bin} depends on the particular bin size Δx_j^{bin} we have chosen. In order to obtain the functional form of the frequency distribution, which should be independent of the binning, we have to divide the number of events by the bin size,

$$N_j = N(x_j) = \frac{N_j^{bin}}{\Delta x_j^{bin}}, \tag{2.3}$$

and can plot the frequency distribution with N_j on the y-axis versus the size x_j on the x-axis. This representation normalizes the distribution to the total number of events n, which we should obtain by integration over the x-axis, or by summing over all (nonequidistant) bins on the x-axis,

$$\int_0^{\infty} N(x)dx = \sum_{j=1}^{n_j} N_j dx_j = \sum_{j=1}^{n_j} N(x_j)\Delta x_j^{bin} = \sum_{j=1}^{n_j} N_j^{bin} = n. \tag{2.4}$$

Instead of expressing the number of occurrences by the actually observed numbers N_j, it is also customary to use a *probability distribution function* $P(x)$, which is simply the number of events in each bin normalized by the total number of events n,

$$P(x) = \frac{N(x)}{n}, \tag{2.5}$$

which has the total integral normalized to unity,

$$\int_{x_{min}}^{x_{max}} P(x)dx = \int_{x_{min}}^{x_{max}} \frac{N(x)}{n} dx = 1. \tag{2.6}$$

Both representations, $N(x)$ or $P(x)$, are called a *differential frequency distribution*, because they express the number of events in a "differential" bin dx.

2.2 Maximum Likelihood Estimator

If the assumption is made that an observed occurrence frequency (or size) distribution can theoretically be described with a power law function, a goodness-of-fit test is in order. Many studies use a chi-square test as a quick and practical method. However, there is a bias involved that can be overcome by using a maximum likelihood estimator (MLE) (Goldstein et al. 2004; Clauset et al. 2007; White et al. 2008; D'Huys et al. 2016). If we order the data in increasing order, with $x_i \geq x_{min}$, $i = 1, \ldots, n$, the MLE for continuous data is given by the power law slope α,

$$\alpha = 1 + n \left[\sum_{i=1}^{n} \ln \left(\frac{x_i}{x_{min}} \right) \right]^{-1}. \tag{2.7}$$

The standard error of the power law slope σ is then

$$\sigma = \frac{\alpha - 1}{\sqrt{n}}. \tag{2.8}$$

Goldstein et al. (2004) showed that the variance of the estimates obtained with the MLE method is significantly lower than that of the estimates using a linear fit on the first five bins in the frequency distribution. It has been shown to be a minimum variance unbiased estimator (White et al. 2008).

2.3 Cumulative Size Distribution

An integrated *differential frequency distribution* $N(x)dx$ is called a *cumulative frequency distribution* $N^{cum}(> x)$, which expresses in each bin the sum of all events that are larger than the size parameter of the bin x,

$$N^{cum}(> x) = \int_x^{x_{max}} N(x)dx, \tag{2.9}$$

which we denote with $N^{cum}(> x)$, in contrast to the differential distribution $N(x)$. The cumulative frequency distribution contains more statistics in the rarer bins at larger sizes, and thus appears to be smoother at the upper end than the differential distribution function. However, the values in each bin are statistically not independent, but always contain information from all other bins on the right-hand side. The particular functional shape at the upper cutoff can dominate the entire distribution function.

If the differential frequency distribution is a power law function with slope α, the cumulative frequency distribution is expected to have a power law slope that is flatter by 1,

$$\begin{aligned} N(x) \quad & \propto x^{-\alpha} \\ N^{cum}(> x) \quad & \propto x^{-\beta} \lesssim x^{-(\alpha-1)}. \end{aligned} \tag{2.10}$$

However, the power law relationship $\beta = \alpha - 1$ is only true when the differential frequency distribution extends to infinite, which is never the case. In reality, there is always a largest event at x_{max}, which causes a sharp cutoff in the differential frequency distribution $N(x)$, but a gradual steepening in the cumulative frequency distribution, because of the missing contributions from $x > x_{max}$. This detail is quite important, because it leads to a significant overestimate of the power law slope when the rule $\alpha = \beta + 1$ is applied. We define a power law distribution function with a sharp cutoff at x_{max},

$$N(x) = (\alpha - 1)x^{-\alpha}, \quad x \leq x_{max}. \tag{2.11}$$

The cumulative frequency distribution function can then be calculated by integrating over the range from x to x_{max},

$$N^{cum}(>x) = n\frac{\int_{x}^{x_{max}} x'^{-\alpha} dx'}{\int_{x_{min}}^{x_{max}} x'^{-\alpha} dx'} = n\frac{(x^{1-\alpha} - x_{max}^{1-\alpha})}{(x_{min}^{1-\alpha} - x_{max}^{1-\alpha})}. \tag{2.12}$$

We see that the second term in the integral, $x_{max}^{1-\alpha}$, steepens the slope and lets the cumulative distribution drop to zero when x approaches x_{max}.

2.4 Rank-Order Plots

If the statistical sample is rather small, in the sense that no reasonable binning of a histogram can be done, either because we do not have multiple events per bin or because the number of bins is too small to represent a distribution function, we can create a rank-order plot. A rank-order plot is essentially an optimum adjustment to minimum statistics that gives every single event a single bin. From an event list of a parameter $x_i, i = 1, \ldots, n$, which is generally not sorted, we have first to generate a rank-ordered list by ordering the events according to increasing size,

$$x_1 \le x_2 \le \ldots \le x_j \le \ldots \le x_n, \quad j = 1, \ldots, n. \tag{2.13}$$

The bins are generally not equidistant, neither on a linear nor logarithmic scale, defined by the difference between subsequent values of the ordered x_j,

$$\Delta x_j^{bin} = x_{j+1}^{bin} - x_j^{bin}. \tag{2.14}$$

In a rank-ordered sequence of n events, the probability for the largest value is $1/n$, for events that are larger than the second-largest event it is $2/n$, and so forth, while events larger than the smallest event occur in this event list with a probability of unity. Thus, the cumulative frequency distribution is simply the reversed rank order,

$$N^{cum}(>x_j) = (n + 1 - j), \qquad j = 1, \ldots, n, \tag{2.15}$$

and the distribution varies from $N^{cum}(>x_1) = n$ for $j = 1$ to $N^{cum}(>x_n) = 1$ for $j = n$. We can plot a cumulative frequency distribution with $N^{cum}(>x_j)$ on the y-axis versus the size x_j on the x-axis. The distribution is normalized to the number of events n,

$$\int_{x_1}^{x_n} N(x)dx = N^{cum}(>x_1) = n. \tag{2.16}$$

The differential frequency distribution function $N(x)$ could in principle be computed from the derivative of the cumulative distribution, but there is usually considerable noise between subsequent events in a rank-ordered histogram, so that deviations from straight power laws are visible, especially on the right-hand side of the rank-ordered histogram.

2.5 Numerical Simulation of Power Law Distributions

For numerical simulations of frequency distributions, for instance, Monte Carlo simulations of SOC models, we need to create randomly distributed values x_i that have a particular prescribed function of their frequency distribution, such as an exponential function for waiting times, or a power law function for energies. Let us prescribe the form of the frequency distribution with a probability function $p(x)$ in the interval $[x, x + dx]$, which has the normalization,

$$\int_0^\infty p(x)dx = 1. \tag{2.17}$$

The total probability $\rho(x)$ to have a value in the range of $[0, x]$ is then the integral,

$$\rho(x) = \int_0^x p(x')dx'. \tag{2.18}$$

If the analytical function $\rho(x)$ can be inverted, say with the analytical inverse function ρ^{-1}, so that

$$x = \rho^{-1}(\rho) = \rho^{-1}(\rho[x]), \tag{2.19}$$

we have a transformation that allows us to generate values x_i from a distribution of probability values ρ_i. There are many numerical random generator algorithms available that produce a random number ρ_i in a unity range of $[0, 1]$, which can then be used to generate values x_i with the mapping transform $x_i = \rho^{-1}(\rho_i)$. The frequency distribution of these values x_i will then fulfill the prescribed function $p(x)$.

As an example we demonstrate the numerical generation of a sample of time scales t that has a frequency distribution function following an exponential function,

$$p(t) = \frac{1}{\tau} \exp\left(-\frac{t}{\tau}\right), \tag{2.20}$$

which fulfills the normalization $\int_0^1 p(t)dt = 1$. The total probability $\rho(t)$ to have a value in the range $[0, t]$ is then the integral function of $p(t)$,

$$\rho(t) = \int_0^t p(t')dt' = \int_0^t \frac{1}{\tau} \exp\left(-\frac{t'}{\tau}\right)dt' = \left[1 - \exp\left(-\frac{t}{\tau}\right)\right]. \tag{2.21}$$

The inverse function $t(\rho)$ of $\rho(t)$ defined by Eq. (2.21) is

$$t(\rho) = -\tau \ln(1 - \rho). \tag{2.22}$$

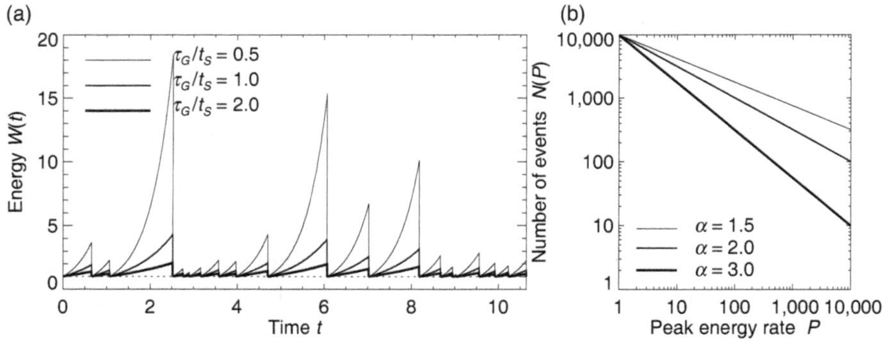

Figure 2.1 (a) An exponential frequency distribution is numerically generated, based on $n = 10,000$ uniformly distributed values ρ_i in the range $[0, \ldots, 1]$ and times $t_i = -\tau \ln(1 - \rho_i)$, leading to the occurrence probability function $p(t)$ as defined in Eq. 2.20. (b) A power law frequency distribution is numerically generated, based on $n = 10,000$ values ρ_i uniformly distributed in the range $[0, \ldots, 1]$, with energies $E_i = (1 + \rho_i)^{1/(1-\alpha)}$ with $\alpha = 1.5$ (Eq. 2.25), leading to the occurrence probability function $p(E)$ as defined in Eq. 2.23. The numerically sampled distributions are shown with log(N)–log(S) histograms, while the theoretically predicted function is rendered with a solid curve (Aschwanden 2011a).

In Figure 2.1a, we use a random generator that produces 10,000 values ρ_i, uniformly distributed in the range of $[0, 1]$, and use the transform Eq. 2.22 to generate values $t_i = -\tau \ln(1 - \rho_i)$ with $\tau = 1$ and sample the frequency distribution of the 10,000 values t_i, which follows the prescribed exponential function $p(t)$ defined in Eq. 2.20.

As a second example, we prescribe a power law function $p(E)$ for the frequency distribution,

$$p(E) = (\alpha - 1)E^{-\alpha}, \tag{2.23}$$

which fulfills the normalization $\int_1^\infty p(E)dE = 1$. The total probability $\rho(E)$ in the range $[0, E]$ is then the integral function of $p(E)$ (Eq. 2.23),

$$\rho(E) = \int_0^E p(\epsilon)d\epsilon = \int_0^E (\alpha - 1)\epsilon^{-\alpha} d\epsilon = \left[1 - E^{(1-\alpha)}\right]. \tag{2.24}$$

The inverse function $E(\rho)$ of $\rho(E)$ (Eq. 2.24) is

$$E(\rho) = [1 - \rho]^{1/(1-\alpha)}. \tag{2.25}$$

In Figure 2.1b, we use a random generator that produces 10,000 values ρ_i uniformly distributed in the range of $[0, 1]$, choose a power law index of $\alpha = 1.5$, and use the transform Eq. 2.25 to generate values $E_i = [1 - \rho_i]^{-2}$ and sample the frequency distribution of the 10,000 values E_i, which follows the prescribed power law function $p(E) = 0.5E^{-1.5}$ as defined in Eq. 2.23, with $\alpha = 1.5$.

2.6 The Exponential-Growth Model

After this basic introduction into the numerical parameterization of the occurrence frequency (or size) distribution, we turn now to theoretical SOC-specific models of the size distributions. One of the first theoretical models that calculates a size distribution of SOC-driven nonlinear energy dissipation events has been pioneered by Rosner and Vaiana (1978), applied to "cosmic flare transients," which has been further developed for logistic avalanche processes (Aschwanden et al. 1998). Avalanches occurring in the state of SOC represent local instabilities that grow explosively (or exponentially) over some time interval. The released energy grows in a nonlinear way above some energy threshold, once triggered, which can be parameterized by some nonlinear function, for instance, by an exponential growth function. We define the time evolution of the energy release rate $W(t)$ of a nonlinear process that starts at a threshold energy of W_0 by

$$W(t) = W_0 \exp\left(\frac{t}{\tau_G}\right), \qquad 0 \le t \le \tau, \tag{2.26}$$

where τ_G represents the exponential (or e-folding) growth time. The process grows exponentially until it saturates at time $t = \tau$ with a saturation energy W_S,

$$W_S = W(t = \tau) = W_0 \exp\left(\frac{\tau}{\tau_G}\right). \tag{2.27}$$

We define a peak energy release rate P which represents the maximum energy release rate W_S, after background subtraction of the threshold energy W_0, that corresponds to the steady-state energy level before the nonlinear growth phase,

$$P = W_S - W_0 = W_0 \left[\exp\left(\frac{\tau}{\tau_G}\right) - 1\right]. \tag{2.28}$$

In the following, we will refer to the peak energy release rate P briefly as "peak energy." For the saturation times τ, which we also call "rise times," we assume a random probability distribution, approximated by an exponential function $N(\tau)$ with e-folding time constant t_S,

$$N(\tau)d\tau = \frac{N_0}{t_S} \exp\left(-\frac{\tau}{t_S}\right)d\tau. \tag{2.29}$$

This probability distribution is normalized to the total number of N_0 events,

$$\int_0^\infty N(\tau)d\tau = N_0. \tag{2.30}$$

In order to derive the probability distribution $N(P)$ of peak energy release rates P, we have to substitute the variable of the peak energy, P, into the function of the rise time $\tau(P)$,

$$N(P)dP = N(\tau)d\tau = N[\tau(P)] \left| \frac{d\tau}{dP} \right| dP. \tag{2.31}$$

This requires the inversion of the evolution function $P(\tau)$ (Eq. 2.28),

$$\tau(P) = \tau_G \ \ln \left(\frac{P}{W_0} + 1 \right), \tag{2.32}$$

and the calculation of its derivative $d\tau/dP$, which is

$$\frac{d\tau}{dP} = \frac{\tau_G}{W_0} \left(\frac{P}{W_0} + 1 \right)^{-1}. \tag{2.33}$$

Inserting the probability distribution of saturation times $N(\tau)$ (Eq. 2.29), the inverted evolution function $\tau(P)$ (Eq. 2.32) and its time derivative $(d\tau/dP)$ from Eq. (2.33) into the frequency distribution $N(P)$ in Eq. 2.31 yields then,

$$N(P)dP = \frac{N_0(\alpha_P - 1)}{W_0} \left(\frac{P}{W_0} + 1 \right)^{-\alpha_P} dP, \tag{2.34}$$

which approaches an exact power law distribution for large peak energies ($P \gg W_0$), with a power law slope α_P of

$$\alpha_P = \left(1 + \frac{\tau_G}{t_S} \right). \tag{2.35}$$

The power law slope thus depends on the ratio of the growth time τ_G to the e-folding saturation time t_s, which is essentially the average number of growth times. Examples of time series with avalanches of different growth times ($\tau_G/t_S = 0.5, 1.0, 2.0$) are shown in Figure (2.1), along with the corresponding power law distributions of energies F. Note that the fastest growing events produce the flattest power law distribution of energies.

This exponential-growth model serves two purposes: (i) The exponential growth curve mimics the typical time evolution of any instability and (ii) it demonstrates that SOC models naturally produce power law functions for the size distribution of dissipated avalanche energies. A caveat, however, is the fact that no theoretical prediction exists for the numerical value of the power law slope α_P (Eq. 2.35).

2.7 The Standard FD-SOC Model

The most frequently observed SOC parameters in astrophysics include: the 1-D spatial length scale L, the 2-D fractal area A, the 3-D fractal volume V, the time duration T, the flux F, and the fluence or energy E, which is approximated with the product of the flux F multiplied with the time duration T, that is, $E = F \times T$. The methodology has been developed in earlier work, using the statistical avalanche model of a slowly driven SOC system (Aschwanden 2012a, 2013a, 2013b, 2014,

2015, 2022a, 2022b), which we call the standard fractal-diffusive (FD-SOC) model here. The main goal of this modeling approach is a deeper physical understanding, and the capability of predicting occurrence frequencies or size distributions $N(x)\,dx \propto x^{-\alpha}\,dx$ of standard SOC parameters $x = (L, A, V, T, F, E)$, based on universal scaling laws. In this section, we derive these theoretical elements analytically from first principles.

In the Section 2.7, we design a generalized SOC model, which has three free parameters, as well as a standard SOC model, which has no free parameters, and can theoretically predict the power law slopes to be observed. We start with the universal scale-free size distribution $N(L)\,dL$ of length scales L given by the scale-free probability conjecture (Aschwanden 2012a, 2014, 2015) (see introduction in Section 1.6),

$$N(L)\,dL \propto L^{-d}\,dL, \tag{2.36}$$

where $d = 1, 2, 3$ are the Euclidean dimensions, and $N(L)dL$ is the number of length scales L observed in the interval $[L, L + dL]$. For most real-world data, the Euclidean space dimension is $d = 3$, but there are exceptions, such as the area A, which obviously has a flat geometry with an Euclidean dimension of $d = 2$.

A next assumption is that SOC parameters x are fractal (or multifractal), see introduction in Section 1.7. In a fractal world, which is an intrinsic property of SOC systems, the Euclidean dimension d represents an upper limit of the fractal dimension D. For the spatial parameters we define the fractal dimensions for 2-D areas A, which is identical to the fractal Hausdorff dimension D_A (Mandelbrot 1977, 1983, 1985),

$$A = L^{D_A}, \tag{2.37}$$

and like-wise for the 3-D volume V, which is identical to the fractal Hausdorff dimension D_V,

$$V = L^{D_V}. \tag{2.38}$$

We can estimate the numerical values of the fractal dimensions D_A and D_V from the mean of the minimum and maximum value in each Euclidean domain,

$$D_A = \frac{(D_{A,min} + D_{A,max})}{2} = \frac{(1+2)}{2} = \frac{3}{2} = 1.50, \tag{2.39}$$

and correspondingly,

$$D_V = \frac{(D_{V,min} + D_{V,max})}{2} = \frac{(2+3)}{2} = \frac{5}{2} = 2.50. \tag{2.40}$$

The range $[D_{min}, D_{max}] = [d-1, d]$ is equivalent to the range of fractals in a multifractal concept. Thus, the range of fractal dimensions is $[D_{min}, D_{max}] = [1, 2]$ in the

2-D Euclidean space, and $[D_{min}, D_{max}] = [2, 3]$ in the 3-D Euclidean space. We can express a generalized fractal dimension D_d as a function of the Euclidean dimension $d = 1, 2, 3$,

$$D_d = d - \frac{1}{2}, \tag{2.41}$$

where $D_2 = D_A$ is the fractal dimension of the area with $d = 2$, and $D_3 = D_V$ is the fractal dimension of the volume with $d = 3$.

In the following equation, we are using universal scaling laws according to the FD-SOC model (Aschwanden 2012a, 2014, 2015), which can all be expressed as a function of the length scale L, such as a diffusive transport distance $L \propto T^{1/2}$ during the time duration T,

$$T \propto L^{2/\beta}, \tag{2.42}$$

which has a coefficient of $\beta = 1$ for classical diffusion, or $\beta \approx 3/2$ for Lévy-flight. The flux F, which is expressed with the power law scaling law $F \propto V^{\gamma}$ (as defined in Eq. 1.11) and the definition of the fractal dimension, that is, $V \propto L^{D_V}$ (Eq. 2.38), yields

$$F \propto V^{\gamma} \propto L^{D_V \gamma}. \tag{2.43}$$

The fluence or energy E, which is expressed by the product of the flux F and the event duration T (for a spatiotemporal SOC event), yields

$$E = F \times T = L^{D_V \gamma + 2/\beta)}. \tag{2.44}$$

Since we have now all scaling laws as a function of the length scale L, it is straightforward to compute the inverse functions $L(x)$ (from Eqs 2.39 to 2.44),

$$L = A^{1/D_A}, \tag{2.45}$$

$$L = V^{1/D_V}, \tag{2.46}$$

$$L \propto T^{\beta/2}, \tag{2.47}$$

$$L \propto F^{1/(D_V \gamma)}, \tag{2.48}$$

$$L \propto E^{1/(D_V \gamma + 2/\beta)}, \tag{2.49}$$

and to calculate the derivatives of Eqs (2.45–2.49),

$$\frac{dL}{dA} = A^{1/D_A - 1}, \tag{2.50}$$

$$\frac{dL}{dV} = V^{1/D_V - 1}, \tag{2.51}$$

$$\frac{dL}{dT} \approx T^{\beta/2-1}, \tag{2.52}$$

$$\frac{dL}{dF} \approx F^{(1/D_V\gamma-1)}, \tag{2.53}$$

$$\frac{dL}{dE} \approx E^{1/(D_V\gamma+2/\beta)-1}. \tag{2.54}$$

We can now substitute the variable L with the SOC parameters $x = [A, V, T, F, E]$ into Eq. 2.36,

$$N(x)dx = N[L(x)] \left| \frac{dL}{dx} \right| dx, \tag{2.55}$$

by inserting $L(x)$ and dL/dx,

$$N(A) = N[L(A)] \left| \frac{dL}{dA} \right| = A^{-\alpha_A}, \tag{2.56}$$

$$N(V) = N[L(V)] \left| \frac{dL}{dV} \right| = V^{-\alpha_V}, \tag{2.57}$$

$$N(T) = N[L(T)] \left| \frac{dL}{dT} \right| = T^{-\alpha_T}, \tag{2.58}$$

$$N(F) = N[L(F)] \left| \frac{dL}{dF} \right| = F^{-\alpha_F}, \tag{2.59}$$

$$N(E) = N[L(E)] \left| \frac{dL}{dE} \right| = E^{-\alpha_E}, \tag{2.60}$$

which defines the power law slopes α_x of the size distributions, where the approximations are calculated in 3-D Euclidean space ($d = 3$). The results of the calculated (or theoretically predicted) power law exponents α_x are listed in Table 2.1. The definitions of the power law slopes $(\alpha_L, \alpha_A, \alpha_V, \alpha_T, \alpha_F, \alpha_E)$ are implicitly defined in Eqs (2.55–2.60), and their numerical values are given in Table (2.1), which are frequently used in the remainder of the book.

The power law slopes α_x of the universal SOC parameters $x = (L, A, V, T, F, E)$ depend on three model parameters: the dimensionality d of the SOC system, the diffusive transport coefficient β, and the flux–volume scaling γ. The (multi)fractal dimensions are $D_A = d - (1/2) = 1.5$ for the area A, and $D_V = d - (1/2) = 2.5$ for the volume V. In the standard SOC model we set $d = 3$, $\beta = 1$, and $\gamma = 1$.

If we measure the two power law slopes of the time duration α_T and the flux α_F, then the physical SOC parameters can directly be inferred from the inversion of the slope $\alpha_T = 1 + (d-1)\beta/2$ (Table 2.1),

Table 2.1 *Theoretically predicted power law slopes α_x of the generalized SOC model with dimensionality d (column 2), the 2-D area SOC model (d = 2, column 3), and the 3-D standard FD-SOC model (d = 3, column 4).*

Parameter	Generalized power law slope	2-D area power law slope	Standard power law slope
Euclidean dimension	$d =$	$(2/1) = 2.00$	$(3/1) = 3.00$
Diffusion type	$\beta =$	$(1/1) = 1.00$	$(1/1) = 1.00$
Fractal dimension	$D_d = d - (1/2) =$	$(3/2) = 1.50$	$(5/2) = 2.50$
Length	$\alpha_L = d =$	$(2/1) = 2.00$	$(3/1) = 3.00$
Area	$\alpha_A = 1 + (d-1)/D_A =$	$(5/3) \approx 1.67$	$(7/3) \approx 2.33$
Volume	$\alpha_V = 1 + (d-1)/D_V =$	$(7/5) = 1.40$	$(9/5) = 1.80$
Duration	$\alpha_T = 1 + (d-1)\beta/2 =$	$(3/2) = 1.50$	$(2/1) = 2.00$
Flux	$\alpha_F = 1 + (d-1)/(D_V\gamma) =$	$(5/3) \approx 1.67$	$(9/5) = 1.80$
Fluence	$\alpha_E = 1 + (d-1)/(D_V\gamma + 2/\beta) =$	$(9/7) \approx 1.29$	$(13/9) \approx 1.44$

$$\beta = 2\frac{(\alpha_T - 1)}{(d-1)}, \tag{2.61}$$

and the inversion of $\alpha_F = 1 + (d-1)/(D_V\gamma)$ (Table 2.1),

$$\gamma = \frac{1}{D_V}\left(\frac{d-1}{\alpha_F - 1}\right), \tag{2.62}$$

These predictions are consistent with those made in previous work (Aschwanden 2012a, 2014, 2015), except for a slightly different assumption in the definition of the minimum fractal dimension, which was assumed to be $D_{V,min} = 1$ earlier, instead of $D_{V,min} = (d-1)$ used here. The derived scaling laws and power law slopes of SOC parameter occurrence frequency distributions should be universally valid for all SOC systems that have fractality $(D_V, D_A) \leq d$, classical diffusion transport ($\beta = 1$), and flux–volume proportionality ($\gamma = 1$). The latter assumption is naturally fulfilled if the data are produced by an incoherent emission mechanism ($\gamma = 1$), while the generalized FD-SOC model also accommodates for coherent emission mechanisms ($\gamma > 1$).

2.8 SOC Scaling Laws

The power law slopes of the size distributions of SOC parameters, which we derived in Table 2.1, can equivalently be expressed by the underlying scaling laws that quantify their correlation. Hence we calculate the correlations among each pair of possibly correlated SOC parameters. For the five SOC parameters, that is, the length scale L, the time duration T, the average flux F, the peak flux P, and the

Table 2.2 *Scaling laws of four SOC parameters, with analytical expressions and numerical exponents for correlated parameter pairs.*

Function		Inverse function	
$L(T) \propto T^{\beta/2}$	$\approx T^{(1/2)}$	$T(L) \propto L^{2/\beta}$	$\approx L^2$
$L(F) \propto F^{1/D_V\gamma}$	$\approx F^{(2/5)}$	$F(L) \propto L^{D_V\gamma}$	$\approx L^{(5/2)}$
$L(E) \propto E^{1/(D_V\gamma+2/\beta)}$	$\approx E^{(2/9)}$	$E(L) \propto L^{D_V\gamma+2/\beta}$	$\approx L^{(9/2)}$
$T(F) \propto F^{2/(\beta D_V\gamma}$	$\approx F^{(4/5)}$	$F(T) \propto T^{\beta D_V\gamma}$	$\approx T^{(5/4)}$
$T(E) \propto E^{D_V\gamma/2+1/\beta}$	$\approx E^{(9/4)}$	$E(T) \propto T^{1/\beta+D_V\gamma/2}$	$\approx T^{(4/9)}$
$F(E) \propto E^{1/(1+\gamma\beta D_V)}$	$\approx E^{(5/9)}$	$E(F) \propto F^{1+2/\beta D_V\gamma/2}$	$\approx F^{(9/5)}$

(spatiotemporal) energy E, the 5 SOC parameters yield then $5 \times 4 = 20$ possibly pair (two-parameter) correlations. In Table 2.2 we calculate the analytical expressions of the scaling low exponents, as well as the numerical values for the standard SOC models ($\gamma = 1$, $\beta = 1$, and $D_V = 5/2 = 2.50$.

2.9 The 2-D Energy Model

The energy of a spatiotemporal SOC event is defined in the standard SOC model by the product of the count rate (F) and the event duration (T) (Eq. 2.44), which is an appropriate measure for nonthermal energies that are quantified by hard X-ray counts (or intensity) in solar and stellar flares. In both solar or stellar flares, down to nanoflares, one can estimate thermal (radiative) energies at the peak time of an event, defined by Rosner et al. (1978),

$$E_A = (3k_B n_e T_e)\ V, \tag{2.63}$$

where k_B is the Boltzmann constant, n_e the electron density, T_e the electron temperature, and V the 3-D volume, all measured at the peak time of an event. The 3-D volume V cannot be measured directly, which led some authors to approximate the volume with a constant height h_0 along the line-of-sight (e.g., Krucker and Benz 1998),

$$V = A\ h_0 = L^{D_A}\ h_0, \tag{2.64}$$

where the fractal area is defined as $A = L^{D_A}$. The fractality is not explicitly mentioned in some of these studies, but every pattern recognition code that measures an area on a pixel-by-pixel basis (at different spatial resolutions) yields approximately the fractal area $A \propto L^{D_A}$ with $D_A < 2$, rather than the encompassing Euclidean area $A = L^2$. Inserting the area fractal dimension $D_A = 3/2$ (Eq. 2.39) into the expression for the thermal energy $E_A \propto L^{D_A}$ (Eq. 2.64), we obtain

$$E_A = (3k_B n_e T_e h_0)\ L^{D_A}. \tag{2.65}$$

In the same way, we substitute the variable L in the size distribution with the energy $x = E$ (Eq. 2.60),

$$N(E_A)\, dE_A = L(E_A)^{-d} \left(\frac{dL}{dE_A}\right) dE_A = E_A^{-\alpha_{E_A}}\, dE_A, \tag{2.66}$$

yielding the power law slope α_{E_A}, for $d = 3$ and $D_A = 3/2$,

$$\alpha_{E_A} = 1 + \frac{(d-1)}{D_A} = \frac{7}{3} \approx 2.33. \tag{2.67}$$

Note that we treat the variables n_e, T_e, and h_0 as constants here, while the scaling law hinges entirely on the correlation between the thermal energy E_A and the length scale L, rendering a first-order approximation to the power law slope α_{E_A}. Since the thermal energy $E_A \propto V \propto L^{D_A}$ (Eq. 2.64) and the fractal area $A \propto L^{D_A}$ (Eq. 2.37) have the same scaling law, the power law index for the size distribution of areas α_A has the same power law value α_{E_A} too,

$$\alpha_{E_A} = \alpha_A = \frac{7}{3} \approx 2.33. \tag{2.68}$$

The definition of the energy made here invokes an isothermal plasma. Note also that the volume of the radiated plasma is approximated with an isotropic and multifractal geometry. Nevertheless, the definition of the thermal energy can accommodate a multithermal formalism, which involves a differential emission measure distribution function $dEM(T_e)\,/dT_e$, characterized by the increase in the emission measure EM, the (mean) electron density n_e, and the volume V,

$$EM = n_e^2\, V, \tag{2.69}$$

which inserted into Eq. 2.63 yields,

$$E_V = (3k_B T_e)\, \sqrt{EM \times V} = (3k_B T_e)\, \sqrt{EM \times A\, h_0}. \tag{2.70}$$

Size distribution of thermal energies, based on emission measure changes EM, yields power law slopes of $\alpha_{E_A} = 2.38 \pm 0.09$ (Krucker and Benz 1998; Parnell and Jupp 2000; Benz and Krucker 2002), while the theoretically expected value is $\alpha_{E_A} = 2.33$ (Table 2.1).

2.10 The 3-D Energy Model

In the 3-D version of the thermal model, which we consider as more realistic than the 2-D model with a constant line-of-sight length h_0, the SOC avalanche volume $V \propto L^{D_V}$ (Eq. 2.38) is defined by the (mean) Hausdorff dimension $D_V = (5/2)$ (Eq. 2.40), which inserted into the expression for the thermal energy is,

$$E_V = (3k_B n_e T_e)\, L^{D_V}. \tag{2.71}$$

We substitute the variable L in the size distribution of the thermal energy E_V (Eq. 2.63),

$$N(E_V)\, dE_V = L(E_V)^{-d} \left| \frac{dL}{dE_V} \right| dE_V = E_V^{-\alpha_{E_V}}\, dE_V, \tag{2.72}$$

yielding the power law slope α_{E_V}, for $d = 3$ and $D_V = 5/2$,

$$\alpha_{E_V} = \alpha_V = 1 + \frac{(d-1)}{D_V} = \frac{9}{5} = 1.80. \tag{2.73}$$

Note that the power law slope is significantly steeper in the 2-D model ($\alpha_{E_A} = 2.33$) than in the 3-D version ($\alpha_{E_V} = 1.80$). Moreover, the two models predict power law slopes below ($\alpha_{E_2} < 2$), which indicates that the size distribution diverges at the high-energy end of the size distribution and thus renders nanoflares less dominant in the energy budget of coronal heating.

Calculations of the multithermal energy using a 3-D model have been performed using Yohkoh, TRACE, SOHO, AIA, and IRIS data (e.g., Shimizu 1995; Berghmans et al. 1998; Berghmans and Clette 1999; Parnell and Jupp 2000; Aschwanden and Parnell 2002; Uritsky et al. 2007; Joulin et al. 2016; Nhalil et al. 2020; Purkhart and Veronig 2022; Kawai and Imada 2022).

2.11 Thresholded Power Law Distributions

The accuracy in the determination of the power law slope α_x can be significantly improved by modeling systematic errors, such as the (astrophysical) background subtraction, incomplete sampling, instrumental sensitivity limitations, and detection thresholds, which affect differential size distributions and require thresholded power law distributions (Section 2.11), finite system-size effects (Section 2.12), and extreme ("Black-Swan" or "Dragon-King") events (Section 2.13).

We start with a data sample $X_i, i = 1, \ldots, n_{ev}$ that contains some size parameter X_i with n_{ev} events. In astrophysical datasets, observables for the size parameter X_i are most often measured in terms of fluxes F_i, time-integrated fluences, intensities, luminosities, or energies E_i, and event time durations T_i. The observable X_i refers to the size of an event, which can be any of these observables. In astrophysical observations, a basic correction that needs to be applied to the size value X_i is the background subtraction

$$x_i = X_i - B_i, \quad i = 1, \ldots, n_{ev}, \tag{2.74}$$

where B_i represents an event-unrelated background flux or fluence, for instance, from the galactic quiescent soft X-ray flux that is irradiated from near the same source direction as the soft X-ray flux of a solar flare. If the flux time profile $F_i(t)$ of an event "i" is available, the background flux can be determined from the preflare flux B_i, and the corrected flux is $F_{i,\mathrm{corr}}(t) = F_i(t) - B_i$. Since the background correction affects the power law slope α_F of the peak flux (see Fig. 7 in Aschwanden 2015), it is advisable to apply a mean estimated background correction at least.

A *size distribution* of events, $N(x)dx$, also called a *differential occurrence frequency distribution*, can ideally be approximated by a power law function as a function of some size parameter x, quantified by three observables (x_1, x_2, and n_0) and one variable (α_x),

$$N(x)dx = n_0\, x^{-\alpha_x} dx, \qquad x_1 \le x \le x_2, \tag{2.75}$$

where x_1 and x_2 are the lower and upper bounds of the power law fitting (inertial) range, n_0 is a normalization constant, and α_x is the power law slope. Uncertainties in previous calculations of the power law slope mostly resulted from the arbitrary choice of fitting ranges $[x_1, x_2]$, which can be largely corrected by a generalized power law function that includes under-sampling at the lower end, and finite system-size effects at the upper end of the inertial range $[x_1, x_2]$.

The straight power law function (Eq. 2.75) can be generalized with a threshold parameter x_0. This generalized function is called *Lomax distribution* (Giles et al. 2011), *the generalized Pareto distribution* (Hosking and Wallis 1987), or the *thresholded power law size distribution* (Aschwanden 2015),

$$N(x)dx = n_0\, (x_0 + x)^{-\alpha_x}\, dx, \tag{2.76}$$

where n_0 is a normalization constant. This threshold parameter x_0 models three different features: such as truncation effects due to incomplete sampling of events below a threshold (if $x_0 > 0$), incomplete sampling due to instrumental sensitivity limitations (if $x_0 > 0$), or subtraction of event-unrelated background (if $x_0 < 0$), as it is common in astrophysical datasets (Aschwanden 2015).

2.12 Finite System-Size Effects

The size distribution can be uniquely approximated with a classical power law function, if the lower bound (x_1) and upper bound (x_2) is well defined. In practice, however, the lower bound is flattened by undersampling, detection thresholds, or instrumental sensitivity limitations, while the upper bound typically shows a gradual steepening due to finite system-size effects (Pruessner 2012). Ignoring these effects leads to power law fits with arbitrary inertial ranges $[x_1, x_2]$, which affects the accuracy of the determined power law slope and its uncertainty. Finite system-size effects are modeled here with an exponential cutoff function, which we can combine with the generalized Pareto distribution (Eq. 2.76),

$$N(x)dx = n_0\, (x_0 + x)^{-\alpha_x} \exp\left(-\frac{x}{x_e}\right) dx, \tag{2.77}$$

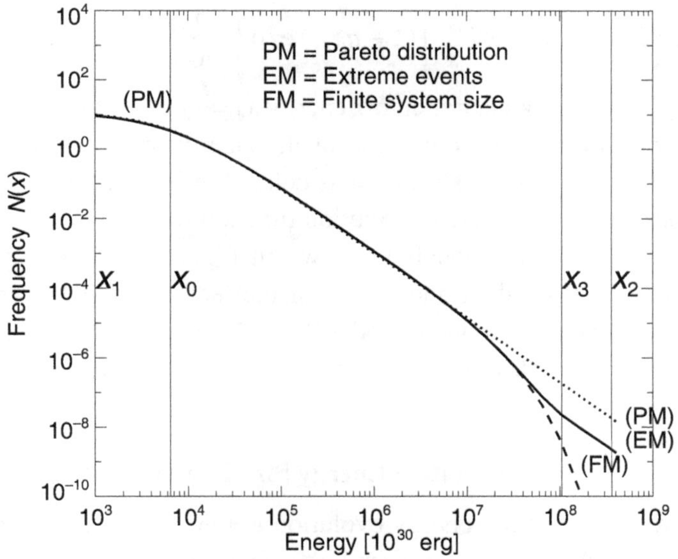

Figure 2.2 Synopsis of three power law models: (PM) = Pareto distribution $[x_1, x_2]$; (EM) = extreme events model $[x_0, x_2]$; and (FM) = finite system-size model $[x_0, x_3]$ (Aschwanden 2021).

which is quantified with an exponential function $\exp(-x/x_e)$ at a characteristic size $x \approx x_e$. The inclusion of finite system-size effects steepens the power law slope asymptotically to infinity at the upper end of the distribution function (Figure 2.2).

2.13 Extreme Events

Sometimes extreme events at the upper end of the size distribution function cannot be fitted with the exponential cutoff function, as expected from finite system-size effects.

These extreme events that deviate from an ideal power law distribution function have also been dubbed as "Black-Swan" and "Dragon-King" events (alluding to their extremely rare appearance) by Sornette (2009) and Sornette and Ouillon (2012), who suggested that they are generated by a different physical mechanism. Detections of such extreme event outliers have occasionally been noted in astrophysical data sets (Aschwanden 2015).

To accommodate these extreme events, we define a size distribution with a second power law component, where the first power law distribution includes the exponential function (Eq. 2.77) with amplitude $(1 - q_{pow})$, while the second power law distribution with amplitude (q_{pow}) extends without a cutoff all the way to the largest event (with an identical power law slope, in order to minimize the number of free parameters),

$$N(x)dx = n_0\,(x_0 + x)^{-\alpha_x} \left[(1 - q_{pow}) \exp\left(-\frac{x}{x_e} \right) + q_{pow} \right] dx, \tag{2.78}$$

where we define the exponential cutoff energy $x_e = x_2\, q_{exp}$, with q_{exp} being a free parameter. This combinatory definition of the size distribution converges to the canonical finite system-size distribution as defined in Eq. (2.78) for $q_{pow} \mapsto 0$, and to the previously defined Pareto distribution function for $q_{pow} \mapsto 1$.

A synopsis of the three models is shown in Figure 2.2, depicting the Pareto distribution model (PM), the finite system-size effect model (FM), and combined with the extreme event component model (EM). Note that the most general model (EM) has three observables (n_0, x_0, x_2) and three variables $(\alpha_x, q_{exp}, q_{pow})$.

2.14 Alternative Energy Size Distributions

An alternative choice of the energy evolution is the logistic equation (Aschwanden et al. 1998), which produces a smoother and more realistic saturation phase of the time evolution $E(t)$ near the peak time of the energy dissipation rate $E_{max} = E(t = t_{max})$, rather than an exponential growth model with a sharp peak in the energy evolution $E(t_{max})$. Comparing these two evolutionary models, one finds that the logistic equation produces ideal power law distributions for the total energy and peak energy dissipation rate (Fig. 3.8 in Aschwanden 2011a), while the exponential growth model features a flattening at the low-end of the size distribution (Fig. 3.3 in Aschwanden 2011a).

From the observational side, energy size distributions are not always ideal power laws. Some sceptics question the existence of power law size distributions at all (Stumpf and Porter 2012). A maximum-likelihood fitting method with goodness-of-fit tests based on the Kolmogorov–Smirnov statistic and likelihood ratios reveals significant deviations from ideal power laws in some cases of empirical, but mostly nonastrophysical, datasets (Clauset et al. 2009). The effect of limited sample sizes has also an influence on the accuracy of the estimated power law slope, which should be quantified by maximum likelihood estimators (D'Huys et al. 2016). Background-corrected data were found to be better characterized by a log-normal function for the peak flux and fluence, rather than by power laws, for some solar flare data sets (Verbeeck et al. 2019). New statistical estimation methods, Independent Approximates (IAs) have been defined and proven to enable closed-form estimation of the parameters of heavy-tailed distributions (Nelson 2020). Alternative models for generating heavy-tailed distributions include Pareto type II functions, percolation, stochastic branching, neuronal avalanches, and so on (Markovic and Gros 2014).

2.15 Summary

The occurrence frequency distributions (or briefly called size distributions) are the most important diagnostics for SOC systems. There are at least three formats for size distributions: (i) the differential size distribution function, (ii) the cumulative size distribution function, and (iii) the rank-order plot, where the latter two methods are preferable for size distributions with a small number of events. Each of the three formats (or methods) has at least three ranges of event sizes:

(i) a range with statistically incomplete sampling (Figure 2.2; $[x_1, x_0]$);
(ii) an inertial range or power law fitting range with statistically complete sampling (Figure 2.2; $[x_0, x_3]$); and
(iii) a range bordering finite system sizes (Figure 2.2; $[x_3, x_2]$).

Only the intermediate range with power law behavior should be used to determine the power law slope from fitting the observed size distributions. An MLE quantifies the uncertainty of the power law fits. The establishment of power law functions in a given observed size distribution depends crucially on the choice of the fitting range, which should have a logarithmic range of at least 2–3 decades. Often the fitted distribution functions exhibit significant deviations from an ideal power law and can be fitted better with alternative functions, such as log-normal distributions, Pareto type-II distributions, and Weibull distributions.

Since we pursue theoretical models of SOC systems in this book, in particular the FD-SOC model, which can be described by an ideal power law distribution function, we conduct tests by comparing the power law slopes of observed astrophysical distributions with the theoretical predictions of a number of SOC parameters, such as the length scale L, the area A, the volume V, the flux F, and the fluence or energy E. The area fractal dimension $D_A = d - 1/2 = 1.50$ and the volume fractal dimension $D_V = d - 1/2 = 2.5$ depend on the dimensionality d. We employ a generalized FD-SOC model that has three free parameters (the Euclidean dimensionality d, the diffusive transport coefficient β, and the flux–volume scaling coefficient γ). Furthermore, we define a standard FD-SOC model that has no free parameters by setting the default parameters ($d = 3$, $\beta = 1$, and $\gamma = 1$), which predicts power law slopes of $\alpha_V = \alpha_F = 1.80$ for the flux size distribution and $\alpha_T = 2.00$ for the time duration size distribution.

3
Waiting Time Distributions

Waiting times are the time differences $\tau_i = t_{i+1} - t_i$ of a chronologically sorted time series $t_i, i = 1, \ldots, n$, which contain information on their statistical distribution function. An analytical treatment of such distribution functions will tell us also whether they are consistent with theoretical power law distribution functions or not, the hallmarks of Self-Organized Criticality (SOC) systems. In Section 3.1, we calculate exact solutions of the waiting time distributions for nonlinear (and nonstationary) flaring rates (Aschwanden et al. 2021; Nurhan et al. 2021; Aschwanden and Johnson 2021; Aschwanden and Nhalil 2022, 2023; Aschwanden 2022a, 2022b).

3.1 Stationary Waiting Time Distributions

Statistical probability distributions $N(\tau)d\tau$ of event waiting times, τ, have distinctly different mathematical functions, depending on whether their generation is produced by a linear or by a nonlinear process. If statistical events are produced by a linear process, the resulting size distribution $N(x)\ dx$ or waiting time distribution $N(\tau)\ d\tau$ follows Poissonian statistics, which can be fitted by an exponential function (on the right-hand side of their distribution),

$$N(\tau)\ d\tau = \lambda_0\ \exp\left[-\lambda_0 \tau\right]\ d\tau, \tag{3.1}$$

where $N(\tau)d\tau$ is the number of events per bin $d\tau$, $\lambda_0 = 1/\tau_0$ is the mean event rate or mean reciprocal waiting time, that is, $\tau_0 = T/N_0$, with T being the total duration of an observed data set and N_0 is the total number of events detected during this time interval. The total number of events is N_0,

$$\int_0^\infty N(\tau)d\tau = N_0. \tag{3.2}$$

By contrast, if a statistical sample is produced by a nonlinear process, rather than by a linear process, the "fat-tail" (or "heavy-tail") of its distribution function becomes important and the resulting size distribution or waiting time distribution $N(\tau)$ fits a power law distribution function,

$$N(\tau)\, d\tau \propto \tau^{-\alpha}\, d\tau, \tag{3.3}$$

with power law slope α_τ. This formalism is valid for a large variety of nonlinear events, including solar and stellar flares, terrestrial catastrophes, also known as avalanche events in the parlance of SOC systems.

How do we distinguish between stationary and nonstationary statistics? Stationary Poisson processes have a constant event rate function λ_0 within the statistical uncertainties in any time interval of a considered data set. Nonstationary data samples have a variable mean event rate $\lambda(t)$ as a function of time t (also called the flare rate function). In the case of solar flares, for instance, the flare event rate varies by factors of $\approx 10^2 - 10^3$ between the solar cycle minimum and maximum (with $T \approx 11$ years). In Section 3.2, we will use the terms *event rate function* and *flare rate function* $\lambda(t)$ interchangeably.

3.2 Nonstationary Waiting Time Distributions

We define nonstationary waiting time distribution functions with a time-dependent flare rate function $\lambda(t)$, which is a generalization from the constant λ_0 (Eq. 3.1) to a time-dependent mean flare rate $\lambda(t)$,

$$N(\tau)\, d\tau = \lambda(t)\ \exp\left[-\lambda(t)\ \tau\right] d\tau. \tag{3.4}$$

Subdividing a waiting time distribution into a piece-wise constant Poisson process with rates λ_i and time intervals τ_i, it can be approximated by Wheatland 2000b, 2001,

$$N(\tau_i) \approx \sum_i \varphi_i \lambda_i \exp\left[-\lambda_i \tau_i\right], \tag{3.5}$$

where

$$\varphi_i = \frac{\lambda_i \tau_i}{\sum_i \lambda_i \tau_i} \tag{3.6}$$

is the fraction of events associated with a given rate λ_i. Inserting the fractions φ_i (Eq. 3.6) into the waiting time probability distribution function (Eq. 3.5) we obtain,

$$N(\tau_i) \approx \frac{\sum_i \lambda_i^2\ \tau_i\ \exp\left[-\lambda_i \tau_i\right]}{\sum_i\ \lambda_i\ \tau_i}, \tag{3.7}$$

which in the asymptotic limit of arbitrary small Bayesian time intervals approaches the integral $N(\tau)$,

$$N(\tau) = \frac{\int_0^T \lambda(t)^2 \tau \exp[-\lambda(t)\tau]\,dt}{\int_0^T \lambda(t)\,\tau\,dt} = \frac{\int_0^T \lambda(t)^2 \exp[-\lambda(t)\tau]\,dt}{\int_0^T \lambda(t)\,dt}. \tag{3.8}$$

The right-hand side equation simplifies somewhat with the cancellation of the variable τ in both the numerator and denominator.

Note that our treatment of nonstationary Poisson processes follows the previous work of Wheatland (2000b, 2001), which is related to the concepts of "mixed Poisson," "compound Poisson," "contagion Poisson," "Poisson auto-regression," and "Poisson point" processes.

3.3 Flare Rate Functions

In nonstationary waiting time distributions, the temporal variability of the mean event rate $\lambda(t)$ needs to be defined, in order to calculate the waiting time distribution $N(\tau)$ (Eq. 3.8). In order to minimize the number of free parameters, following the principle of Occam's razor, we define here nonstationary event rate functions $\lambda(t)$ that have three parameters only (T, λ_0, p). These flare rate functions have: (i) a total duration T and a mean event rate λ_0, (ii) are symmetric in the time interval before and after the peak time $t = T/2$, and (iii) vary the shape of the (flare) event rate function (as a function of time) by a single exponent p (also called degree or order of polynomial), to which we refer as nonlinearity parameter also. We define three different flare rate function models, including the polynomial (Figure 3.1), sinusoidal, and Gaussian function,

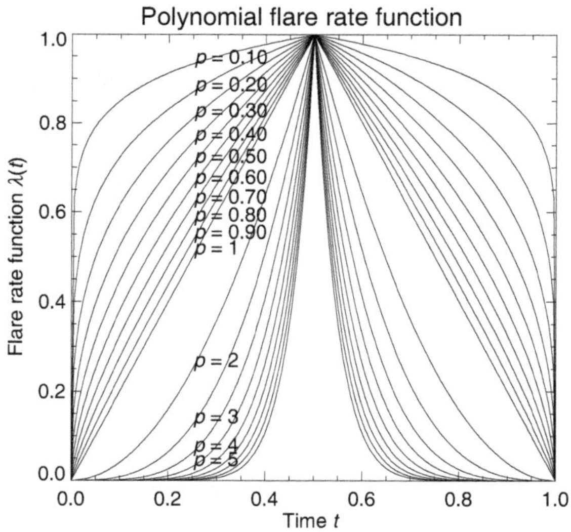

Figure 3.1 Parameter range of polynomial flare rate functions $\lambda(t)$ for exponents $p = 0.1, \ldots, 10$, including the linear case $p = 1$ for $\lambda(t) \propto t$, or the nonlinear (quadratic) case $p = 2$ for $\lambda(t) \propto t^2$ (Aschwanden et al. 2021).

$$\lambda(t) = \lambda_0 \begin{cases} \left(\frac{T/2-|t-T/2|}{T/2}\right)^p & \text{Polynomial model} \\ \left(\sin\left[\pi\frac{t}{T}\right]\right)^p & \text{Sinusoidal model} \\ \left(\exp\left[-\frac{(T/2-t)^2}{2\sigma^2}\right]\right)^p & \text{Gaussian model} \end{cases} . \quad (3.9)$$

These time-dependent functions $\lambda(t)$ are suitable to represent the time variation of a cycle, such as the solar cycle of $T \approx 11$ years. For the parameter $p = 1$, the polynomial model mimics a simple triangular function with linear rise time and decay time, for $p = 2$ a quadratic function, for $p = 1/2$ a square root function, and for large exponents $p \gg 1$ it asymptotically approaches the Dirac delta function. The parameter space of t and $\lambda(t)$ is shown in Figure 3.1 for the polynomial flare rate function, which illustrates that functions $\lambda(t)$ with degree $p = 0.1, \ldots, 10$ cover most of the $\lambda(t)$ parameter space.

Similar time profiles are formulated for the sinusoidal and Gaussian function (Eq. 3.9). The Gaussian width σ is defined from the full width at half maximum (FWHM) being equal to the half duration, that is, FWHM = T/2, and $\sigma = (T/4)/\sqrt{2\ln 2}$. The selection of three different functional forms (polynomial, sinusoidal, and Gaussian) allows us to explore systematic errors in the power law slope of waiting time distributions, which are generally larger than the formal errors obtained from least-square fitting. Another motivation for the choice of our selection of flare rate functions $\lambda(t)$ is the availability of analytical exact solutions, for the polynomial model (see Section 3.5), for the sinusoidal flare rate model (Nurhan et al. 2021), for the Gaussian flare rate model (Section 3.7), as well as for the Pareto type-II distribution function (Section 3.6).

3.4 Exact Solution of Waiting Time Distribution

We derive here an analytical exact solution of the waiting time distribution $N(\tau)d\tau$ (Eq. 3.8),

$$N(\tau) = \frac{2\int_0^{T/2} \lambda(t)^2 \exp\left[-\lambda(t)\tau\right] dt}{2\int_0^{T/2} \lambda(t)\, dt}, \quad (3.10)$$

for the general case of a flare rate function $\lambda(t)$, chosen symmetrically about the peak time $t = T/2$ in the time range $0 \le t \le T$, in terms of a polynomial with degree p (Eq. 3.9),

$$\lambda(t) = \lambda_0 \left(\frac{t}{T/2}\right)^p . \quad (3.11)$$

Inserting Eq. 3.11 into Eq. 3.10 leads to,

$$N(\tau,p) = \frac{\int_0^{T/2} \lambda_0^2 \, [t/(T/2)]^{2p} \, \exp\,[-\lambda_0 \, [t/(T/2)]^p \, \tau] \, dt}{\int_0^{T/2} \lambda_0 \, [t/(T/2)]^p \, dt}. \tag{3.12}$$

Substituting the variable $x = t/(T/2)$, with $dx/dt = 1/(T/2)$ and $dt = dx(T/2)$, renders Eq. 3.12 as,

$$N(\tau,p) = \frac{\int_0^1 \lambda_0^2 \, x^{2p} \, \exp\,[-\lambda_0 \, x^p \tau] \, dx}{\int_0^1 \lambda_0 \, x^p \, dx}, \tag{3.13}$$

where the integral limits change from $t = [0, T/2]$ to $x = [0, 1]$. The integral in the denominator of Eq. 3.13 is simply,

$$\int_0^1 \lambda_0 \, x^p \, dx = \frac{\lambda_0}{(p+1)}, \tag{3.14}$$

while the integral $I(\tau,p)$ in the numerator of Eq. (3.10) is,

$$I(\tau,p) = \int_0^1 \lambda_0^2 \, x^{2p} \, \exp\,[-\lambda_0 \, x^p \tau] \, dx. \tag{3.15}$$

We change the variable $y = x^p$, or $x = y^{1/p}$, with the differential $dx = (1/p) y^{(1-p)/p}$ dy, which yields,

$$I(\tau,p) = \int_0^1 \frac{\lambda_0^2}{p} y^{(1+p)/p} \, \exp\,[-\lambda_0 \, \tau \, y] \, dy. \tag{3.16}$$

In the limit of $\tau \mapsto \infty$, the exponential factor drops very fast, so that the upper integral limit is of order $O(e^{-\beta})$, where β is defined by

$$\beta = \lambda_0 \tau, \tag{3.17}$$

yielding,

$$I(\tau,p) = \int_0^1 \frac{\lambda_0^2}{p} y^{(1+p)/p} \, \exp\,[-\beta \, y] \, dy. \tag{3.18}$$

Substituting β with $q = \beta \, y$ and $dq = \beta \, dy$,

$$I(\tau,p) \approx \frac{\lambda_0^2}{p \, \beta^{2+1/p}} \int_0^\infty q^{(1+p)/p} \, \exp\,[-q] \, dy, \tag{3.19}$$

leads to the gamma function Γ (Bronstein and Semendjajew 1960),

$$\Gamma(\alpha) = \int_0^\infty q^{(\alpha-1)} \exp[-q]\, dq, \tag{3.20}$$

and the integral over the range $q = [0, \infty]$ is, in the asymptotic expansion of the exact solution,

$$I(\tau, \alpha) \approx \frac{\lambda_0^2}{p\, \beta^\alpha}\, \Gamma(\alpha), \tag{3.21}$$

with the power law slope α,

$$\alpha = 2 + \frac{1}{p}. \tag{3.22}$$

This result has also been derived in an extension of the waiting time distribution model applied to stellar X-ray bursts (Wand et al. 2017). Combining this expression with the denominator integral $\lambda_0/(p+1)$ (Eq. 3.14), we obtain an approximation of the waiting time distribution function,

$$N(\tau, \alpha) \approx \lambda_0\, (\alpha - 1)\; \Gamma(\alpha)\; (\lambda_0\, \tau)^{-\alpha}. \tag{3.23}$$

The exact solution of this integral is the incomplete gamma function (Bronstein and Semendjajew 1960),

$$\gamma(\alpha, \beta) = \int_0^\beta q^{(\alpha-1)} \exp[-q]\, dq, \tag{3.24}$$

where the integral boundaries are $q = [0, \beta]$, in contrast to $q = [0, \infty]$ in the gamma function. The exact solution of the waiting time distribution as a function of the waiting time τ and nonlinearity parameter p of the flare rate function $\lambda(t) \propto t$ reads then in the final form as,

$$N(\tau, \alpha) = \lambda_0\, (\alpha - 1)\, \gamma(\alpha, \lambda_0 \tau)\; (\lambda_0 \tau)^{-\alpha}. \tag{3.25}$$

The power law slope $\alpha = (2 + 1/p)$ can approximately be expressed in terms of the gamma function Γ, but the exact solution has an additional dependence of the waiting time τ in the incomplete gamma function γ (Eq. 3.25).

Alternative solutions of the waiting time distribution $N(\tau)$ can be obtained also for Pareto type-II distribution functions, which is presented in Section 3.6. All solutions, the Gamma function Γ solution (Eq. 3.21), the incomplete Gamma function γ solution (Eq. 3.25), and the Pareto type-II distribution exhibit the same characteristics: The steepest power law slope is in the asymptotic limit of the largest waiting times, and a gradual rollover and flattening to a constant value occurs at the shortest waiting times, as illustrated in Figure 3.2.

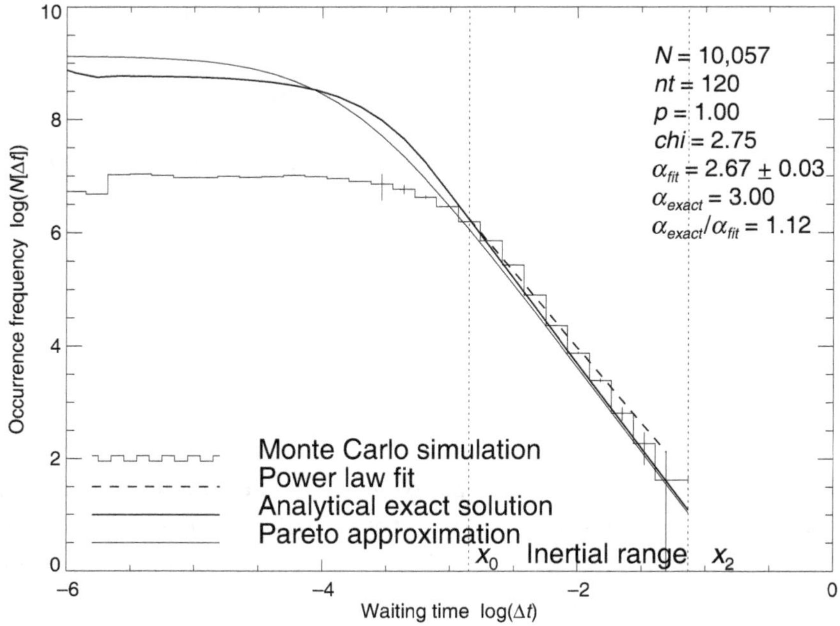

Figure 3.2 Comparison of three different calculation methods of the waiting time distribution function: Monte Carlo simulation (histogram) with power law fit (dashed curve), analytical exact solution (thick solid curve), and Pareto approximation (thin solid curve). The fitting range is defined by the inertial range $[x_0, x_2]$. The ratio of the slopes $\alpha_{exact}/\alpha_{fit} = 1.12$ is attributed to the Poissonian weighting bias (Aschwanden et al. 2021).

3.5 Exact Solution of Waiting Time Distribution for Linear Case ($p = 1$)

The waiting time distribution $N(\tau)\, dt$ of a Poissonian random process can be represented by an exponential distribution (Eq. 3.1) for the case of a stationary flare rate function $\lambda(t) = \lambda_0$. For a linearly increasing flare rate function $\lambda(t) = t/(T/2)$ (with $p = 1$ in Eq. 3.9), the probability function is (Wheatland et al. 1998; Wheatland and Litvinenko 2002; Aschwanden and McTiernan 2010),

$$N(\tau) = \frac{\int_0^T \lambda(t)^2 \exp{[-\lambda(t)\tau]} dt}{\int_0^T \lambda(t)\, dt}. \tag{3.26}$$

Inserting the flare rate function $\lambda(t) = \lambda_0 (t/(T/2))^1$, we obtain the integral,

$$N(\tau) = \frac{\int_0^{T/2} \lambda_0^2 \left(\frac{t}{T/2}\right)^2 \exp{[-\lambda_0\, \tau\, (t/(T/2))]} dt}{\int_0^{T/2} \lambda_0 \left(\frac{t}{T/2}\right) dt}. \tag{3.27}$$

Substituting the variables $x = t/(T/2)$, $dt = (T/2)dx$, and $a = -\lambda_0\tau$, we obtain the integral,

$$N(\tau) = \frac{\int_0^1 \lambda_0^2\, x^2 \exp^{ax}\, dx}{\int_0^1 x\, dx}, \tag{3.28}$$

which has the analytical solution (Bronstein and Semendjajew 1960),

$$N(\tau) = \lambda_0 \; \exp^{ax} \left(\frac{x^2}{a} - \frac{2x}{a^2} + \frac{2}{a^3} \right)_{x=0}^{x=1} \tag{3.29}$$

$$N(\tau) = 2\lambda_0 \exp^{a} \left[\left(\frac{1}{a} - \frac{2}{a^2} + \frac{2}{a^3} \right) - \frac{2}{a^3} \right], \tag{3.30}$$

which can be expressed with $a = -\lambda_0\tau$ to show the explicit function of the waiting time τ,

$$N(\tau) = 2\lambda_0 \left[\frac{2}{\lambda_0^3\tau^3} - \exp^{-\lambda_0\tau} \left(\frac{1}{\lambda_0\tau} + \frac{2}{\lambda_0^2\tau^2} + \frac{2}{\lambda_0^3\tau^3} \right) \right]. \tag{3.31}$$

3.6 Pareto Type-II Approximation

An alternative generalization of a simple straight power law distribution function $N(\tau) \propto \tau^{-\alpha}$ is the Pareto type-II distribution function, $N(\tau) \propto (1+\tau)^{-\alpha}$, which exhibits a power law-like function at large waiting times and becomes constant at small waiting times. Such a Pareto type-II function has been used previously to fit waiting time distributions (Aschwanden and McTiernan 2010; Aschwanden and Freeland 2012; Aschwanden and Dudok de Wit 2021). Here we present an analytical approximation of the waiting time distribution function.

We start with the expression of the waiting time distribution in terms of the gamma function (Eq. 3.23). In the asymptotic limit, we have

$$N(\tau) \sim \lambda_0 \; (\alpha - 1)\Gamma(\alpha) \; (\lambda_0\tau)^{-\alpha}, \tag{3.32}$$

where the slope α of the power law is (Eq. 3.22),

$$\alpha = 2 + \frac{1}{p}. \tag{3.33}$$

This is also asymptotic to a Pareto type-II distribution of the form,

$$N_P(\tau) = \frac{N_0}{\left(1 + \frac{c\tau}{\alpha}\right)^{\alpha}}. \tag{3.34}$$

Equating the two expression (Eq. 3.32) and (Eq. 3.33) yields,

$$N_0 \left(\frac{\alpha^\alpha}{c^\alpha}\right) = \lambda_0 \frac{(\alpha - 1)\Gamma(\alpha)}{(\lambda_0)^\alpha}. \tag{3.35}$$

Also, for a uniform approximation, we require that

$$N_P(0) = N_0(1 - c\tau + \ldots) \tag{3.36}$$

so that the constant N_0 is,

$$N_0 = \lambda_0 \left(\frac{p+1}{2p+1}\right) = \lambda_0 \left(\frac{\alpha - 1}{\alpha}\right), \tag{3.37}$$

which then yields the equivalence,

$$\lambda_0 \left(\frac{\alpha - 1}{\alpha}\right)\left(\frac{\alpha^\alpha}{c^\alpha}\right) = \lambda_0 \frac{(\alpha - 1)\Gamma(\alpha)}{(\lambda_0)^\alpha}, \tag{3.38}$$

and the constant c,

$$c = \frac{\alpha\lambda_0}{(\alpha\Gamma(\alpha))^{1/\alpha}}. \tag{3.39}$$

This theoretical solution has been checked numerically and compared with the exact solution based on the incomplete gamma function solution (Eq. 3.25), the asymptotic approximation, the Pareto distribution obtained from the power series, and the Pareto distribution obtained from the Padé approximation. The power series solution is inaccurate when $\lambda_0\tau > 1$. The Padé approximation is reasonably accurate for all $\lambda_0\tau$ and matches the power law slope accurately.

3.7 Solution for Gaussian Event Rate Function

After we calculated exact solutions for the polynomial flare rate function $\lambda(t)$ (Section 3.4), while exact solutions for the sinusoidal flare rate are given in Nurhan et al. (2021), we derive here an approximate solution for the Gaussian flare rate also. The definition of the waiting time integral is (Eq. 3.10),

$$N(\tau) = \frac{\int_0^{T/2} \lambda^2(t)\exp(-\lambda(t)\tau)dt}{\int_0^{T/2} \lambda(t)dt}, \tag{3.40}$$

where the Gaussian rate function is (Eq. 3.9),

$$\lambda(t) = \lambda_0 \exp(-(t - T/2)^2/2\sigma^2). \tag{3.41}$$

We change the variables

$$y = (T/2 - t)/\sqrt{2}\sigma, \tag{3.42}$$

$$dy = -dt/\sqrt{2}\sigma, \tag{3.43}$$

$$\lambda = \lambda_0 e^{-y^2}, \tag{3.44}$$

$$\beta = \frac{T}{2^{3/2}\sigma}, \tag{3.45}$$

and obtain the waiting time integral in terms of the error function erf(β),

$$N(\tau) = \lambda_0 \frac{\frac{2}{\sqrt{\pi}} \int_0^\beta e^{-2y^2} \exp(-\lambda_0 e^{-y^2}\tau) dy}{\frac{2}{\sqrt{\pi}} \int_0^\beta e^{-y^2} dy} = \lambda_0 \frac{I(\tau, \beta)}{\mathrm{erf}(\beta)}, \tag{3.46}$$

where

$$I(\tau, \beta) = \frac{2}{\sqrt{\pi}} \int_0^\beta e^{-y^2} \exp(-\lambda_0 e^{-y^2}\tau) dy, \tag{3.47}$$

$$\mathrm{erf}(\beta) = \frac{2}{\sqrt{\pi}} \int_0^\beta \exp(-x^2) dx. \tag{3.48}$$

Now we have,

$$I(\tau, \beta) = \frac{1}{\lambda_0^2} \frac{d^2}{d\tau^2} \frac{2}{\sqrt{\pi}} \int_0^\beta \exp(-\lambda_0 e^{-y^2}\tau) dy. \tag{3.49}$$

We expand the exponentially around $y = \beta$, as this will be the smallest value in the exponential and will dominate asymptotically.

$$e^{-y^2} = e^{-\beta^2}(1 - 2\beta(y - \beta) + \ldots). \tag{3.50}$$

Consequently we obtain,

$$\int_0^\beta \exp(-\lambda_0 e^{-y^2}\tau) dy \sim \int_{\beta-\epsilon}^\beta \exp(-\lambda_0 \tau e^{-\beta^2}(1 + 2\beta(\beta - y)) dy \tag{3.51}$$

$$= \exp(-\lambda_0 \tau e^{-\beta^2}) \int_0^\epsilon \exp(-2\lambda_0 \tau \beta e^{-\beta^2} q) dq, \tag{3.52}$$

which introduces only an exponentially small error, so that we can now take $\epsilon \to \infty$,

$$\int_0^\beta \exp(-\lambda_0 e^{-y^2}\tau) dy \sim \frac{\exp(-\lambda_0 \tau e^{-\beta^2})}{2\lambda_0 \tau \beta e^{-\beta^2}}. \tag{3.53}$$

Taking the leading order terms,

$$-\frac{d}{d\tau} \frac{\exp(-\lambda_0 \tau e^{-\beta^2})}{2\lambda_0 \tau \beta e^{-\beta^2}} = \frac{\exp(-\lambda_0 \tau e^{-\beta^2})}{2\beta\tau} \left(1 + \frac{1}{\lambda_0 \tau e^{-\beta^2}}\right), \tag{3.54}$$

$$\frac{d^2}{d\tau^2}\frac{\exp(-\lambda_0 \tau e^{-\beta^2})}{2\lambda_0 \tau \beta e^{-\beta^2}} \tag{3.55}$$

$$= \frac{\lambda_0 e^{-\beta^2} \exp(-\lambda_0 \tau e^{-\beta^2})}{2\beta\tau}\left(1 + \frac{2}{\lambda_0 \tau e^{-\beta^2}} + O\left(\frac{1}{\lambda_0 \tau}\right)^2\right), \tag{3.56}$$

and then obtain,

$$N(\tau) \sim \lambda_0 \frac{e^{-\beta^2} \exp(-\lambda_0 \tau e^{-\beta^2})}{\sqrt{\pi}\beta\lambda_0 \tau \operatorname{erf}(\beta)}\left(1 + \frac{2}{\lambda_0 \tau e^{-\beta^2}}\right). \tag{3.57}$$

In the δ-function limit, we have $\sigma \to 0$ and $\beta \to \infty$, so that the second term in the expansion will dominate and $\alpha = 2$, consistent with the δ-function limit we have seen as $p \to \infty$ in the polynomial case. Comparing with the numerical solution, it is apparent that a power-like form is found for $\lambda_0 \tau \approx 5 - 10$ and does not apply at larger values of τ. Note that power laws are generally substantially different as σ varies.

3.8 Do Power Laws Exist?

In waiting time studies, numerical tools are used that produce true random distributions of flare events (which should follow exponential functions with Poisson statistics), as well as nonstationary Poissonian distributions. The latter group produces more or less power law-like functions, but it has never been demonstrated whether we should expect an exact power law function. Since numerous deviations from exactly straight power law functions (in a log(N)-log(S) histogram) have been noted before (Wheatland et al. 1998; Aschwanden and McTiernan 2010), some criticism has been raised whether power laws exist at all (Stumpf and Porter 2012). Here we have an answer that is summarized in Figure 3.2. We calculated the analytical exact solution of a waiting time distribution for the linear case (with time evolution $\lambda(t) \propto t^p = t^1$, or $p = 1$), which is derived in Section 3.5, Eq. 3.31. The solution contains terms of τ, τ^2, τ^3, and $e^{-\tau}$, and therefore is of the type of an ideal power law function $\tau^{-\alpha}$ in the asymptotic limit only ($\lambda_0 \tau \mapsto \infty$), where all other terms are exponentially small. A Pareto type-II function $(1 + \tau)^{-\alpha}$ is a good approximation (Aschwanden and Dudok de Wit 2021; Aschwanden and McTiernan 2010; Aschwanden and Freeland 2012) that is accurate to first-order terms, but shows a gradual change in the power law slope with respect to the exact analytical solution. Moreover, in our numerical simulation of waiting time distributions we obtain a power law-like function too (Figure 3.2), but the Poissonian weighting of a power law fit is biased toward the lower end (where the error bars are smallest and thus have the largest weight there). Thus, if a power law fit is obtained over the full inertial range, we predict that the average slope is slightly flatter than the asymptotic limit at the upper end, which is evident in the example shown in Figure 3.2 (dashed line). Other examples are shown in Figure 3.2, where deviations from ideal power laws show up clearly, since the nonstationary Poissonian model (with large time variability of the mean flare rate) predicts that a power law-like convolution consists of a superposition of multiple (but finite number) of exponential distributions. In summary, deviations from ideal power law functions are not significant for relatively small samples (with $N_0 \approx 10^1 - 10^3$) but become significant for larger samples (with $N_0 \gtrsim 10^4$), as demonstrated in other works (Aschwanden 2015, 2019a; Aschwanden and Dudok de Wit 2021), which are theoretically expected and thus should be modeled correspondingly.

Let us investigate physical models that possibly could explain the observed power law slopes α of waiting time distributions. Naively, to first order, one would expect that the distribution $N(\tau)$ is reciprocal to the mean waiting time, that is, $N(\tau) \propto \tau^{-1}$, simply because the number of fragments is reciprocal to the length of a fragment in a one-dimensional "fragmentation" process. In other words, the product of the number N multiplied with the duration of a time scale T is invariant, that is, $N_T \times T = \text{const}$. However, this simplest model that predicts a scaling of τ^{-1} is not consistent with observations, which show distributions of $\approx \tau^{-2}, \ldots, \tau^{-3}$. It is likely that the 1-D dimensionality in the time domain represents an oversimplified model.

However, if we consider the generation of waiting times in the spatiotemporal domain, a spatial scale with length L and time scale T are required in a minimal model, which are coupled by the length scale distribution $N(L)dL$ and the transport equation $L(T)$. We can find such a model in the self-organized criticality concept. Both relationships can be expressed with power laws, the scale-free size distribution of avalanches, $N(L)\, dL \propto L^{-d} dL$ (Eq. 2.36), and the diffusive transport mechanism, $L \propto T^{\beta/2}$ (Eq. 2.42), with $d = 1, 2, 3$ the Euclidean dimension, and $\beta = 1$ the classical transport coefficient. Taken these two relationships together, our SOC model predicts a power law distribution of flare durations T, that is, $N(T) \propto T^{-\alpha_T}$, where the power law slope is defined as $\alpha_T = 1 + (d-1)\beta/2$ (Table 2.1). Thus, the SOC model predicts a value of $\alpha_T = 2$ for the 3-D space geometry, while the values of $\alpha_T = 1$ for 1-D and $\alpha_T = 1.5$ for 2-D geometry are below the observed values, and thus can be ruled out. Interestingly, the predicted value of $\alpha_T = 2.0$ for the 3-D world agrees with the independently calculated approximation of the Pareto type-II model (Eq. 3.33) in the asymptotic limit of $\tau \mapsto \infty$ (Eq. 3.34), which corroborates the SOC theory and the Pareto distribution function to some extent.

3.9 Interval–Size Relationship

This simple analytical model in terms of an exponential growth phase with saturation after a random time interval goes back to Willis and Yule (1922) who applied it to geographical distributions of plants and animals. Yule's model was applied to cosmic rays (Fermi 1949), to cosmic transients and solar flares (Rosner and Vaiana 1978; Aschwanden et al. 1998), to the growth dynamics of the world wide web (Huberman and Adamic 1999), as well as to the distribution of the sizes of incomes, cities, internet files, biological taxa, and in gene family and protein family frequencies (Reed and Hughes 2002). For the application to solar flares, Rosner and Vaiana (1978) interpreted the time interval between two subsequent events in terms of an energy storage time. However, it was argued that there is no observational evidence for a correlation between the energy storage time and the magnitude of a solar flare peak (Lu 1995; 1995a; 1995b; Crosby 1996; Wheatland 2000b; Georgoulis et al. 2001). By contrast, the exponential rise time τ in our model here corresponds to the energy release time during an instability (Aschwanden et al. 1998).

3.10 Waiting Times and Turbulence

Boffetta et al. (1999) measured the statistics of quiescent (waiting) times between subsequent bursts over 20 years and found a power law distribution with an exponent of $\alpha_T \approx 2.4$, which they interpret as a failure of the SOC model, since it does not exhibit an exponential size distribution that is expected for a stationary waiting time distribution. Consequently, a number of authors (Boffetta et al. 1999; Giuliani et al. 2000; Carbone et al. 2002) interpret power law distributions of solar flares in terms of turbulence. However, this interpretation is not conclusive, since nonstationary waiting time distributions produce power law distribution functions. Our predicted values of power law slopes is $\alpha_T \approx (2 + 1/p)$ (Eq. 3.22) matches the observed slopes of $\alpha_T \approx 2.3 - 2.5$ for nonlinear power law slopes in the range of $p = 2 - 3$.

The main diagnostic capability of waiting time distribution functions is the discrimination between random Poisson processes and processes with memory, which can be inferred even from stellar flares (Li et al. 2018; Lei et al. 2020).

An alternative time-dependent model of the energy of a stochastic flaring solar active region has been analytically calculated for the *jump-transition model* (Kanazir and Wheatland 2010).

3.11 Summary

The size distribution of waiting times τ_i, which one obtains from chronological sorting of time intervals ($\tau_i = [t_{i+1} - t_i]$), is found to have an exponential distribution in the case of a stationary Poissonian process. In reality, however, the waiting time distributions reveal power law-like distribution functions, which can be modeled in terms of nonstationary Poisson processes by a superposition of Poissonian distribution functions with time-varying event rates. We model the time evolution of such waiting time distributions by polynomial, sinusoidal, and Gaussian functions, which have exact analytical solutions in terms of the incomplete Gamma function, as well as in terms of the Pareto type-II approximation, which has a power law slope of $\alpha_\tau = 2 + 1/p$, where $p = 1$ represents the linear time evolution, or with $p \gtrsim 2$ representing nonlinear growth rates, which have a power law slope of $\alpha_\tau = 2 + 1/p \approx 2.5$. Our mathematical modeling confirms the existence of significant deviations from ideal power law size distributions (of waiting times), but no correlation or significant interval–size relationship exists, as would be expected for a simple (linear) energy storage-dissipation model.

Part II

Astrophysical SOC Phenomena

4

Solar Flare Hard X-Rays

4.1 Solar Flare Hard X-Ray Frequency Distributions

The application of self-organized criticality (SOC) theory to solar phenomena outnumbers all other astrophysical applications. Therefore we break the subject down into observational statistics from different wavelengths in the following sections (hard X-rays in Chapter 4, soft X-rays in Chapter 5, EUV in Chapter 6, radio in Chapter 8, etc.), and test the theoretically predicted size distributions, waiting time distributions, and fractality for each wavelength group separately. This subdivision into different wavelength groups is even more justified by the fact that we know that different physical emission mechanisms are at work at different wavelengths.

Solar flares provide the energy source for acceleration of nonthermal particles, which emit bremsstrahlung in hard X-ray wavelengths, when the nonthermal particles interact with a high-density plasma via Coulomb collisions. Most solar flares display an impulsive component in hard X-rays, produced by accelerated coronal electrons that precipitate toward the chromosphere, and produce intense hard X-ray emission at the footpoints of flare loops via the electron (thick-target) bremsstrahlung process (Brown 1971). Therefore, hard X-ray pulses are a reliable signature of solar flares, often detected at energies $\gtrsim 20$ keV, and for smaller flares all the way down to $\gtrsim 5$ keV (Lin et al. 1984).

Solar flare event catalogs, containing the peak photon count rate (P), fluence or energy (E), and flare duration (T), have been compiled from a number of spacecraft or balloon-borne hard X-ray detectors over the last three decades, such as from the Orbiting Solar Observatory (OSO-7) (Datlowe et al. 1974), a University of Berkeley balloon flight (Lin et al. 1984), the Hard X-Ray Burst Spectrometer (HXRBS) onboard the Solar Maximum Mission (SMM) (Dennis 1985; Schwartz et al. 1992; Crosby et al. 1993), the Burst and Transient Source Experiment (BATSE) onboard the Compton Gamma Ray Observatory (CGRO) (Schwartz et al. 1992; Biesecker et al. 1993, 1994; Biesecker 1994), the Wide Angle Telescope for Cosmic Hard

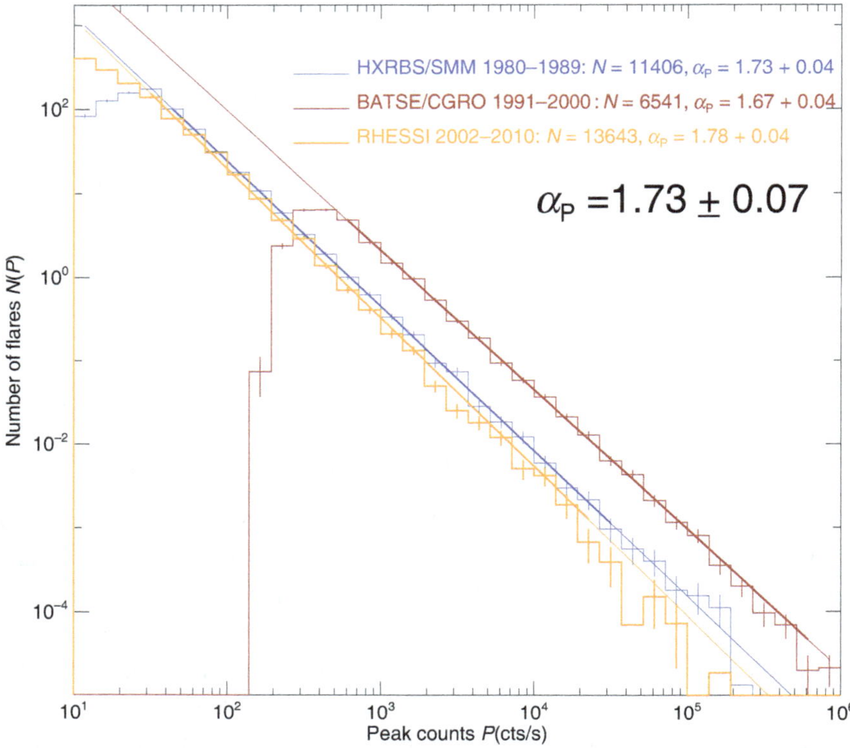

Figure 4.1 Occurrence frequency distributions of hard X-ray peak count rates P [cts s^{-1}] observed with HXRBS/SMM (1980–1989), $P \lesssim 2 \times 10^5$, BATSE/CGRO (1991–2000), $P \lesssim 10^6$, and RHESSI (2002–2010), $P \lesssim 10^5$, with power law fits. An average preflare background of 40 [cts s^{-1}] was subtracted from the HXRBS count rates. Note that BATSE/CGRO has larger detector areas and thus records higher count rates (Aschwanden 2011a).

X-rays (WATCH) onboard the GRANAT spacecraft (Crosby 1996, 1998; Georgoulis et al. 2001); the International Sun/Earth Explorer (ISEE-3) (Lu et al. 1993; Lee et al. 1993; Bromund et al. 1995); the Payload for High Energy Burst Spectroscopy (PHEBUS) onboard GRANAT (Perez-Enriquez and Miroshnichenko 1999), the Ramaty High Energy Solar Spectroscopic Imager (RHESSI) spacecraft (Su et al. 2006; Christe et al. 2008; Lin et al. 2001), and the ULYSSES spacecraft (Tranquille et al. 2009). Three examples of hard X-ray peak flux distributions are shown in Figure 4.1. Note that the size distribution of peak counts exhibit a sharp cutoff at the low end of the histogram, due to a fixed count rate threshold that is generally used in the compilation of hard X-ray flare catalogs, and thus the power law slope can be determined with high accuracy. From the histograms shown in Figure 4.1, we can estimate peak flux thresholds of $P \gtrsim 40$ cts s^{-1} for HXRBS/SMM and RHESSI, and $P \gtrsim 500$ cts s^{-1} for BATSE/CGRO, which represent instrument-dependent peak flux ranges where statistical sampling is complete and power law fitting can be applied. Other parameters have generally a gradual rollover at the left side of the histogram due to incomplete sampling and finite-resolution effects, which causes truncation effects in the histogram and hampers the accuracy of the power law fit. The size distribution of solar flare hard X-ray counts, which has already been pointed out before the SOC concept came along (Dennis 1985), is still one of the "cleanest" power law size distributions measured in astrophysics (Figure 4.1).

A compilation of occurrence frequency distribution power law slopes of solar hard X-ray flare peak fluxes (α_P), fluences or energies (α_E), and flare durations (α_T) is listed in Table 4.1. In this table, we combine both the power law slopes α_E from the fluences (which is the time-integrated or total number of hard X-ray counts per flare) and from nonthermal energies (which are computed from the hard X-ray energy spectrum assuming a low-energy cutoff at 10 or 25 keV), both representing a physical quantity that is proportional to the energy. In Table 4.1, we indicate also the number of events, which constrains the accuracy of the fitted power law slopes. Synthesizing the datasets with the largest statistics (HXRBS/SMM, BATSE/CGRO, and RHESSI), the following median values of the power law slopes were found: $\alpha_P = 1.74$ for the peak fluxes (Table 4.1), $\alpha_E = 1.62$ for the fluences or energies, and $\alpha_T = 2.20$ for the flare durations (Aschwanden 2011b). The uncertainties of the power law slope quoted in literature generally include the formal fitting error only, while the standard deviations given here reflect methodical and systematic uncertainties also, since every dataset has been analyzed from different instruments, different time epochs, and with different analysis methods. One of the largest systematic uncertainty results from the (flare-unrelated) preflare background subtraction, because the preflare flux is often not specified in solar flare catalogs. Nevertheless, given these systematic uncertainties, the observed values are consistent with the theoretical predictions of the basic fractal-diffusive (FD) SOC model (Section 2.7), based on an Euclidean space dimension of $d = 3$, a mean fractal dimension of $D_V = 2.50$, and classical diffusion $\beta = 1$, which yields $\alpha_P = 1.80$ for peak fluxes, $\alpha_E = 1.44$ for spatiotemporal (see Table 2.1) energies, and $\alpha_T = 2.00$ for durations (Table 4.1). As a footnote, the definition of the minimum fractal dimension has been changed from $D_{V,min} = 1$ (in earlier work) to $D_{V,min} = (d - 1)$ in the present work, which changes the power law slope of the energy slightly, from $\alpha_E = 1.50$ to $\alpha_E = 1.44$. Nevertheless, the basic FD-SOC model predicts the correct power law slopes within the uncertainties of hard X-ray measurements.

Frequency-size distributions of solar flares are generally sampled from the entire Sun, and thus from multiple active regions that are present on the visible

Table 4.1 *Frequency distributions measured from solar flares in hard X-rays and γ-rays, the median of the observed slopes (second-last line), and the theoretical prediction based on the FD-SOC model (last line).*

Power law slope of peak flux α_P	Power Law slope of fluence α_E	Power law slope of durations α_T	Number of events n	Instrument and threshold energy	References
1.8			123	OSO–7 (>20 keV)	Datlowe et al. (1974)
2.0			25	UCB (>20 keV)	Lin et al. (1984)
1.8			6775	HXRBS (>20 keV)	Dennis (1985)
1.73 ± 0.01			12,500	HXRBS (>25 keV)	Schwartz et al. (1992)
1.73 ± 0.01	1.53 ± 0.02	2.17 ± 0.05	7045	HXRBS (>25 keV)	Crosby et al. (1993)
1.71 ± 0.04	1.51 ± 0.04	1.95 ± 0.09	1008	HXRBS (>25 keV)	Crosby et al. (1993)
1.68 ± 0.07	1.48 ± 0.02	2.22 ± 0.13	545	HXRBS (>25 keV)	Crosby et al. (1993)
1.67 ± 0.03	1.53 ± 0.02	1.99 ± 0.06	3874	HXRBS (>25 keV)	Crosby et al. (1993)
1.61 ± 0.03			1263	BATSE (>25 keV)	Schwartz et al. (1992)
1.75 ± 0.02			2156	BATSE (>25 keV)	Biesecker et al. (1993)
1.79 ± 0.04			1358	BATSE (>25 keV)	Biesecker et al. (1994)
1.59 ± 0.02		2.28 ± 0.08	1546	WATCH (>10 keV)	Crosby (1996)
1.86	1.51	1.88	4356	ISEE–3 (>25 keV)	Lu et al. (1993)
1.75	1.62	2.73	4356	ISEE–3 (>25 keV)	Lee et al. (1993)
1.86 ± 0.01	1.74 ± 0.04	2.40 ± 0.04	3468	ISEE–3 (>25 keV)	Bromund et al. (1995)
1.80 ± 0.01	1.39 ± 0.01		110	PHEBUS (>100 keV)	Perez-Enriquez and Miroshnichenko (1999)
1.80 ± 0.02		2.2 ± 1.4	2759	RHESSI (>12 keV)	Su et al. (2006)
1.58 ± 0.02	1.7 ± 0.1	2.2 ± 0.2	4241	RHESSI (>12 keV)	Christe et al. (2008)
1.6			243	BATSE (>8 keV)	Lin et al. (2001)
1.61 ± 0.04			59	ULYSSES (>25 keV)	Tranquille et al. (2009)
1.74	**1.62**	**2.20**		Median	All HXR observations
1.80	**1.44**	**2.00**		FD-SOC prediction	Aschwanden (2012a)

hemisphere at a given time. This configuration corresponds to a multi-sandpile situation, and the resulting power law distribution is composed of different individual active regions, which may have different physical conditions and sizes. In particular, different sizes may cause an exponential cutoff at the upper end of the size distribution due to finite system-size effects. A study of flare statistics on individual active regions, however, did not reveal significant differences in their size distributions, and thus the size distributions of individual active regions seem to follow the universal power law slopes that are invariant, individually as well as in a superimposed ensemble (Wheatland 2000c), except for one particular active region (Wheatland 2010).

4.2 Solar Flare Hard X-Ray Waiting Times

The theory of waiting time distribution functions is presented in Chapter 3. Observationally, we show in Figure 4.2 a representative sample of six waiting time distribution functions (Aschwanden and McTiernan 2010), using data from HXRBS/SMM Figure 4.2a and b, WATCH/GRANAT (Figure 4.2c), BATSE/CGRO (Figure 4.2d), ICE/ISEE-3 spacecraft (Figure 4.2e), and RHESSI (Figure 4.2f).

One of the first waiting time distributions of solar flares was published by Pearce et al. (1993), using 8319 hard X-ray flare events detected with HXRBS/SMM, finding a power law distribution $N(\tau) \propto \tau^{-0.8 \pm 0.1}$ in the range of $\tau = 1-100$ min, which could not be explained (Figure 4.2a). Obviously, the fitted range of waiting times was chosen to small, as well as within a to short waiting time range (Figure 4.2a). Datasets covering a larger range of waiting times, especially over 4–6 orders of magnitude (ISEE-3, HXRBS/SMM, BATSE/CGRO, and RHESSI), can all be fitted by the same waiting-time distribution with a power law function or Pareto-type distribution function with a power law slope of $\alpha_\tau \approx 2$ (Aschwanden and McTiernan 2010).

For a dataset of 182 hard X-ray flares detected by WATCH/Granat (Figure 4.2(c)), a waiting time distribution with a power law slope of $p = 0.78 \pm 0.13$ was found in the range of $\tau = 10-300$ min, with an exponential rollover at the lower end of the waiting time distribution (Crosby 1996), which is similar to the result of Pearce et al. (1993) and covers a to small waiting time range for a reliable determination of the power law slope.

A similar dataset observed with the more sensitive BATSE/CGRO was analyzed by Biesecker (1994), which shows a power law function at the high end of waiting times, and a rollover at the low end of waiting times, similar to the HXRBS/SMM (Figure 4.2(b)) or RHESSI (Figure 4.2(e)).

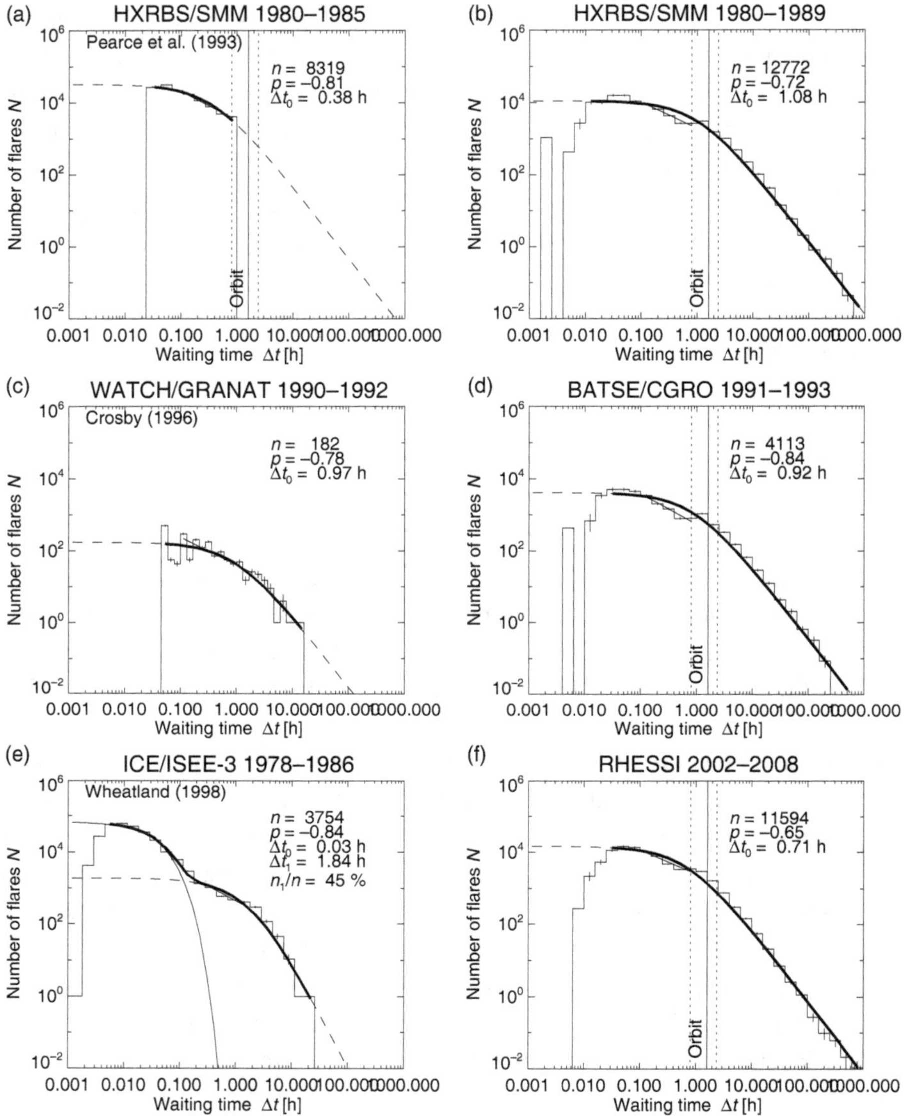

Figure 4.2 Waiting time distributions of six different datasets: HXRBS/SMM (a and b), WATCH/Granat (c), ISEE-3/ICE (e), BATSE/CGRO (d), and RHESSI (f). The distribution of the observed waiting times is shown with histograms, and the fits of nonstationary Poisson processes with dashed curves (with a power law slope tail of $\alpha_\tau = 2$) (Aschwanden and McTiernan 2010).

A dataset of 6,919 hard X-ray bursts at energies ≥ 30 keV detected by ICE/ISEE-3 (Figure 4.2(e)) was sampled by Wheatland et al. (1998), showing a double-hump distribution of waiting times, which indicates an overabundance of short waiting times ($\tau \approx 10$ s–10 min) compared with a nonstationary Poisson process fitted to the longer waiting time intervals of $\tau \approx 10$–1,000 min (Wheatland et al. 1998; Wheatland and Eddy 1998). Modeling of the waiting time distribution in terms of a nonstationary Poisson process with Bayesian blocks was attempted but no suitable fit was found. The overabundance of short waiting times was interpreted as clustering of multiple hard X-ray bursts per flare event (Wheatland et al. 1998).

In Figure 4.2(f), we show waiting time statistics of 11,594 hard X-ray flare events observed with (RHESSI) during 2002–2008. The frequency distribution of waiting times can be fitted with a power law function for waiting times in the time range of $\tau \approx 2-1{,}000$ h (Figure 4.2f), so the longer waiting times $\tau \gtrsim 2$ h are consistent with a nonstationary Poisson random process in a SOC system. At shorter time intervals, there is an excess of events at waiting times near the orbital period of $\tau \lesssim t_{orbit} \approx 1.6$ h (Figure 4.2e), which is caused by instrumental effects.

A first reason of this instrumental effect is that events that start at spacecraft night cannot be detected until the spacecraft comes out of the night part of the orbit, which causes a clustering that systematically increases from one half to one full orbital period. A second reason is that large events that extend over more than one full orbital period are double-counted in each subsequent orbit. Additional data gaps result from irregular switch on/off periods during some parts of the orbit, such as when the spacecraft passes through the South Atlantic Anomaly (SAA). Such instrumental biases have been modeled with Monte Carlo simulations in previous waiting time studies (e.g., Biesecker 1994). Thus, the main observational result of these hard X-ray studies is that the waiting time distribution of solar flares is consistent with a power law function for waiting times $\tau \gtrsim 1$ h, consistent with a nonstationary Poissonian random process. The fact that the driver process for solar flares is nonstationary does not contradict with models of self-organized criticality (Lu and Hamilton 1991; Charbonneau et al. 2001), except that the input rate is highly variable. Alternative interpretations, such as intermittent turbulence have been considered too (Boffetta et al. 1999; Lepreti et al. 2001).

The six waiting time distributions shown in Figure 4.2 can all be fitted by a Pareto-type distribution (Aschwanden and McTiernan 2010), which amounts to a power law distribution at the high end, and a rollover to a constant value at the low end (Figure 4.2),

$$N(\tau) \propto N_{\tau,0} \left(1 + \left[\frac{\tau}{\tau_0}\right]^{-\alpha_\tau}\right), \qquad (4.1)$$

where the function converges to $N(\tau \mapsto 0) \mapsto N_{\tau,0}$ at the low end, and $N(\tau \mapsto \infty) = 0$ at the high end. A common interpretation of this rollover (or deficit of short waiting times) is incomplete sampling, caused by limited instrumental sensitivity. Alternatively, an explanation in terms of a low count cutoff plus a noise component for the frequency distribution of the hard X-ray count rates of solar flares has been proposed also (Li et al. 2013).

4.3 Solar Flare Hard X-Ray Time Profiles

The time profile of a SOC parameter, for instance, the flux $F(t)$, generally exhibits an exponential growth phase, followed by a saturation or decay phase (Figure 1.2), in the case of a simple flare with a single energy dissipation pulse but can display much more complex behavior in the case of clusters with multiple overlapping peaks. Time series analysis of solar hard X-ray sub-bursts has been performed with wavelet analysis down to sub-second intervals (Figure 4.3), which reveal that the size distributions of (elementary) hard X-ray sub-burst fluxes range from power law functions to exponential functions (Aschwanden et al. 1998; Cheng et al. 2012).

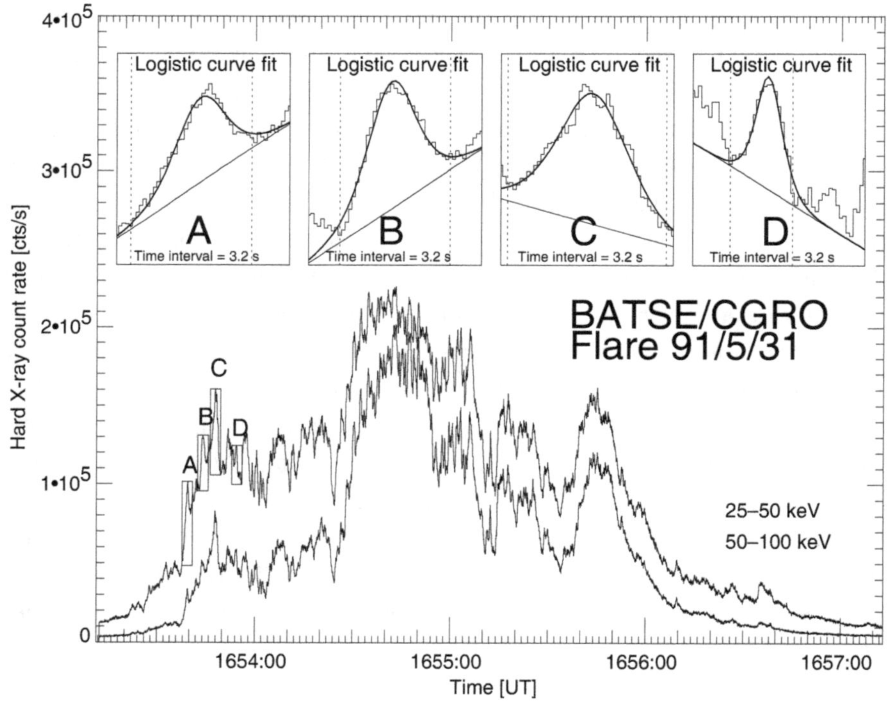

Figure 4.3 Examples of >25 and >50 keV hard X-ray time profiles, observed with BATSE/CGRO with 64-ms time resolution. Four sub-bursts at the beginning of the flare are enlarged (marked with the labels A, B, C, and D. Note that the solar flare (1991, May 31) consists of significant sub-bursts with sub-second time scales, while the sub-bursts themselves show no significant substructures, and thus may represent time-resolved elementary structures of particle acceleration processes (Aschwanden et al. 1998).

However, the size distributions $N(T)$ of hard X-ray sub-burst durations during a flare were found to be mostly exponential (Aschwanden et al. 1998), probably due to finite system-size effects in each flaring region. Hard X-ray spikes are produced during the rise time of flare hard X-ray emission in 90% and during the peak times in 70% (Cheng et al. 2012). Hard X-ray kernels detected in Hα and C IV reveal power law size distributions with slopes of $\alpha_P \approx 1.5$ for peaks, $\alpha_T \approx 2.3$ for time durations, and $\alpha_\tau \approx 1.8$ for waiting times τ, which all are close to the prediction of the FD-SOC model (Table 2.1), and thus can be considered as evidence for SOC avalanche behavior of a single flare event, or for fractal current sheets in the impulsive reconnection region (Nishizuka et al. 2009). The Hölder exponent and multifractal characteristics of a hard X-ray time profile indicates the "bursty nature" of solar flare X-ray emission, with each reconnection event producing a different and uncorrelated nonthermal particle source (McAteer et al. 2007, 2013b). A novel wavelet method, called the local intermittency measure (LIM) analysis, has been introduced (Dinkelaker and MacKinnon 2013a, 2013b), which can reveal scale-invariant time evolutions, such as the fragmentation of the energy release cascading from large to smaller structures (the "top-down" scenario), or a small flare event that is avalanching into a larger structure (the "bottom-up" scenario), but it was found that neither of the two extremes captures the totality of flare time profiles (Dinkelaker and MacKinnon 2013a, 2013b).

4.4 Solar Flare Finite System-Size Effects

Deviations from "straight" power law size distributions are expected due to the finite system-size effects of the largest events. We search for outliers in extreme events of statistical size distributions of solar flare hard X-ray datasets (Aschwanden 2019b, 2021), motivated by the *Dragon-King hypothesis* of Sornette (2009), which suggests that the most extreme events in a statistical distribution may belong to a different population, and thus may be generated by a different physical mechanism, in contrast to the strict power law behavior of SOC models produced by a single physical mechanism. Identifying such disparate outliers is important for space weather predictions, because it might indicate underestimates of the largest events. Possible physical mechanisms to produce such outliers could be generated by sympathetic flaring (i.e., multiple flares happening almost at the same time). However, we find that Dragon-King events are not common in solar and stellar flares, identified in 4 out of 25 solar and stellar flare datasets only (Aschwanden 2019b). The cumulative size distributions for solar hard X-ray flares are shown in Figure (4.4), for three different hard X-ray instruments (HXRBS/SMM, BATSE/CGRO, RHESSI), three different solar cycles, and for three different SOC parameters (flare peaks P, total counts C, and total durations T). Most of these

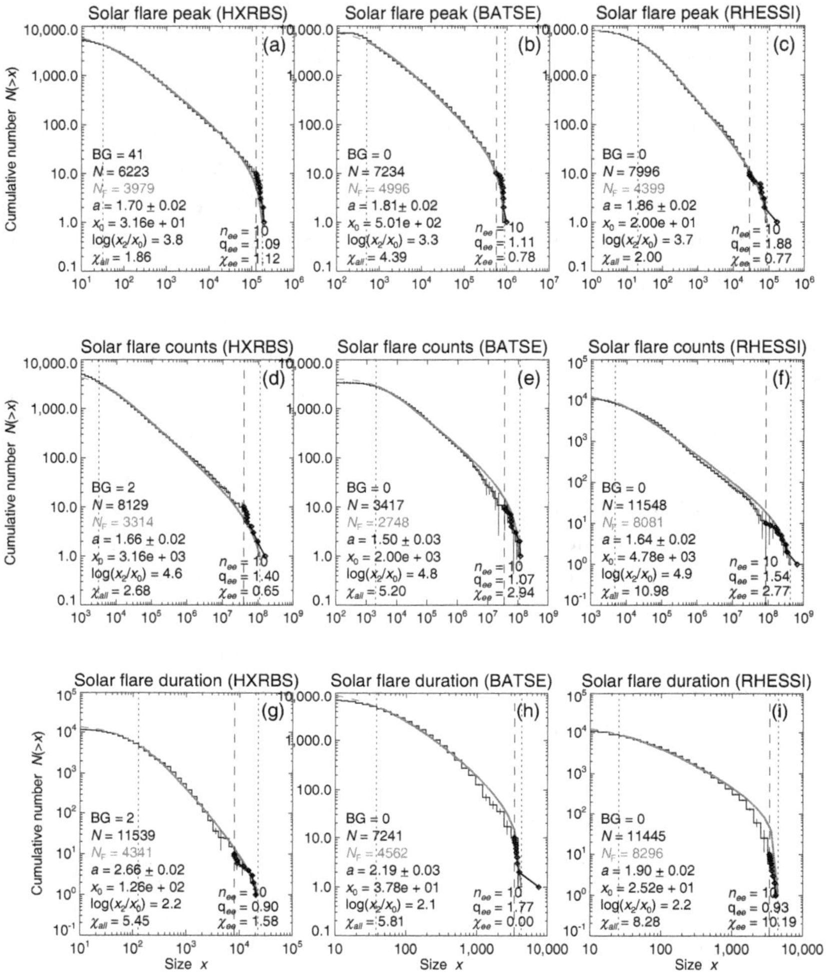

Figure 4.4 Solar flare hard X-ray peak photon rates (a–c), fluences or counts (d–f), and flare durations (g–i) are histogrammed from HXRBS observations (a, d, g), BATSE/CGRO observations (b, e, h), and RHESSI observations (c, f, i) (shown in form of histograms with uncertainties), and in form of the best-fit cumulative distribution functions (smooth curves). The chi-square is given for all events above the threshold x_0, $\chi_{all}(x > x_0)$, and for the 10 most extreme events separately, $\chi_{ee}(x > x_3)$ (Aschwanden 2019b).

nine cumulative size distributions do not show significant deviations of the best-fit cumulative size distributions for the largest events at the high end. Consequently, small, large, and extreme flares are essentially scale-free and can be modeled with a single physical mechanism. In very large datasets ($N \gtrsim 10^4$), we find significant deviations from ideal power laws in almost all datasets. Nevertheless, the fitted power law slopes constrain physical scaling laws in terms of flare areas and volumes, which have the highest nonlinearity in their scaling laws.

In a complementary study (Aschwanden 2021), we generalize the idealized "straight" power low size distribution and Pareto distribution functions in order to accommodate for incomplete sampling, limited instrumental sensitivity, finite system-size effects, "Black-Swan," and "Dragon-King" extreme events. The findings are: (i) solar flares show no finite system-size limits up to $L \lesssim 200$ Mm, which is about the size of the largest active region, but solar flare durations reveal an upper flare duration limit of $T \lesssim 6$ h. Schrijver et al. (2012), estimated the energy of the largest solar flare event to $E_{max} \lesssim 10^{33}$ erg, amounting to an X30 GOES-class event, which may occur once in 30 years with a probability of $\approx 10\%$. From this maximum value E_{max}, which represents the amount of dissipated magnetic (nonpotential) energy, an energy partition of nonthermal energies $(51 \pm 17)\%$, energies in accelerated ions $(6 \pm 14)\%$, CME energies $(7 \pm 14)\%$, and energy in direct heating $(7 \pm 17)\%$ was found (Aschwanden et al. 2017). In the statistical mean, only a fraction $(7 \pm 3)\%$ of the potential magnetic energy is dissipated during a flare, which amounts to a fraction of $(60 \pm 26)\%$ of the free energy (e.g., the difference between the nonpotential and the potential magnetic energy), is dissipated in solar flares (Aschwanden 2019).

4.5 Summary

Solar flare hard X-ray events are produced by the electron thick-target bremsstrahlung process at electron energies of $\gtrsim 20$ keV. Large statistical samples of hard X-ray fluxes, fluences, energies, flare durations, and waiting times have been observed with instruments from three different spacecraft (HXRBS/SMM, BATSE/CGRO, RHESSI, etc.) from three different solar cycles and analyzed with different automated event detection methods. Despite of this large variety of data, all datasets reveal self-consistent results, for instance, power law peak fluxes with a slope of $\alpha_P = 1.74$, which match the theoretical prediction of the fractal-diffusive SOC model, that, $\alpha_P = 9/5 = 1.80$. Systematic errors and uncertainties of these datasets include insufficient fitting ranges, spacecraft orbital data gaps, finite-size effects, south Atlantic anomaly data gaps, instrumental sensitivity, incomplete samples, thresholded event selection, and background subtraction.

5

Solar Flare Soft X-Rays

5.1 Size Distributions of Solar Flare Soft X-Ray Fluxes

Solar flares display signatures of thermal emission in soft X-ray wavelengths, besides the nonthermal emission detected in hard X-rays (Chapter 4). The emission in both wavelength regimes is triggered by the same instability, which is part of the *chromospheric evaporation model* (e.g., Fisher et al. 1985), but are produced by different parts of the physical processes. While hard X-rays are mostly produced by collisional bremsstrahlung of nonthermal electrons precipitating down into the dense chromosphere, soft X-ray line emission and continuum emission is excited by impulsive heating of the chromospheric plasma. The precipitating electrons essentially control the heating rate of the chromospheric plasma, while the energy emitted from the heated thermal plasma (with typical flare temperatures of $T_e \approx 10 - 35$ MK) follows approximately the time integral of the hard X-ray-driven heating rate, a relationship that has been dubbed the *Neupert effect* (Dennis and Zarro 1993). Because of this intimate relationship between soft X-rays and hard X-rays in solar flares, similar energy or size distributions are expected in both wavelength regimes, which is indeed the case, as the compilations in Tables 4.1 and 5.1 show.

Size distributions of soft X-ray peak fluxes, fluences, and durations were mostly obtained from flare (or microflare) detections with the OSO-3 spacecraft (Hudson et al. 1969), the Explorer (Drake 1971), Yohkoh /SXT (Shimizu 1995; Shimojo and Shibata 1999), the SMM/BCS (Lee et al. 1993, 1995), and the GOES spacecraft (Feldman et al. 1997; Veronig et al. 2002a; Yashiro et al. 2006; Aschwanden and Freeland 2012; Martin et al. 2010; Li et al. 2012), and the SOHO/SUMER instrument (Wang et al. 2006).

Interestingly, the power law slope of the peak count rate distribution in the range of $\alpha_{P,SXR} \approx 1.6 - 2.0$ (Table 5.1) is similar to that of the hard X-rays, that is,

Table 5.1 *Frequency distributions measured from solar flares in soft X-rays. Measurements with no preflare background subtraction are marked with parentheses. The acronyms for the instruments and spacecraft are: OSO = Orbiting Solar Observatory; SXT = Soft X-ray Telescope (aboard Yohkoh); BCS = Bent-Crystal Spectrometer (aboard SMM = Solar Maximum Mission); GOES = Geostationary Operational Environmental Satellite; SUMER = Solar Ultraviolet Measurements of Emitted Radiation (onboard the SOHO = Solar and Heliospheric Observatory).*

Power law slope of peak flux α_P	Power law slope of fluence α_E	Power law slope of durations α_T	Log range	Instrument	References
1.8			1	OSO-3	Hudson et al. (1969)
1.75 ± 0.1	1.44		2	Explorer	Drake (1971)
1.63–1.89	1.5–1.6		2	SXT	Shimizu (1995)
1.79			2	BCS	Lee et al. (1995)
1.86			2	GOES	Lee et al. (1995)
1.88 ± 0.21			3	GOES	Feldman et al. (1997)
1.7 ± 0.4			2	SXT	Shimojo and Shibata (1999)
1.98	1.88		3	GOES	Veronig et al. (2002a)
(2.11 ± 0.13)	(2.03 ± 0.09)	(2.93 ± 0.12)	3	GOES	Veronig et al. (2002a)
(2.16 ± 0.03)	(2.01 ± 0.03)	(2.87 ± 0.09)	3	GOES	Yashiro et al. (2006)
	1.75 ± 0.05		1	SUMER	Wang et al. (2006)
(2.3)			3	GOES	Martin et al. (2010)
2.04 ± 0.04	2.35		2	GOES	Li et al. (2012)
1.88 ± 0.12			1	GOES	Wheatland (2010)
1.98 ± 0.11		2.02 ± 0.04	5	GOES	Aschwanden and Freeland (2012)
1.83	**1.75**	**2.02**		Median	All SXR observations
1.80	**1.44**	**2.00**		FD-SOC	Aschwanden (2012a)

$\alpha_{P,HXR} \approx 1.6 - 2.0$ (Table 4.1) and thus implies proportionality between the hard X-ray fluences and the soft X-ray fluxes. Since the Neupert effect predicts that the time profile of soft X-rays approximately follows the time integral of the impulsive hard X-rays, one would, according to the Neupert effect, expect that the soft X-ray peak flux distribution should be equal to the hard X-ray fluences, which is however not always the case (Lee et al. 1993, 1995). Moreover, the soft X-ray emission generally lasts longer than hard X-ray emission (Feldman et al. 1997).

The compilation of power law slopes in soft X-ray parameters (Table 5.1) is based on event detections obtained from imaging data (Yohkoh /SXT, SMM/BCS), as well as from time profiles (OSO-3, Explorer, and GOES). Despite of the different dimensionality of detected SOC avalanches (i.e., 1-D for time profile fluxes, and 2-D for image fluxes), the resulting size distributions in hard X-rays and soft X-rays appear to be largely invariant (Table 5.1), ignoring those cases with no background subtraction.

However, there are discrepancies visible due to inappropriate data analysis. Some of the size distributions of soft X-ray peak fluxes have been found to have values steeper than $\alpha_P = 2.0$ (Veronig et al. 2002a; Yashiro et al. 2006; Martin et al. 2010), which we can understand now, in hindsight, to be a consequence of neglecting the subtraction of the preflare background fluxes, which make up for a substantial amount of the total flux for small flares. Feldman et al. (1997) investigated the power law slope for five different thresholds (GOES class A1, A4, B1, B4, C1) and confirmed that the flare occurrence rates are a function of the background-subtracted flare fluxes. Yashiro et al. (2006) subdivided the detected flare events into datasets with and without coronal mass ejections (CME) and found different power law slopes in the fluence, peak flux, and duration parameters. The subdivision of a size distribution into partial subsets, however, violates the basic assumption of completeness of the size distribution (above some suitable threshold) in our standard SOC model.

For detection of SOC events in imaging data, the data preparation includes also analysis of data compression features, dark-current subtraction, spacecraft attitude jitter removal, exposure time normalization, and truncation effects, as demonstrated in the analysis of Yohkoh/SXT data (Shimizu 1995). Truncation effects produce deviations from ideal power law slopes mostly for small events, as demonstrated by Monte Carlo type simulations (Lee et al. 1993, 1995).

The largest dataset (with over 300,000 detected events) has been obtained from GOES light curves (Aschwanden and Freeland 2012), using a five times higher detection sensitivity than the NOAA flare catalog, from which a size distribution with a power law slope of $\alpha_P = 1.98 \pm 0.11$ for the background-subtracted soft X-ray peak fluxes P is found, which agrees closely with the theoretical prediction of the standard FD-SOC model, $\alpha_P = 1.80$ (Table 2.1). Small differences between the two GOES wavebands are identified (Figure 5.1), $\alpha_{F,\mathrm{low}} = 2.04 \pm 0.04$ for the low $\lambda = 1 - 8$ Å channel, and $\alpha_{F,\mathrm{high}} = 1.85 \pm 0.04$ for the high $\lambda = 0.5 - 4$ Å channel (Li et al. 2012). The higher energy channel has a flatter distribution than the lower energy channel, which challenges the universality of the power law distribution (Li et al. 2016b). Other cases with deviations from ideal power law size distributions were also attributed to arbitrary threshold definitions and insufficient degradation corrections (due to the degradation of EUV detectors over time spans of years)

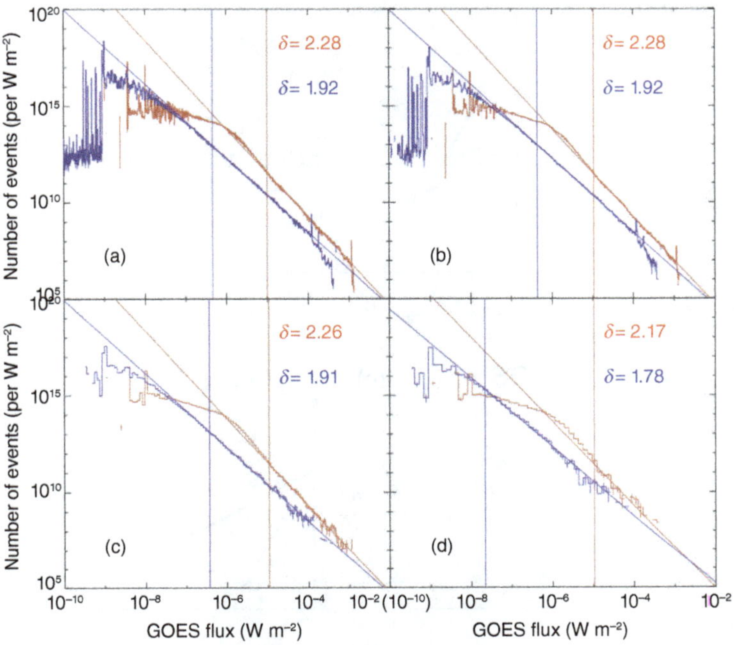

Figure 5.1 Size distributions of the soft X-ray flux in the GOES 1-8 Å (upper curve) and the GOES 0.5-4 Å (lower curve) (Li et al. 2016a). Note the difference of the power law slopes in the low-energy and high-energy channels, which indicate some deviation from universality.

(Ryan et al. 2016; Verbeeck et al. 2019). However, even when flare event detection in GOES time profiles is background-corrected, some of the size distributions were found not to be well-fitted, while log-normal distribution functions appear to produce better fits in a few cases (Verbeeck et al. 2019).

Factors that determine the power law index of an energy distribution of solar flares, down to nanoflares, include the variable driver of the solar activity, the type of coronal phenomena (such as Quiet regions, active regions, coronal holes, or off-limb events), the energy range, the magnetic flux, the magnetic nonpotentiality (twist), and the magnetic reconnection type (Kawai and Imada 2022). It was even suggested that the SOC dimensionality could modify the power law slope (Kawai and Imada 2022): for instance Parker's nanoflare heating scenario (Parker 1988) could be controlled by 1-D granule length scales, and by 2-D sunspot length scales (Figure 5.2), according to the Carmichael-Sturrock-Hirayama-Kopp-Pneuman (CSHKP) reconnection scenario.

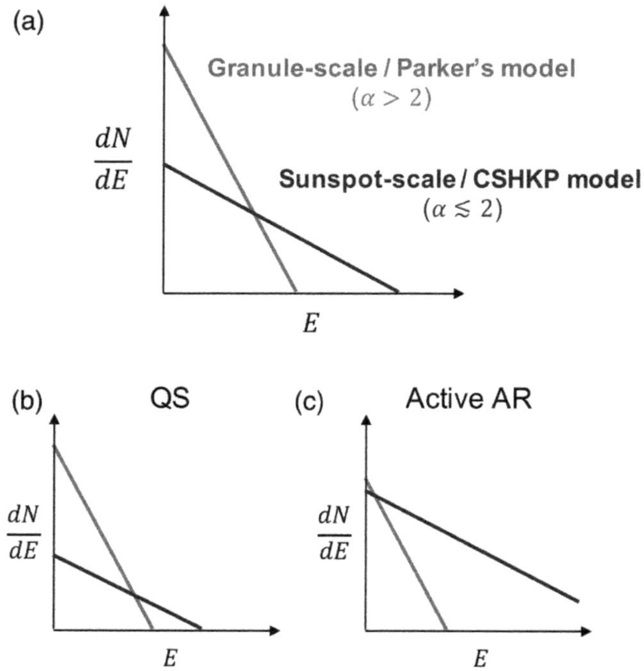

Figure 5.2 Energy distributions of CSHKP flare models and Parker nanoflare models. In the Quiet Sun, the distribution of granule-scale events (Parker nanoflare model) stands out owing to the absence of sunspot-scale events (b). In active regions, the distribution of granule-scale events is hidden by that of CSHKP model events (c) (Kawai and Imada 2022).

5.2 Solar Flare Soft X-Ray Waiting Times

The statistics of inter-occurrence times between soft X-ray-emitting flare events has been tested using the GOES flare catalog. Theoretical modeling of the waiting time distributions $N(\tau)$ is presented in Chapter 3, which includes the stationary waiting time function (Section 3.1), the nonstationary waiting time distribution function (Section 3.2), and the Pareto type-II waiting time distribution function (Section 3.6). An example of time scale definitions is given in Figure 5.3. The time profile shown in Figure 5.3 contains one waiting time $\tau = t_2 - t_1$, defined by a start time t_1 and an end time t_2, beginning and ending when the soft X-ray rise time exceeds a fixed flux threshold.

The stationary waiting time distribution function is expected to be an exponential function, since each flare represents a statistically independent event, occurring at random times during an observational time span, with a constant flare rate. The assumption of a constant flare rate, however, is a very unrealistic model, at least for solar flare events, which show variability over three orders of magnitude from solar minimum to solar maximum (Figure 5.8). If one abandons the inadequate

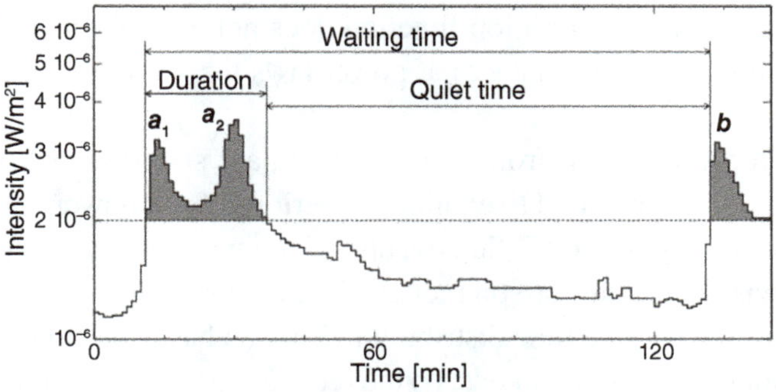

Figure 5.3 Definitions of time scales are shown: the waiting time, the quiet time, the event duration, and the absolute time (x-axis). Using a threshold of $I = 2 \times 10^{-6}$ W/m^2, the first event consists of two peaks (a_1, a_2), while the second event consists of one peak (b) (Baiesi et al. 2006).

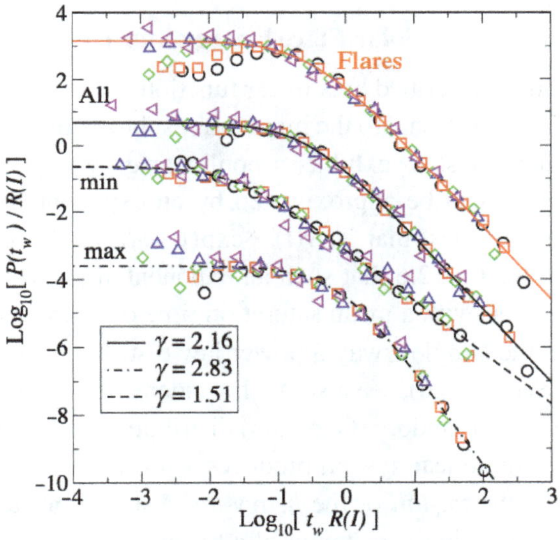

Figure 5.4 The waiting time distribution functions of soft X-ray flare intervals are shown from GOES data, for the solar minimum, the solar maximum, the entire catalog (all), and the flare catalog, arbitrarily shifted in vertical direction, with the power law slopes indicated, where γ represents here the waiting time α_τ (Baiesi et al. 2006).

model of a stationary Poisson process and takes nonstationary Poisson processes into account, one obtains power law-like size distributions of waiting times, which match the observations, for example, Baiesi et al. (2006). The variability of soft X-ray fluxes during different phases of the solar cycle affects even the power law slopes: the power law slope of waiting time distributions is flattest ($\alpha_\tau = 1.51$) during the solar minimum, and steepest ($\alpha_\tau = 2.83$) during the solar maximum, while the sum of all yields a mean value of $\alpha_\tau = (1.51 + 2.83)/2 = 2.16$ (Figure 5.4).

Thus the waiting time distribution function does not exhibit universality. We discuss the existence (or nonexistence) of power laws for waiting time distributions in Section 3.8.

Evidence for a departure from a power law flare size distribution for a single small active region was found to exhibit a superior fit for a power law function plus a roll-over at the largest GOES flux events, which was considered to be consistent with a piecewise-constant Poisson model (Wheatland 2010).

Variation of the waiting time distribution function has been simulated with three different cellular automaton models: (i) the reduced Lu and Hamilton (1991) model, (ii) the anisotropic optimized model, and (iii) the isotropic optimized model, where the latter was found to reproduce the straight power law function best (Farhang et al. 2019).

5.3 Solar Flare Energy Storage

A linear system can be described by a linear function, such as $y(t) = y_b + y_0(t - t_0)$, where the output is proportional to the input (after subtraction of the pre-event level y_b), while a nonlinear system exhibits a nonlinear growth phase, such as $y(t) = y_b + y_0(t - t_0)^p$, which can be approximated by an exponential function during a few e-folding growth times, that is, $y(t) \propto \exp(t - t_0)$, as depicted in Figure 5.5. We have shown in Section 2.6 that such an exponential-growth model, subject to random time durations (with a mean saturation time of t_S and an e-folding growth time τ_G), yields in the simplest way a power law distribution function, with slope $\alpha_P = (1 + \tau_G/t_S)$ (Section 2.6), see also Aschwanden et al. (1998). Consequently, a linear system produces a random (Gaussian) distribution of (output) sizes that falls off steeply, while a nonlinear system produces a power law-like size distribution, which is also called the *fat-tail* or the *heavy-tail* distribution function. This is the most essential difference in the behavior of a linear and a nonlinear system.

A pioneering early study explored the linear type of dynamic models, which assumes a correlation $y \propto t_S$, between the saturation time scale t_S and the dissipated energy y (after background subtraction y_b). In particular, attributing an energy to the output parameter y, a model of linearly growing energy storage in the solar corona is assumed, where the dissipated energy flux is applied to a solar flare, or to some other cosmic transient phenomenon (Rosner and Vaiana 1978). The linear model, however, has two major problems: (i) linear growth of any flare parameter has not been observed during the rise time of an energy dissipation event and (ii) a linear correlation between dissipated energies and waiting time intervals has not been observed either.

Specifically, the relation between the time intervals of individual microflares was investigated, but no correlation between the time intervals and soft X-ray

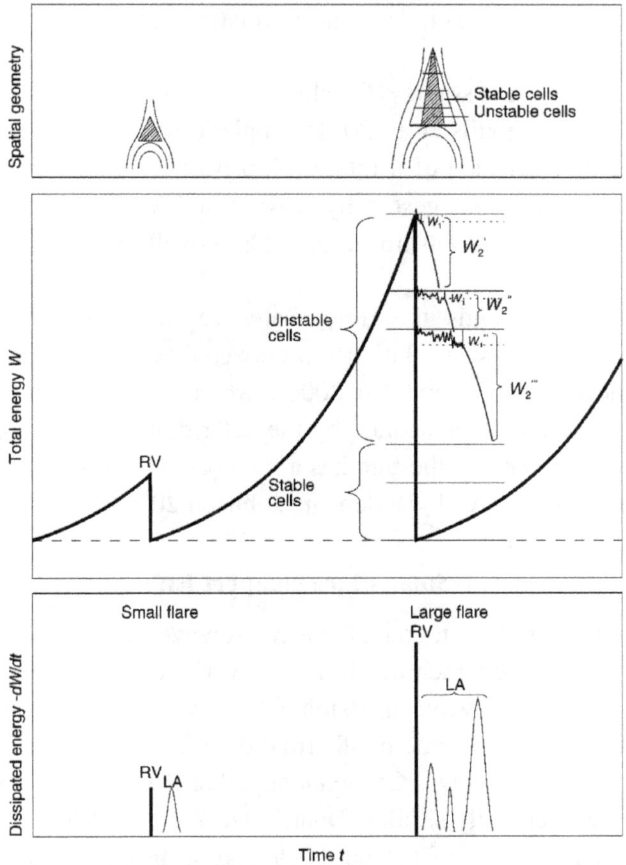

Figure 5.5 Cartoon illustrating the relationship between the Rosner and Vaiana (1978) model (RV) and the logistic avalanche (LA) model. The RV model assumes that energy storage grows monotonically between two subsequent flares, linearly or exponentially (thick linestyle). In the LA model, flare instabilities are triggered randomly, once their dissipated energy exceeds a threshold (Aschwanden et al. 1998).

intensity was found (Shimojo and Shibata 1999). Scatterplots between the waiting time since previous bursts and the peak count rate, obtained from WATCH/GRANAT observations, do not reveal significant correlations (Crosby et al. 1998; Georgoulis et al. 2001). Wheatland (2000) examined whether solar flares exhibit an interval–size relationship, using events selected from the same active region, and concluded also that the data do not support an interval–size relationship, not even in a single active region. Correlations between energies and waiting times are believed to be a physical property and are not an effect of obscuration by overlapping waiting time intervals (Lippiello et al. 2008). Hudson (2020) investigated relationships between solar flare magnitude and time intervals before the flare peak, as well as after the flare peak, and concluded that correlations appear to be nonexistent or intermittent.

In summary, all of these studies (Crosby et al. 1998; Shimojo and Shibata 1999; Wheatland 2000; Georgoulis et al. 2001; Lippiello et al. 2008; Hudson 2020) do not support the original idea of a relationship between flux and waiting time in solar flares, as originally suggested by Rosner and Vaiana (1978). Such a lack of correlation defies a relationship in general, as well as a linear relationship in particular.

However, if one extends the time scales to the length of a solar cycle (≈ 11 years), an interval–size relationship between the solar cycle amplitude and descending time exists (Du and Du 2006), which represents the most extreme time scale only, ultimately controlled by the solar dynamo. Indeed, including the solar cycle demonstrates that the Sun has a memory from hours (in solar flares) to decades (in solar cycles) (Aschwanden and Johnson 2021).

5.4 Solar Flare Neupert Effect

The variation of the time integral of the microwave or hard X-ray intensity in flares often matches closely the time derivative variation of the soft X-ray intensity (Neupert 1968). The microwave emission is believed to be produced by gyrosynchrotron radiation from the same nonthermal distribution of accelerated electrons that produces the hard X-rays. Consequently, hard X-ray and microwave light curves of flares are generally similar (Dennis and Zarro 1993). The Neupert effect can be formulated in terms of the hard X-ray time profile $F_{\mathrm{HXR}}(t)$ and the soft X-ray time profile $F_{\mathrm{SXR}}(t)$,

$$F_{\mathrm{SXR}}(t) = \int_{t_0}^{t_0+T} F_{\mathrm{HXR}}(t')\, dt' \qquad t_0 \le t \le (t_0 + T). \tag{5.1}$$

coupled by an integral relationship, where $t = t_0$ is the start time of the flare, and T is the flare duration. The inverse relationship is,

$$F_{\mathrm{HXR}}(t) = \left(\frac{dF_{\mathrm{SXR}}(t)}{dt} \right) \qquad t_0 \le t \le (t_0 + T). \tag{5.2}$$

Obviously the two light curves $F_{\mathrm{HXR}}(t)$ and $F_{\mathrm{SXR}}(t)$ cannot peak at the same time, and thus the universality of their size distribution slopes $\alpha_{\mathrm{F,HXR}}$ and $\alpha_{\mathrm{F,SXR}}$ appears to be violated in single-peaked time profiles. The study of Dennis and Zarro (1993) demonstrated that 71% of the analyzed flares showed a good correlation between the impulsive hard X-ray and soft X-ray impulsive peaks (for an example, see Figure 5.6) and 50% showed a poor correlation with the gradual hard X-ray peaks. About 50% show a coincident timing between soft X-ray and hard X-ray peak times that is consistent with the expectation of the Neupert effect (Veronig et al. 2002b),

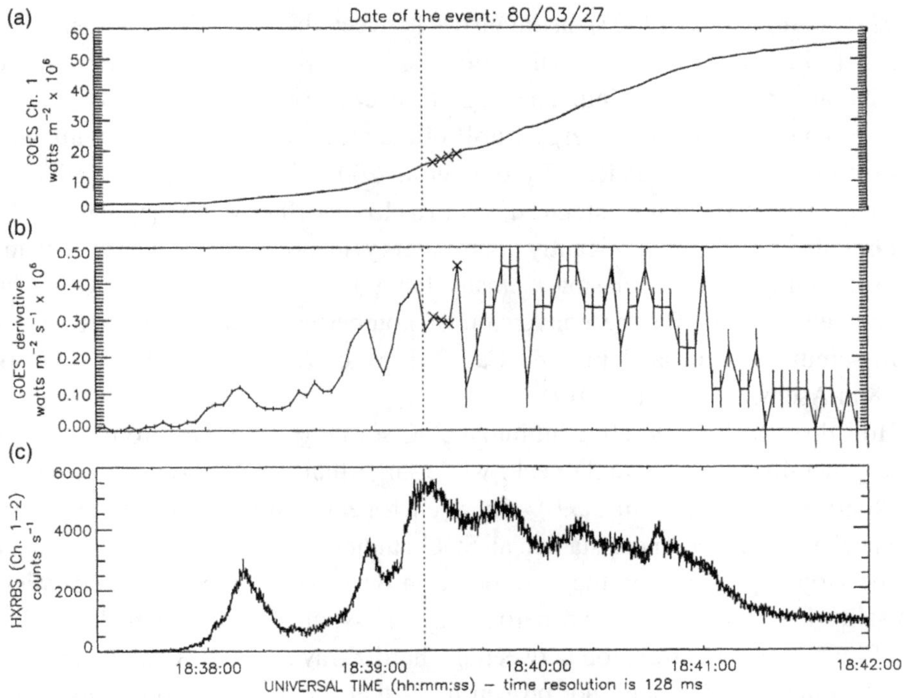

Figure 5.6 (a) GOES 1-8 Å soft X-ray time profile of the 1980, March 27, solar flare. (b) The time derivative of the top plot, which predicts the hard X-ray profile according to the Neupert effect. (c) HXRBS/SMM 26-51 keV hard X-ray time profile. Note the good overall correlation of the multipeak structure of the flare (Dennis and Zarro 1993).

while a significant fraction deviates from the Neupert relationship and thus requires an additional energy transport mechanism than the nonthermal electron population (Veronig et al. 2002a, 2002b, 2002c, 2005).

However, the physical processes of conductive cooling and radiative loss are neglected in the approximative formalism of the Neupert effect (Eqs 5.1–5.2), while a temperature–density scaling law is required (Lee et al. 1995). High-temperature plasma (> 16.5 MK) is more likely than low-temperature plasma to exhibit the Neupert effect (McTiernan et al. 1999).

Studies of the spatial Neupert effect became possible with Yohkoh imagery (Shimizu 1995), which localizes the sites of nonthermal electron precipitation as the source of the Neupert effect (Tomczak 1999; Ning and Cao 2010b; Li et al. 2009). Any deviation from the Neupert effect relationship (Eq. 5.1–5.2) basically means that the hot plasma giving rise to the SXR emission is not heated exclusively by the thermalization of the accelerated electrons that are responsible for the hard X-ray emission (Dennis and Zarro 1993; Lee et al. 1993; Veronig et al.

2002a). Veronig et al. (2005) distinguishes between the "empirical Neupert effect" (Eq. 5.1–5.2) and the "theoretical Neupert effect," which includes effects of conductive and radiative cooling. The largest uncertainty in calculating the Neupert effect comes from the low-energy cutoff of the electron injection spectrum, which is found to be time dependent (Veronig et al. 2005). Early RHESSI results show that the electron spectrum extends down to as low as 10 keV in many flares, which implies that there is significantly more energy in nonthermal electrons than in the soft X-ray emitting plasma in many flares (Dennis et al. 2003). Other estimates yield 2–20% for the conversion of nonthermal energy into thermal (soft X-ray-emitting) plasma (Ning and Cao 2010a), or $E_{th}/E_{nth,e} = 12\% \pm 11\%$ (see Eq. 8 in Aschwanden et al. 2017).

How does the Neupert effect fit into a SOC scenario? The time profile of a small flare generally shows a single peak, while large flares generally exhibit multiple peak structures (e.g., Figure 5.6), where each peak can be interpreted as a SOC avalanche. However, most statistical SOC studies measure the SOC parameters of the largest peak during the flare duration only. The Neupert effect predicts a causality between the soft and hard X-ray emission of a flare, which manifests itself by a positive correlation between the hard X-ray and soft X-ray fluxes or fluences. It appears (although not proven here) that this positive correlation, which implies universality of the power law-like size distributions and their power law slopes, strengthens the proportionality $\alpha_{F,\mathrm{HXR}} \propto \alpha_{F,\mathrm{SXR}}$ in the cross-correlation $CCC = F_{\mathrm{HXR}} \propto F_{\mathrm{SXR}}$, while the cases that are not consistent with the Neupert effect indicate a poor cross-correlation coefficient. Thus the Neupert effect can be used to identify subsets of SOC-consistent (well-correlated) cases.

5.5 Solar Flare Soft X-Ray Energetics

In the following, we derive some basic physical relationships that are related to the SXR-emitting plasma in solar flares. The observation of solar flares or transient brightenings in soft X-rays, as described in Shimizu (1995), yields a volume emission measure $EM = n_e^2 V$ and an electron temperature T_e, which defines the mean electron density n_e,

$$n_e = \sqrt{\frac{EM}{V}} \tag{5.3}$$

in terms of the volume V, approximated by a single-loop geometry with loop length L and cross-sectional width w,

$$V = L\, w^2, \tag{5.4}$$

where the line-of-sight depth is set equal to the cross-sectional width w, and a filling factor of unity is assumed. The gas pressure p in the brightening or flare loops is given by,

$$p = 2n_e k_B T_e, \tag{5.5}$$

where k_B is the Boltzmann constant and T_e is the electron temperature at the flare (or brightening) peak time. For a confined flare plasma, the magnetic pressure p_m is

$$p_m = \left(\frac{B_G^2}{8\pi} \right), \tag{5.6}$$

which corresponds to a magnetic energy E_m of (in units of Gauss),

$$E_m = \left(\frac{B_G^2}{8\pi} \right) V = \left(\frac{B_G^2}{8\pi} \right) L w^2. \tag{5.7}$$

Shimizu (1995) estimated the radiative loss F_r using the radiative loss function $\Lambda(T)$ specified by Rosner et al. (1978),

$$F_r = n_e^2 \; \Lambda(T) \; V = EM \; \Lambda(T). \tag{5.8}$$

In addition, the conductive loss F_c was then estimated by,

$$F_c = \kappa T^{5/2} \nabla T \; A = 2\kappa T^{7/2} \frac{w^2}{L}, \tag{5.9}$$

using the approximations $\nabla T = 2T/L$, the area $A = w^2$, and κ the classical Spitzer conductivity. By combining both the radiative loss rate F_r and the conductive loss rate F_c, Shimizu (1995) arrived at the total energy loss rate of,

$$E_{loss} = (F_r + F_c)\tau_d, \tag{5.10}$$

where τ_d is the duration of a flare or transient brightening. The resulting size distribution of the different energies was found to be in the range of $E_{tot} \approx 10^{27}-10^{29}$ erg with a power law slope of $\alpha_E \approx 1.5 - 1.6$ (Figure 5.7).

The results of Shimizu (1995) have two significant implications: First, the observed power law slopes have a median of $\alpha_F = 1.83$ for the peak flux, and $\alpha_E = 1.75$ for the energy or fluence, which is close to the theoretical values predicted by the standard FD-SOC model, $\alpha_F = 1.80$ for the peak flux, and $\alpha_E = 1.44$ for the energy or fluence (Table 5.1). Second, the energy distribution is not diverging ($\alpha_E < 2$), as would be expected for nanoflare heating scenarios ($\alpha_E > 2$). The total energy supplied by transient brightenings and flares, assuming that the power law continues to lower energy, is estimated to be at most a factor 5 smaller than the

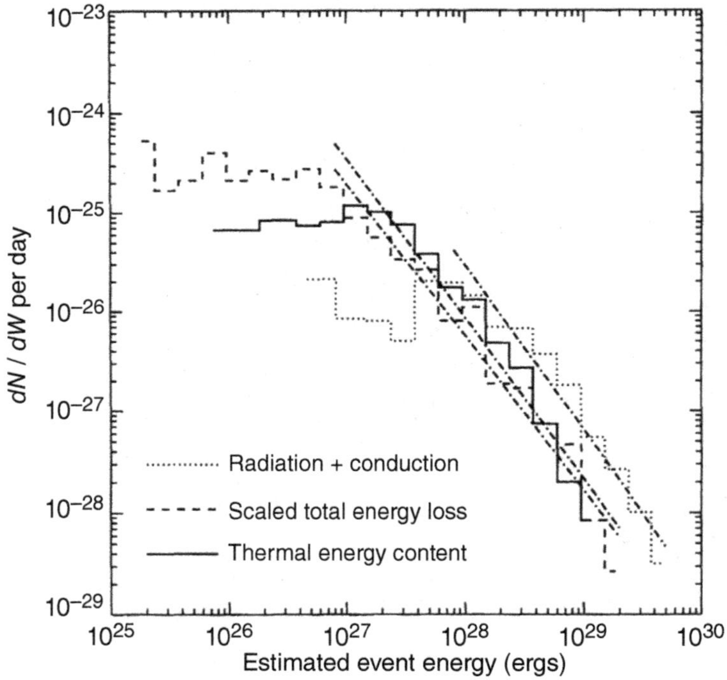

Figure 5.7 Occurrence frequency distributions of transient brightenings as function of the total energy, observed with Yohkoh/SXT: radiation plus conductive loss (dotted line), scaled total energy loss (dashed line), and thermal energy content (histogram in solid line style). Each distribution can be represented by a single power law with the index $\alpha_E \approx 1.5-1.6$ (dash dotted lines), although the distribution deviates from a "straight" power law function at the lower end of the energy range (Shimizu 1995).

heating rate required for the active region corona, and thus is insufficient to supply the coronal heating rate in active regions (Shimizu 1995). The observed energy distribution (Figure 5.7) reveals also a rollover at energies of $E \lesssim 10^{27}$ erg, which is caused by instrument sensitivity limits, instrument dead time, and/or obscuration by bright coronal features (Shimizu 1995).

We should be aware that the size distributions of different forms of energy do not need to be identical, because they all have different definitions. Some forms of energies are given by the standard SOC model in terms of the (time-integrated) spatiotemporal energy (Table 2.1), while three additional definitions in terms of radiative loss (Eq. 5.8), conductive loss (Eq. 5.9), and total energy loss (Eq. 5.10) are given by Shimizu (1995). Each energy size distribution can be tested separately. A list of physical parameter ranges observed during flares and transient soft X-ray brightenings is given in Table 5.2.

Table 5.2 *Physical parameter ranges of transient brightenings observed with Yohkoh/SXT (Shimizu 1995).*

Physical parameters	Parameter range
Electron temperature	$T_e = 4 - 8$ MK
Volume emission measure	$EM = 10^{44.5} - 10^{47.5}$ cm^{-3}
Electron density	$n_e = 2 \times 10^9 - 2 \times 10^{10}$ cm^{-3}
Gas pressure	$p = 5 - 20$ dyne cm^{-2}
Loop length	$L = 5 \times 10^3 - 4 \times 10^4$ km
Loop width	$w = 2 \times 10^3 - 7 \times 10^3$ km
Duration	$t_d = 2 - 7$ min
Total loss energy (radiative and conductive)	$E_{tot,1} = 5 \times 10^{26} - 5 \times 10^{29}$ erg
Scaled total energy loss	$E_{tot,2} = 10^{25} - 10^{29}$ erg
Thermal energy content	$E_{th} = 10^{26} - 10^{29}$ erg

Regarding the completeness in the detection of transient brightenings or flares, imaging observations are generally superior to nonimaging data, because the background flux from the rest of the Sun (and its variability) can simply be eliminated by extracting the flux from flare areas only. On the other side, nonimaging data may have the advantage of higher temporal resolution, down to $dt = 2$ s for GOES time profiles.

5.6 Variability of SOC Driver

Flare statistics from the GOES satellite have been sampled over a period of 48 years by now (1975–2023), which covers four solar cycles. Since the soft X-ray flux from the Sun varies by about three orders of magnitude during each solar cycle, due to the variability of emerging magnetic fields and the resulting coronal plasma heating rate, which is ultimately driven by the solar magnetic dynamo, the Sun is an ideal system to study SOC systems with variable drivers. It was already noted for flare hard X-ray fluxes that the size distribution is steeper during the solar maximum years than during the declining solar cycle years (Bai 1993). While the power law of the soft X-ray peak rate was found to be invariant during different solar cycles, having a roughly constant value of $\alpha_F = 1.98 \pm 0.11$, the time durations were found to have a variable slope from $\alpha_T \approx 2.0$ during solar minima to $\alpha_T \approx 2 - 5$ during solar maxima (Figure 5.8), which was explained in terms of a flare pile-up effect (Aschwanden and Freeland 2012). In other words, the *separation of time scales*, that is, the waiting times and flare durations, is violated during the busy periods of the solar cycle maximum. By contrast, an opposite trend has been reported for a 158-day modulation of the flare rate (Bai 1993).

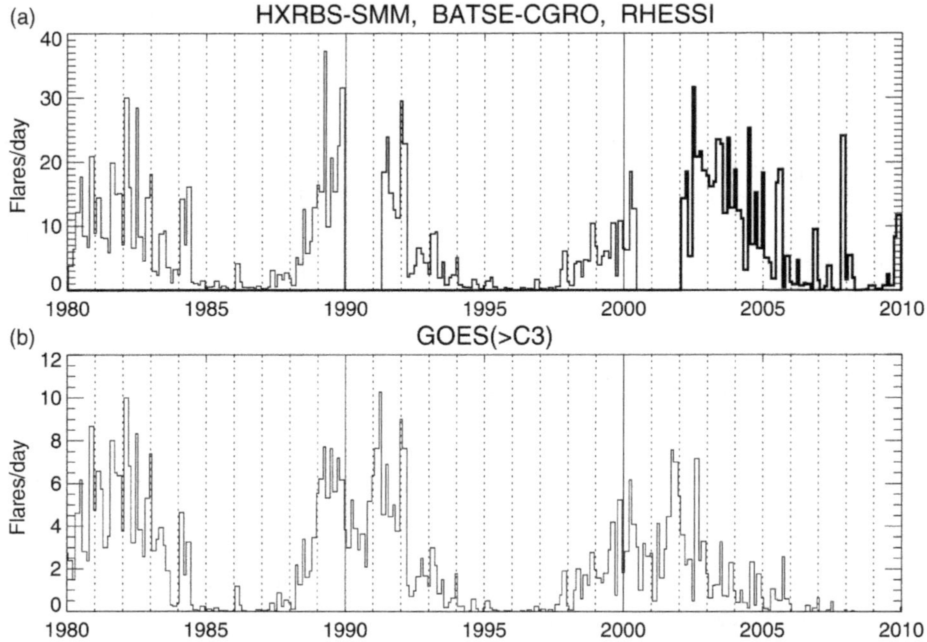

Figure 5.8 (a) Monthly averages of solar-flare rates observed during the last three solar cycles in hard X-rays with SMM/HXRBS (1980–1989), CGRO/BATSE (1991–2000), and RHESSI (2002–2010). (b) Monthly averages of solar-flare rate observed in soft X-rays with GOES, including events above the C3-class threshold level (Aschwanden 2011b).

The stationary Poisson distribution assumes a constant flaring rate, which obviously is in stark contrast to the observed time evolution of the mean flaring rate, which varies by about three orders of magnitude from solar cycle minimum to solar cycle maximum, as shown in Figure 5.8. This raises the question which SOC parameters are invariant or universal.

According to the study of Aschwanden and Freeland (2012), the mean power law slope α_F of the flux size distribution is indeed found to be invariant during each solar cycle, and thus exhibits universality. On the other side, the power law slope α_τ of the waiting time τ size distribution varies in synchrony with a solar activity index, such as the sunspot number or flare rate (Aschwanden et al. 2011b, 2011c) and thus is not universal (Figure 5.8). This can easily be understood. Since the mean flare rate λ is the total duration of all observations T_{obs}, divided by the total number N_{ev} of all detected events, that is, $\lambda = T_{obs}/N_{ev}$, the flare rate has to drop when the number of events increases during the solar cycle maximum. Consequently the flare rate λ cannot be an universal value.

Nevertheless, the observed waiting distribution exhibits a power law, in agreement with the theoretically predicted models (Section 2), such as the stationary

Poisson model, the nonstationary Poisson model, the Pareto type-II model, the reduced Lu and Hamilton model, the anisotropic, and isotropic optimized model (see Section 5.2).

5.7 Solar Flare SOC or Turbulence

As an alternative to interpret the observed size distributions of solar flares in terms of the Neupert effect, a number of studies suggested that *magneto-hydrodynamic (MHD)* turbulence could explain the observed power law size distributions (Lee et al. 1995; DeMenech and Stella 2002; Buchlin et al. 2006; Uritsky et al. 2007, 2010; Watkins et al. 2009; Georgoulis 2011; Morales et al. 2020). It was argued that the lack of a well-defined temperature-density scaling law requires that additional heating by an agent other than accelerated electrons, such as turbulence, is necessary to explain the observations (Lee et al. 1995). The prototype sandpile model of a SOC time series that is obtained by decomposing into waves of toppling show intermittent fluctuations, so that the sandpile model provides a qualitative analog of many features that are common in turbulent phenomena (De Menech and Stella 2002). Evidence for the existence of small scales, created by turbulence,

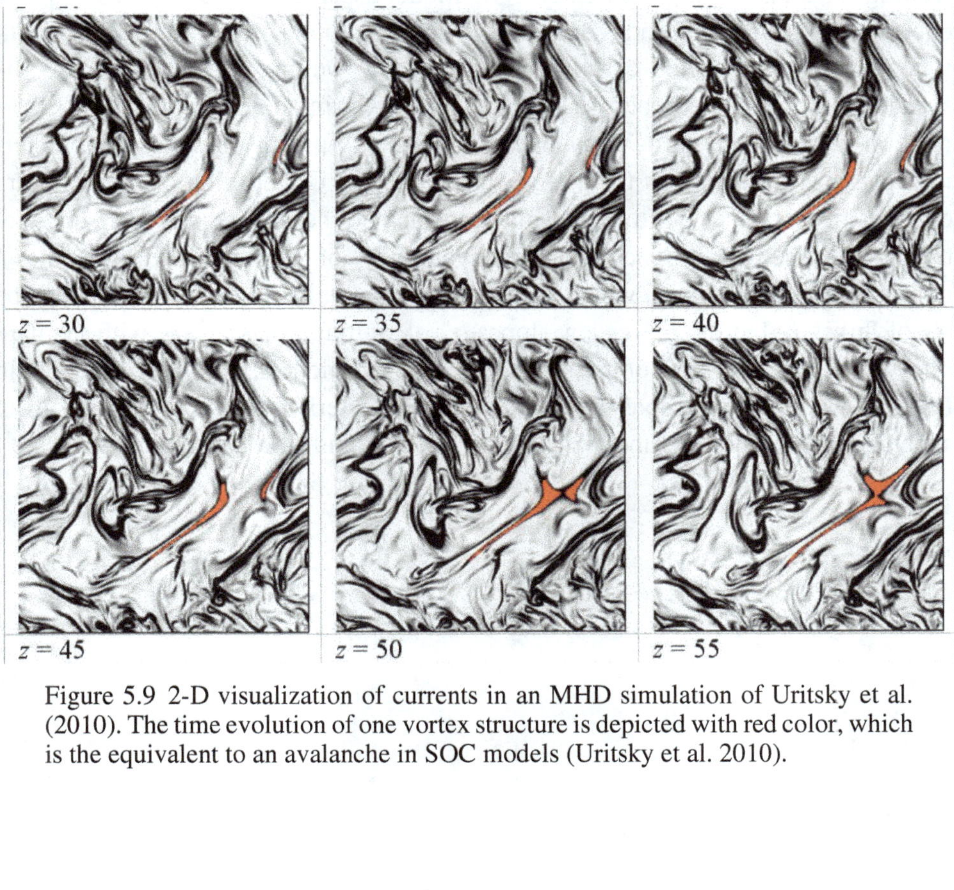

Figure 5.9 2-D visualization of currents in an MHD simulation of Uritsky et al. (2010). The time evolution of one vortex structure is depicted with red color, which is the equivalent to an avalanche in SOC models (Uritsky et al. 2010).

is inferred from their observed size distributions, Fourier spectra, and structure functions (Buchlin et al. 2006). SOC and intermittent turbulence are considered as complementary aspects of a common complex dynamical process (Uritsky et al. 2007; Morales et al. 2020), where SOC avalanches of magnetic energy dissipation is conveyed by turbulent particle flows (Uritsky et al. 2007), but it can also be explained by intermittent turbulence alone, without SOC (Watkins et al. 2009). MHD simulations of 3-D MHD turbulence (Figure 5.9) allows us to analyze the evolution of helical vortices and MHD-driven currents, and it is suggested that this process should be called turbulence dynamics, rather than turbulence criticality, and the resulting intermittency appears to be driven by local instabilities (Uritsky et al. 2010). Some major flares were quantified by their fractality, multifractality, and Kolmogorov turbulence, but these scale-free and multiscale characteristics were found not to be optimal tools for prediction of solar flares (and eruptions) (Georgoulis 2012).

5.8 Summary

In this chapter, we focus on the statistics of SOC-related solar flare parameters in soft X-ray wavelengths, which primarily includes their size distribution and their waiting time distribution. An early SOC model assumed a linear increase of the energy storage (Rosner and Vaiana 1978), but this pioneering model is not consistent with the expected correlation between the waiting time interval and the subsequently dissipated energy. The Neupert effect in solar flares implies a correlation between the hard X-ray fluence and the soft X-ray flux, which predicts identical size distributions for these two parameters. Quantifying of thermal flare energies in soft X-ray emitting plasma needs also to include radiative and conductive losses. The intermittency and bursty variability of the solar dynamo implies a nonstationary SOC driver, which yields a universal value for the power law slope α_F of fluxes F, but the power law slopes α_τ of waiting times τ vary with the flare rate. While our focus encompasses primarily SOC models, alternative models in terms of MHD turbulence can explain some characteristics of SOC features also, such as size distribution functions, Fourier spectra, and structure functions.

6

Solar EUV Nanoflares

6.1 Solar EUV Nanoflares: Size Distributions

Large solar flares (with energies of $E \approx 10^{30} - 10^{33}$ erg) exhibit heated plasma with peak temperatures of $T_e \approx 10 - 35$ MK, most conspicuously detected in soft X-rays, which cools down to temperatures of $T_e \approx 1 - 2$ MK and is readily detected in the postflare phase in extreme ultraviolet (EUV) wavelengths. Also small flares, microflares, and nanoflares (with energies of $E \approx 10^{24} - 10^{27}$ erg) radiate mostly in the EUV temperature range. Combining these wavelengths, one can obtain statistics of solar flare energies extending over up to nine orders of magnitude (Figure 6.1), hence the term "nanoflares." Therefore, gathering flare statistics in EUV is expected to complement the low-energy part of the size distribution sampled in soft X-rays and hard X-rays.

A compilation of occurrence frequency distributions of flare datasets observed in EUV is given in Table 6.1. The theoretically expected values of the power law slopes, according to the standard FD-SOC model (Section 2.7; Table 2.1), are

$\alpha_L = 3.00$ for the length scale L,
$\alpha_A = 2.33$ for the event area A,
$\alpha_V = 1.80$ for the event volume V,
$\alpha_F = 1.80$ for the flux F,
$\alpha_T = 2.00$ for the event duration T, and
$\alpha_E = 1.44$ for the fluence E.

We can test three energy models (Table 2.1): the fluence that is defined by the product of the flux multiplied with the event duration ($E = F \times T$); the 3-D energy that is proportional to the volume V, and the 2-D energy that is proportional to the area A. The observational values match the theoretical predictions (Table 2.1) in a number of previous studies (Aschwanden et al. 2000a, 2000b).

Some of the first EUV power law slopes were pioneered by Krucker and Benz (1998) and Benz and Krucker (2002), assuming a constant depth of the

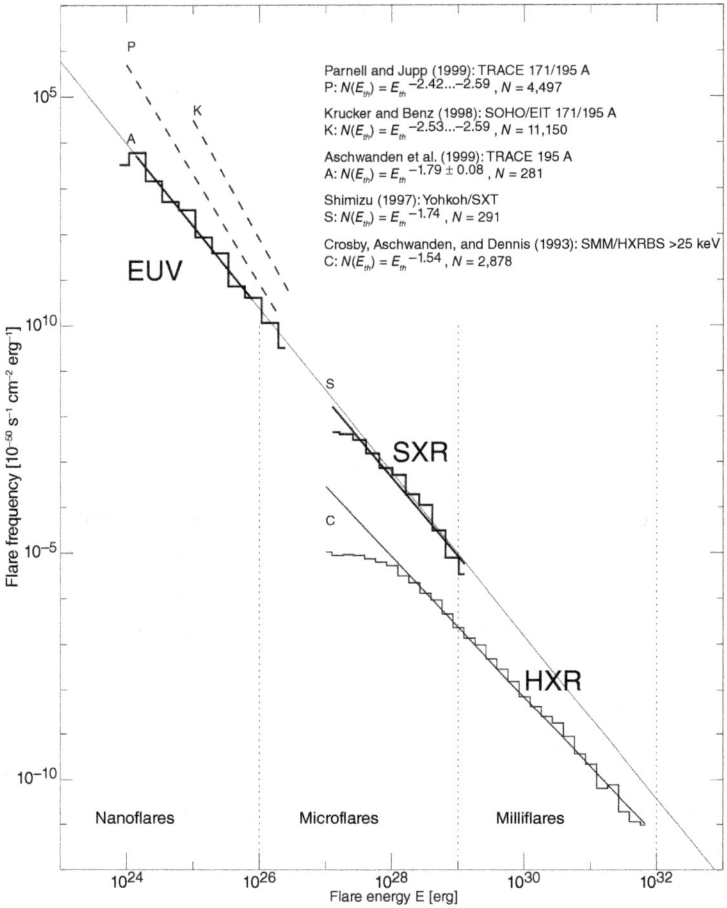

Figure 6.1 Composite flare frequency distribution on a normalized scale in units of 10^{-50} flares per time unit (s^{-1}), area unit (cm^{-2}), and energy unit (erg^{-1}). The diagram includes EUV flares analyzed in Aschwanden et al. (2000b), from Krucker and Benz (1998), from Parnell and Jupp (2000), transient brightenings in soft X-rays (Shimizu 1995), and hard X-ray flares (HXR) (Crosby et al. 1993). All distributions are specified in terms of thermal energy $E_{th} = 3n_e k_B T_e V$, except for the case of HXR flares, which is specified in terms of nonthermal energies in > 25 keV electrons. The slope of $\alpha_F = 1.8$ is extended over the entire energy domain of $10^{24} - 10^{32}$ erg (Aschwanden et al. 2000b).

EUV-emitting volumes, so that the energy is proportional to the 2-D flare area A. For those studies where the nanoflare energy is proportional to the flare area, we expect a power law slope of $\alpha_A \approx 5/3 \approx 2.33$, which indeed matches the observed range of $\alpha_A \approx 2.0 - 2.7$ (column 1 in Table 6.1). For other studies where the nanoflare energy is proportional to the flare volume, we expect a power law slope of $\alpha_F \approx 9/5 = 1.80$, which is indeed matched by the observed median value of $\alpha_F \approx 1.74$ (column 2 in Table 6.1). Hence it is very important to discriminate

Table 6.1 *Frequency distributions measured in small-scale (nanoflare) events in EUV, UV, and Hα wavelengths. The theoretically predicted power law slopes of the flare areas, $\alpha_A = 2.33$, the flare flux, $\alpha_F = 1.80$, and the time duration ($\alpha_T \geq 2.00$), see Table 2.1.*

Power Law slope of flare area α_A	Power Law slope of flare flux α_F	Power Law slope of duration α_T	Waveband λ (Å)	References
2.3 − 2.6			EIT,171,195	Krucker and Benz (1998)
2.0 − 2.6			EIT,171,195	Parnell and Jupp (2000)
(1.19 ± 1.13)			EIT 195	Aletti et al. (2000)
2.31 − 2.59			EIT,171,195	Benz and Krucker (2002)
2.04 − 2.52			EIT,171,195	Benz and Krucker (2002)
2.56 ± 0.23	1.94 ± 0.09		TRACE,171,195	Aschwanden et al. (2000a,b)
	1.93 ± 0.06		TRACE,171	Aschwanden and Parnell (2002)
	1.75 ± 0.07		TRACE,195	Aschwanden and Parnell (2002)
	1.52 ± 0.10		AlMg	Aschwanden and Parnell (2002)
2.12 ± 0.05			6563	Georgoulis et al. (2002)
	1.5 − 3.0		1 − 500	Greenhough et al. (2003)
	1.4 − 2.0		EIT 171,195,284	McIntosh and Gurman 2005
	1.57 ± 0.03		SUMER/S IV 933 Å	Buchlin et al. (2006)
	1.44±0.01		SUMER/Lyϵ	Buchlin et al. (2006)
	1.74 ± 0.01		SUMER/S IV 944 Å	Buchlin et al. (2006)
2.51 − 2.53	1.66 − 1.70	1.96 − 2.02	STEREO/EUVI	Uritsky et al. (2007)
	1.5	2.3	1550	Nishizuka et al. (2009)
2.03 − 2.99	1.64 − 1.91	2.12 − 2.31	STEREO/EUVI	Uritsky et al. (2013)
2.42 − 2.52		2.02 − 2.66	STEREO 171	Aschwanden et al. (2014)
2.66 − 2.69		2.50 − 2.52	STEREO 195	Aschwanden et al. (2014)
2.14 − 2.18		2.15 − 2.24	STEREO 284	Aschwanden et al. (2014)
2.58 − 2.70		2.61 − 2.74	STEREO 304	Aschwanden et al. (2014)
	1.6 ± 0.07		AIA 193	Alipour and Safari (2015)
	1.35 − 2.52			Verbeeck et al. (2019)
2.2 − 2.9				Ulyanov et al. (2019)
2.49	**1.74**	**2.20**	Median	All EUV observations
2.33	**1.80**	**2.00**	FD-SOC prediction	Aschwanden (2012a)

between the two methods of nanoflare detection: If the nanoflare energies are derived from the flare area A, we expect a steeper slope ($\alpha_A \approx 2.33$) than in the case of nanoflare energies derived from 3-D volumes ($\alpha_F \approx 1.80$). The scaling of the volume–area relationship is approximately $V \propto A^{3/2} = A^{1.5}$, if the fractal dimensions are neglected, but $V \propto A^{5/3} = A^{1.67}$, if the fractal dimensions are included (as defined in Eqs 2.37–2.38).

The range of power law slopes measured in EUV appears to be self-consistent with those measured in hard X-rays (Table 4.1) and soft X-rays (Table 5.1), according to the FD-SOC model. Some inconsistencies can be attributed to different event detection methods, incomplete sampling, arbitrary choices of fitting ranges, or inadequate choices of fitting functions (see, for instance, Aletti et al. 2000).

An important bias is the incomplete temperature coverage when statistics of nanoflares is obtained with a single (narrowband) EUV filter, which results into relatively steep power law slopes, while synthesized energy statistics combined from a broader range of EUV, soft X-ray, and hard X-ray filters yield flatter power law slopes (Aschwanden and Parnell 2002).

Equally important is the scaling law used in the definition of flare energy. The classical approach is to estimate the thermal flare energy $E_{th} = 3k_B n_e T_e V$ from the peak electron density n_e, the flare peak temperature T_e, and the flare volume V. However, the flare volume V cannot directly be measured, but only the flare area A instead. The scaling of the thermal energy depends crucially on the used geometric scaling law. Some authors used a cylindrical "pill-box" model $V = Az_0$ with a constant column depth z_0, which corresponds to a geometric scaling law $A \propto L^{D_A}$ (Eq. 2.37), while a spherical volume scales as $V \propto L^{D_V}$ (Eq. 2.38). In Section 2.7, we derived a distribution $N(A) \propto A^{-\alpha_A}$ for 2-D areas, and $N(V) \propto \alpha_F \propto V^{-\alpha_V}$ for 3-D volumes. We will see that event areas A are used for automated event detection in imaging data, while event volumes V are indirectly used to measure the total flux from a volume in the case of (nonimaging) time series data. Since the volume–area relationship is nonlinear, that is, $V \propto A^{5/3} = A^{1.67}$, the power law slopes are different for size distributions of flare volumes ($\alpha_F = 1.80$) and flare areas ($\alpha_A = 2.33$).

Other factors that play a role are the geometric scaling of fractal volumes, the flare selection, the flare detection algorithm, the detection threshold, the synchrony in different temperature filters, the completeness of sampling, truncation effects in small samples, the power law fitting method, see discussion in Benz and Krucker (2002). For the temperature bias, which occurs for single-temperature data in EUV wavelengths, it has been demonstrated that the power law of nanoflare energies (with a typical value of $\alpha_F \approx 1.8$) lowers to a flatter slope $\alpha_E = 1.4$ in an unbiased, complete sample (combining EUV-emitting, 171, 195 and soft X-ray-emitting, AlMg) temperatures.

Nonetheless, flare statistics from different wavelength regimes start to converge, as shown in Figure 6.1. What is still needed is a unified detection method that uniformly samples events from the largest giant flare down to the smallest nanoflares. Good agreement of the power law slopes (α_A, α_V, and α_T) in the observed and theoretically predicted EUV size distributions is found in the studies of Aschwanden et al. (2000a, 2000b) and Uritsky et al. (2007, 2013). Note that both the slope of the fluence ($\alpha_E = 1.44$) and the slope of the volume ($\alpha_F = 1.80$) are less than the critical value of $\alpha_{crit} = 2$, so that the energy integral does not diverge at finite nanoflare values, which makes nanoflaring energetically less important. However, the size distribution of the durations of spatiotemporal energy release events in the Quiet Sun is consistent with random (exponential) behavior, rather than with power law behavior (Uritsky et al. 2013; Uritsky and Davila 2014).

The largest dataset of any astrophysical size distribution has been collected over a time span of nine years with the EIT/SOHO instrument (McIntosh and Gurman 2005), which amounts to the staggering size of $n_{EUV} = 1.3 \times 10^8$ bright point events. Although we expect that large statistics translates into a high accuracy of their power law slope, this is not the case generally, because the power law slope of the time durations ($\alpha_T \approx 1.4 - 2.0$) exhibits significant deviations from an ideal power law (at the high-energy side of the size distribution).

While the size distributions of fluxes, energies, and durations generally exhibit power laws in solar flares, sub-structures of flares, called *flare kernels* or *elementary bursts*, show power law distributions as well, especially in the C IV ≈ 1550 Å wavelength (Nishizuka et al. 2009), or in ≥ 20 keV hard X-rays (Aschwanden et al. 1998).

The energy distribution of thermal energies in nanoflares was also measured with the TESIS/CORONAS-PHOTON spacecraft and the AIA /SDO, yielding power law slopes of $\alpha_E = 2.2 - 2.9$ (Ulyanov et al. 2019). However, the total energy of nanoflares in the range of $E \approx 10^{23} - 10^{26}$ has turned out to be lower than the energy losses of the solar corona through radiation, by a factor of ≈ 30. In order to explain coronal heating by nanoflares, their distribution index must extend at least down to $E \lesssim 10^{21}$ erg, with the same power law slope (Ulyanov et al. 2019).

As an alternative to power law-like size distributions, a log-normal function was fitted to AIA/SDO and XRS/GOES data (Verbeek et al. 2019; Alipour and Safari 2015). Interestingly, GOES/XRS peak fluxes were better described with a power law, while AIA/SDO data are better described by a log-normal distribution. However, the exponential-like drop-off at flare energies is not satisfactorily fitted by neither distribution function. The choice of a power law-like function that represents an acceptable fit to the size distribution is of more fundamental importance for physical SOC modeling than deviations from ideal power laws. Consequently, formal uncertainties (such as chi-squares) are generally second-order effects, compared with the effects of incomplete sampling or finite-size effects, in the derivation of scaling parameters of power law-distributed data (D'Huys et al. 2016). Deviations from an ideal power law function are even more conspicuous in cumulative size distributions due to finite system-size effects, and should be fitted with adequate distribution functions (e.g., see examples in Alipour et al. 2022).

6.2 Solar EUV Nanoflares: Waiting Times

After we examined the waiting time distributions $N(\tau) \propto \tau^{-\alpha_\tau}$ in hard X-rays (Section 4.2; Figure 4.2), in soft X-rays (Section 5.2; Figure 5.4), and the theoretical derivations (Section 3), we expect similar power law slopes for flare events observed in EUV, namely (Eq. 3.22),

$$\alpha_\tau = 2 + \frac{1}{p}, \tag{6.1}$$

where p is the nonlinearity index, the degree (or order) of the polynomial in the mean time evolution of the flare rate $\lambda(t) \propto t^p$, that is ($p = 1$) for linear models, or ($p \gtrsim 2$) for nonlinear models.

Interestingly, events with a linear time evolution ($p = 1$) predict power law slopes of $\alpha_\tau = 3.0$, while events with a nonlinear time evolution ($2 < p < \infty$) predict power law slopes in the range of $\alpha_\tau = (2.0 - 2.5)$. The nonlinearity index p was found to vary in the following ranges (Figure 6.2): $p = 2.2 \pm 0.1$ for clustering of solar flares, $p = 1.8 \pm 0.2$ for partially occulted flares (due to the solar rotation), and $p = 2.8 \pm 0.5$ for the solar dynamo (due to the solar cycle), which constrains the solar memory over a huge time range from hours to decades (Aschwanden and Johnson 2021).

In order to minimize the complexity of multiple nonlinear systems, one study focuses on the statistics of flares by restricting to the same active region. The flare

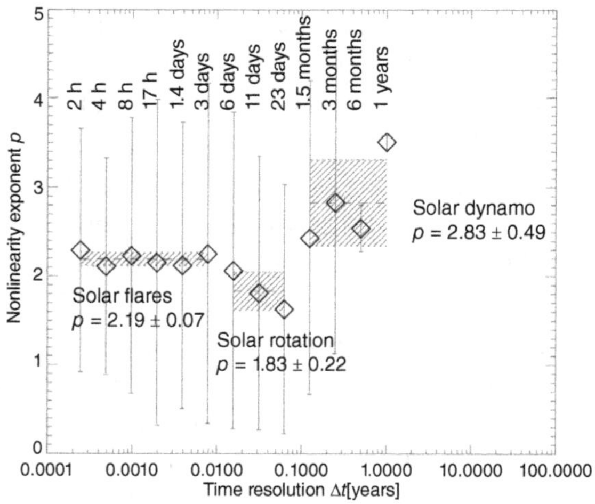

Figure 6.2 The nonlinearity index $p(\tau)$ as a function of the time resolution τ is calculated from an automated GOES flare detection algorithm. Note the triple regime of solar flares (in the time range of $\tau \approx 2$ h–3 days), the solar rotation (in the time range of $\tau \approx 6$–45 days), and the solar dynamo (in the time range of $\tau \approx 1.5$ month–1 year) (Aschwanden and Johnson 2021).

rate was approximated by a piece-wise constant over the duration of a flare. Nevertheless, deviations from a simple power law function were still detected in the observations (Wheatland 2010).

Studies of the statistical properties of the EUV solar flux sudden variation have also been carried out by using data from the dual-frequency Global Navigation Satellite Systems (GPS) (Monte-Moreno and Hernandez-Pajares 2014). The solar flux variation time series was characterized by a fractional Brownian model for the long-term dependence, and a power law distribution for the extreme values in the time series.

6.3 Solar EUV Nanoflares: Cellular Automatons

The first cellular automaton (CA) simulations applied to solar flares were first carried out in hard X-ray wavelengths (Lu and Hamilton 1991; Lu et al. 1993). These first CA simulations reproduce the simulated power laws of $N(E) \propto E^{-1.4}$ and $N(F) \propto F^{-1.8}$, in agreement with the predictions of the (fractal-diffusive) FD-SOC model (i.e., $\alpha_E = 1.44$ and $\alpha_F = 1.80$, see Table 2.1), which is consistent with a nonstationary Poissonian distribution, rather than with a stationary (exponential) distribution function. Follow-up CA simulations investigated the dependence of the power law slope on the grid size, as well as the dimensionality of CA systems (Lu et al. 1993).

Motivated by the recent interest in nanoflares, which contribute to the overall magnetic energy dissipation mostly when the power law slope is steeper than the critical value of $\alpha \geq \alpha_{crit} = 2$ (Hudson 1991), hybrid avalanche models were constructed with the observed power law slopes of $\alpha_E \approx 1.8$ at the high-energy side of the size distribution, and a secondary component with a (hypothesized) steeper slope of $\alpha_E \approx 3.5$ at the low-energy side of the size distribution (Vlahos et al. 1995). Similar broken power laws appear when WATCH/GRANAT data were subdivided by time duration (Georgoulis et al. 2001). The so-called anisotropic statistical flare model of Vlahos et al. (1995) appears to be similar to the avalanche CA model of Bak et al. (1988) and Lu and Hamilton (1991) at first glance, but in contrast, produces artificially enhanced nanoflare event rates. Moreover, we notice that inertial (power law) ranges extend over a single decade only in CA simulations of active regions driven by observed magnetograms (Dimitropoulou et al. 2011), while typical SOC systems display size distributions over 3–4 decades. Deviations from straight power law slopes are noted in magnetogram-based CA simulations also (Dimitropoulou et al. 2013).

The CA simulations of solar flare SOC models have led to the following physical picture: Flares are the result of an avalanche of small-scale reconnection events cascading through a highly stressed coronal magnetic structure, driven to a critical state by random photospheric motions of its magnetic footpoints, and this way provide a natural and convenient computational framework to examine Parker's hypothesis of coronal heating by nanoflares (Charbonneau et al. 2001).

Magnetic discontinuities are formed and evolve either when the footpoints of thin magnetic fibrils are shuffled by the continuous velocity field on the photosphere, or when new magnetic flux emerges from the convection zone. Linear force-free extrapolations of the photospheric magnetic field are calculated in the study of Vlahos and Georgoulis (2004), based on magnetic free energies,

$$E = \frac{\lambda^2 \delta h}{8\pi} \sum_{l=1^N} (\mathbf{B}_{ff,l} - \mathbf{B}_{p,l})^2, \tag{6.2}$$

where $\mathbf{B}_{\mathbf{ff,l}}$ is the linear force-free field, and $\mathbf{B}_{p,l}$ is the potential field at location l. Power law slopes of $\alpha_E \approx 1.35 - 1.60$ are calculated, which are consistent with the theoretical prediction of the FD-SOC model, that is, $\alpha_E = 1.44$ (Section 2.7).

However, there is ample room for variants of the Bak-Tang-Wiesenfeld (BTW) model: isotropic CA models, anisotropic CA models, discretized MHD CA models, divergence-free field braiding models, or branching process models (Aschwanden 2011a, p. 63).

As an example of an anisotropic CA model, the basic lattice structure of the cellular automaton model of Morales and Charbonneau (2008, 2009), based on a 2-D lattice with anisotropic connectivity, is shown in Figure 6.3, consisting of initially parallel field lines that have their footpoints rooted in a horizontal 1-D lattice grid (x), while their lengths extend over the vertical axis (y), so that every field line position has a unique coordinate (x, y). This initial setup is periodically disturbed by a random displacement of node (x, y) to position $(x + \Delta x, y)$, which is measured

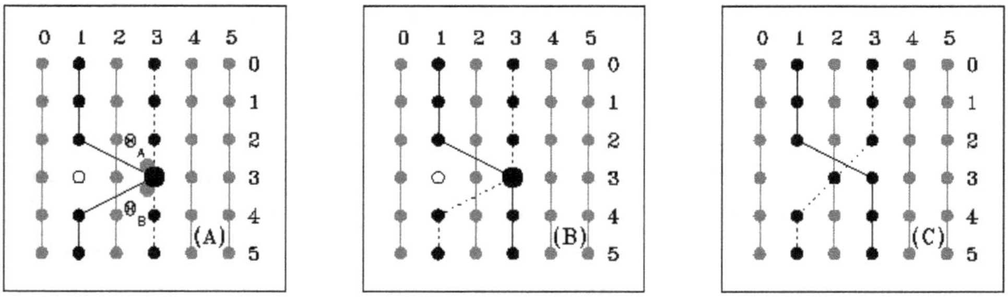

Figure 6.3 Sequence of a two-step scheme in the divergence-free field braiding cellular automaton model of Morales and Charbonneau (2008). Node $j = 3$ of field $i = 1$ is displaced by two units, causing misalignment angles θ_a and θ_b above a critical value (left frame A), triggering a new connectivity between $i = 1$ and $i = 3$ (middle frame B), which is still not stable and triggers another new connectivity between line 1 and 3 (dashed: right frame C) (Morales and Charbonneau 2008).

by a misalignment angle $\theta(x,y)$. If the misalignment angle exceeds a critical angle θ_c, the intersecting field lines can reconnect into an alternative configuration by a "cut-and-splice" operation, until a subcritical misalignment angle is reached. The redistribution rule involves a new connectivity for crossing field lines that has a smaller misalignment angle. The example shown in Figure 6.3 shows a two-step redistribution scheme where field lines 1 and 3 exchange their connectivity in two time steps and end up with a sub-critical angle everywhere. Multiple reconnection steps are often required to reach a new stable state. Such chains of reconnection steps represent the avalanches in sandpiles. Note that the avalanches are highly anisotropic, in contrast to isotropic avalanches in the BTW sandpile model.

Further CA models applied to multistranded coronal loops were developed in 1-D (Lopez Fuentes and Klimchuk 2010), as well as in 2-D (Lopez Fuentes and Klimchuk 2015, 2016) (Figure 6.4), using the Enthalpy-Based Thermal Evolution of Loops (EBTEL) model, which is a 0-D hydrodynamic code. The EBTEL code is based on scaling laws under equilibrium conditions between the heating rate and the conductive cooling rate, and/or the radiative cooling rate (Cargill et al. 2012). Since the EBTEL code simulates the time evolution of avalanches, size distributions of avalanche energies can be inferred. The energy released in an EBTEL avalanche event corresponds to the difference in the magnetic energy before and after reconnection,

$$\Delta E = (E' - E) = \frac{(\Delta B_h)^2}{4}, \tag{6.3}$$

where B_h is the horizontal magnetic field. The mean nanoflare heating rate $< Q >$ (averaged both in time and along the strand) is,

$$< Q > \approx \frac{v B_v^2 \tan\theta}{L}, \tag{6.4}$$

with L the length of a nanoflare flux tube. The resulting energy size distribution is found to have a slope of $\alpha_E = 2.47$ (Lopez Fuentes and Klimchuk 2010), which appears to correspond to the slope of the 2-D nanoflare areas, $\alpha_A = 2.33$ (Table 2.1).

A similar approach to explain the structure of coronal loops in terms of self-organized braiding has been proposed (Berger and Asgari-Targhi 2009; Berger et al. 2015), but the simulated size distributions come out significantly steeper ($\alpha_E \approx 3.0$) in the numerical simulations than in the theoretical predictions (i.e., $\alpha_E \approx 1.44$, and $\alpha_F = 1.80$, similar to the result of Lopez Fuentes and Klimchuk 2010).

A more detailed approach that demonstrates how an MHD avalanche might occur in a multithreaded coronal loop was simulated by Hood et al. (2016). It was found that only one (out of 23) threads needs to be unstable, in order to start

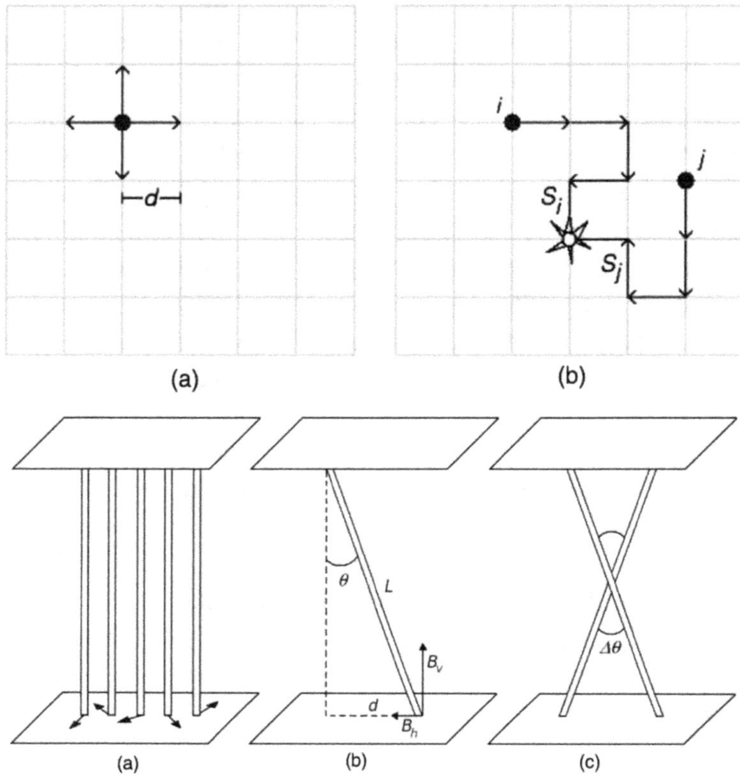

Figure 6.4 *Top:* Footpoint motion (trajectory S_i) and loop top motion (trajectory S_j) in a Cartesian grid is depicted. *Bottom:* A schematic description of a CA model is shown. (a) Initially, all strands are parallel and there is no magnetic stress between neighbors. The horizontal planes represent two distant portions of the photosphere connected by the strands. (b) Photospheric motions displace the strand footpoints, increasing the mutual inclinations and producing magnetic stress. (c) The inclination of the strands (θ) with the appearance of a horizontal component of the magnetic field (Lopez Fuentes and Klimchuk 2010).

an avalanche event when the others are below marginal stability, which produces episodic (avalanche-like) heating events, rather than constant heating of the solar corona (Hood et al. 2016).

Attempts have been undertaken to detect and extract small-scale heating events and their underlying loop geometries from 3-D magneto-hydrodynamic (MHD) simulations (Guerreiro et al. 2015, 2017). About 10^4 small-scale heating events have been detected at any instant above the simulated area of 128 Mm^2, at heights of $h \approx 1.5 - 23.0$ Mm. Volumes and energy distributions of the events were found to resemble power laws (Guerreiro et al. 2015, 2017). The temporal and spatial distribution of the Ohmic heating in a 3-D MHD model was found to follow a power law and can be considered as a SOC system (Bingert and Peter 2013).

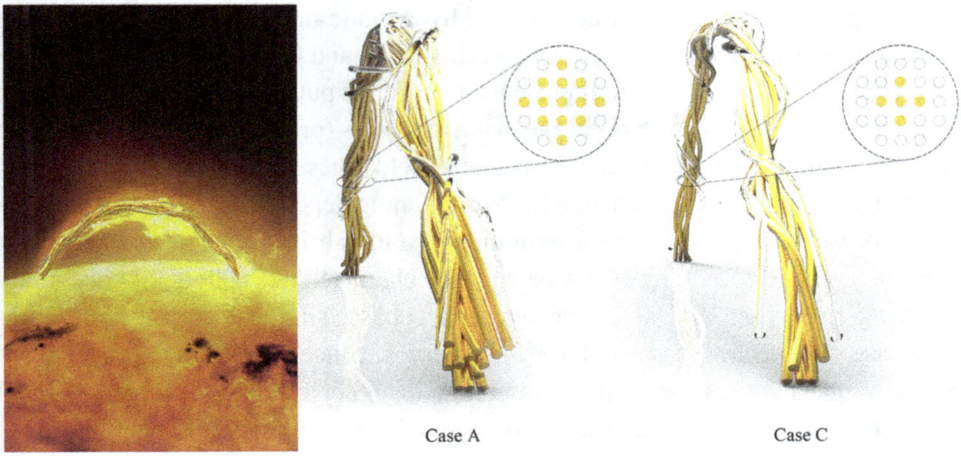

Figure 6.5 Cellular automaton simulations of a coronal loop, consisting of numerous loop strands. The cross-sectional structure of a coronal loop is rendered with a bundle of 13 circular structures (Case A) and 5 circular structures (Case C) (Farhang et al. 2018).

A different approach to obtain statistics of nanoflare energies is the identification of coronal heating episodes in 3-D MHD simulations, using the *Bifrost* code (Kanella and Gudisksen 2017). In one run with 4,136 identified events in a volume with $768 \times 786 \times 331$ grid cells, power law size distributions with power law slopes of $\alpha_F = 1.5 \pm 0.02$ (at low energies) and $\alpha_F = 1.53 \pm 0.03$ (at high energies) and $\alpha_F = 2.53 \pm 0.22$ were found, which differs from the theoretical prediction of the FD-SOC Model because it displays a broken (or double) power law size distribution (see Figure 4 in Kanella and Gudisksen 2017).

The concept of turbulent photospheric flows that twist the coronal field was simulated with a statistical model based on the principle of minimum energy in a coronal loop undergoing magnetic reconnection, which is described as an avalanche process (Farhang et al. 2018), see Figure 6.5. The result of a power law index, $\alpha_F = 1.84 \pm 0.02$, is close to the theoretical prediction of the FD-SOC model, $\alpha_F = 1.80$ (Table 2.1). Additional simulations with an anisotropic optimization model are given in Farhang et al. (2019). CAs based on twisted magnetic flux ropes are described in Wang et al. (2022).

6.4 Solar EUV Nanoflares: Coronal Loops

The most conspicuous structures that are observed in solar EUV images are the so-called *coronal loops*, which appear as semi-circular bundles of magnetic field lines, having typical widths of $w \approx 1$ Mm and lengths of $L \approx 10 - 100$ Mm (Cirtain et al. 2013; Aschwanden and Peter 2017). The apparent length of coronal loops has been explained and modeled in terms of hydrostatic and super-hydrostatic scale heights (Aschwanden and Nitta 2000; Aschwanden and Charbonneau 2002). Since every observed loop structure represents a subset of putative magnetic field lines, the question arises what physical mechanism selects (or fills) bright loops, among a volume-filling complete set of magnetic field lines in a given field-of-view. Although the physical process of a loop brightening depends on the assumed radiative process, we can consider the evolution of a loop brightening as an elementary avalanche process, as long it can be suitably characterized with an exponentially growing rise phase and a decreasing decay phase, similarly to the more complex rise and decay phase of flare light curves. Since bright emission in EUV wavelengths requires hotter temperatures and higher electron densities, coronal loops can be considered as tracers of coronal heating processes near the footpoints of EUV loops. The cross-sectional width of coronal loops observed in EUV wavelengths may thus reveal the geometry of the coronal heating function, which operates in flaring active regions, as well as in the quiet-Sun network. While the geometry of solar flares is quantified by an isotropic and fractal 3-D structure in the FD-SOC model, coronal loops are 1-D structures, due to the action of the magnetic field, which is a strongly anisotropic structure.

There has been a controversy whether coronal loops display a uniform thickness along their length ($w(s)$) or whether they expand with height, as expected from a linear $(\nabla \times \mathbf{B}) = \alpha \mathbf{B}$, or nonlinear magnetic field model (Lopez Fuentes and Klimchuk 2006). An implication of loop width variations $w(s)$ is the geometric correction of loop volumes, which are often approximated by cylindrical flux tubes, that is, $V_{loop} \approx L w^2$, which can be used to characterize the (multi)fractal geometry of loops. The highest resolution observations reveal that active region loop widths are fully resolved with the Hi-C instrument, with a mean value of $w \approx 550$ km, while the same loops are marginally resolved with the AIA/SDO instrument (Aschwanden and Peter 2017). This result is highly relevant to quantify the filling factor or the fractal dimension of EUV emission.

6.5 Solar EUV Nanoflares: Field Line Braiding

Coronal heating models are traditionally modeled in terms of direct (electric) current (DC) models, alternating current (AC) models (Alfvén waves), or MHD turbulence (Zirker 1993; Uritsky et al. 2007). For the interpretation of nanoflare mechanisms, however, a controversy arose whether twisting or braiding is more consistent with the observed size distributions (Zirker and Cleveland 1993a, 1993b). The mean helical twisting number and the braiding linkage number of solar coronal loops in active regions has been determined to be about a half turn (Aschwanden 2019). The example simulated in Figure 6.7 illustrates a possible cascading magnetic reconnection process.

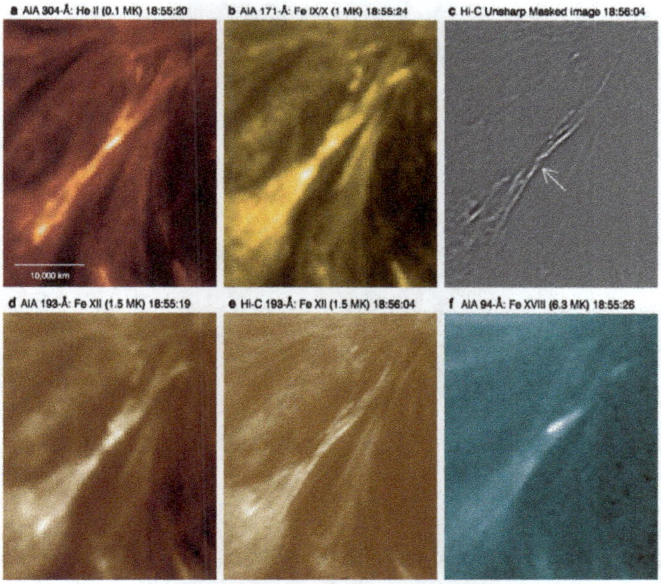

Figure 6.6 Observations of braided filament structures from the Hi-C rocket flight and the AIA/SDO spacecraft. The six panels are obtained from six EUV different wavelengths, 304 Å , 171 Å , Hi-C unsharp masked, AIA 193 Å , Hi-C 193 Å , and AIA 94 Å (Cirtain et al. 2013).

The Parker model for heating of the solar corona involves magnetic reconnection of braided magnetic field lines (Berger and Asgari-Targhi 2009). In alternating current (AC) models, photospheric footpoint motion generates MHD waves that propagate upward along magnetic flux tubes and dissipate their energy in the corona (Berger et al. 2015). Reconnected field lines evolve into a self-organized critical state (Berger et al. 2015).

A direct confirmation of the braided geometry (Figure 6.6) is difficult, since the spatial resolution of relevant observations may be insufficient. Only very few cases with braided geometries have been reported so far (Figure 6.6), such as with the Hi-C instrument (Cirtain et al. 2013). A reanalysis of Hi-C data revealed a 100 times higher magnetic energy associated with the field line braiding, where the overall twist of the flux rope increased by about a half turn within 12 min (Thalmann et al. 2014).

Numerical 3-D simulations of magnetic flux braiding were performed, driven by footpoint motion, which relax into nonlinear force-free equilibria (Longbottom et al. 1998).

The inevitable consequence of the magnetic braiding process is a loss of equilibrium of the magnetic field, probably via magnetic reconnection processes (Wilmot-Smith et al. 2009a). The process of spinning and braiding are found to be driven by a kink instability in a submerged flux tube (Longcope et al. 2007).

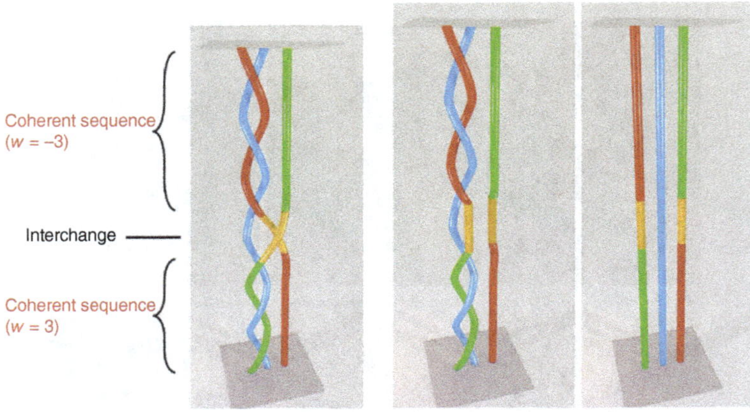

Figure 6.7 Simulation of a cascading magnetic reconnection process (Berger and Asgari-Targhi 2009).

If the degree of braiding of the magnetic field is increased, the so-called *squashing factor* increases with an exponentially function (Wilmot-Smith et al. 2009b). A resistive 3-D MHD simulation code mimics the dynamics of braided coronal loops, which starts from a near force-free state in the beginning, leading to the formation of two thin current layers, and results in an untwisting of the global field structure (Wilmot-Smith et al. 2009a; 2009b). A threshold level of the braided loop contains $10^{26} - 10^{28}$ ergs of available free magnetic energy, which is sufficient to heat large nanoflares (Pontin and Hornig 2015). Estimating the rate of field line braiding in the solar corona and photospheric flows is obtained with magneto-convection simulations (Candelaresi et al. 2018).

6.6 Solar EUV Nanoflares: Magnetic Reconnection

How can we understand the power law distribution function of electromagnetic emission in solar flares in general? On a microscopic level, SOC models operate via next-neighbor interactions, while magnetic reconnection scenarios describe a physical process on a macroscopic level. Moreover, there is some ambiguity in the interpretation of a size distribution or a power spectrum, since different physical mechanisms (such as SOC versus fluid turbulence) can produce the same power law spectrum (for instance, Bershadskii and Sreenivasan 2003).

A number of specific magnetic reconnection processes have been reformulated in order to accommodate the SOC rules, with the goal to reproduce the power law distribution functions and the resulting power law exponents. For instance, one model employs the conservation of the magnetic helicity during the avalanche transport in a lattice grid, which yields values in the range of $\alpha = 1.25 - 1.35$ (Chou 1999, 2001), which is not consistent with the FD-SOC model that predicts $\alpha_F = 1.80$ for the mean flux F. Another magnetic reconnection model applies separator reconnection in a minimum current corona model, where stress produces currents flowing along the magnetic field's network (Longcope and Noonan 2000). This model produces power law exponents of $\alpha_F = 1.48 \pm 0.13$ for the peak flux, $\alpha_E = 1.34 \pm 0.05$ for the energy, and $\alpha_T = 1.53 \pm 0.11$ for the time duration, all being significantly lower than the FD-SOC prediction ($\alpha_F = 1.80$, $\alpha_E = 1.44$, $\alpha_T = 2.0$). This discrepancy may be a consequence of the MHD approximation in the minimum current corona model, which implies that the magnetic field remains in its state of minimum energy at all times, and thus may not be appropriate to describe the dissipated energies of SOC avalanches as they occur in solar flares (Longcope and Noonan 2000).

McIntosh and Charbonneau (2001) simulated the size distribution of flare areas A, yielding a mean value of $\gamma = 1.41 \pm 0.04$ for the ratio γ of the 2-D and 3-D fractal dimensions. In the FD-SOC model (Section 2.7), the mean fractal dimensions $D_A = 1.5$ and $D_V = 2.5$, defined by the Hausdorff dimension, $A = L^{D_A}$ and $V = L^{D_V}$, yield a ratio of $\gamma = D_V/D_A = 5/3 = 1.67$. The discrepancy between the observed and theoretically predicted value of γ underscores the importance of spatial modeling of SOC avalanches (McIntosh and Charbonneau 2001; McIntosh et al. 2002).

A magnetic reconnection model that produces cascades of reconnecting loops has been simulated by injection of small scales with opposite magnetic polarities, which aggregate at the loop footpoints and trigger magnetic reconnection whenever two loops collide or annihilate (Hughes et al. 2003). Size distributions for the number of loops n emerging from each footpoint with slopes of $\alpha_n = 2.0 \pm 0.1$ are found, and slopes of $\alpha_E = 3.0$ for flare energies, which is substantially larger than predicted from the FD-SOC model ($\alpha_E = 1.44$). However, it can be reconciled for energy definitions that are proportional to the length scale, that is, $E \propto L$, since $N(E) \propto E^{-\alpha_E}$ with $\alpha_E = \alpha_L = 3$. This is one more example of a CA model that fits a plausible physical model, but does not match the observed power law slopes.

Instead of deriving the size distribution from a sample of solar flare events, statistics was also gathered from samples of "elementary flare sub-bursts," also called "flare kernels," observed in ultra-violet (C IV) wavelengths, which exhibit numerous flare kernels that are localized in the chromospheric flare ribbons (Nishizuka et al. 2009). The time profiles of the C IV kernels showed intermittent bursts, with peak intensity slopes of $\alpha_F \approx 1.5$, time duration slopes of $\alpha_T = 2.3$, and waiting time slopes of $\alpha_\tau = 1.8$, which agree roughly with predictions of the FD-SOC model ($\alpha_F = 1.80$, $\alpha_T = 2.0$, and $\alpha_\tau \gtrsim 2.0$).

Numerical simulations of the solar corona driven by convective motions at the photospheric level were performed by Knizhnik et al. (2018). The underlying physical model uses temperature changes, current distributions, and Poynting fluxes as proxies for heating, which yield power law frequency distributions with slopes of $\alpha_A = 2.49 \pm 0.03$ for flare areas A, $\alpha_V = 1.90 \pm 0.01$ for flare volumes V, $\alpha_T = 2.53 \pm 0.06$ for flare durations T, and $\alpha_F = 1.76 \pm 0.02$ for flare intensities T, which agree well with the theoretical predictions of the FD-SOC model: $\alpha_A = 2.33 \pm 0.03$ for flare areas A, $\alpha_V = 1.80 \pm 0.01$ for flare volumes V, $\alpha_T = 2.00 \pm 0.06$ for flare durations T, and $\alpha_F = 1.80 \pm 0.02$ for flare intensities F. Testing three different quantities, the obtained size distributions appear to follow a power law that is better matched for strictly positive incremental temperature changes (above the mean), than for horizontal Poynting flux fluctuations, or for current density fluctuations (Knizhnik et al. 2018).

The energy of a SOC avalanche has been quantified by thermal energies (in SXR, EUV, and UV wavelengths), as well as by nonthermal energies (in HXR and radio wavelengths), where thermal energies are associated with incoherent radiation, and nonthermal energies could be produced by both incoherent or coherent emission. For an example, nonthermal particles can be accelerated by quasi-static electric fields, such as a convective super-Dreicer electric field. Litvinenko (1996) calculated a convective electric field in a reconnecting current sheet using the Ampére law. Anastasiadis et al. (2004) calculated then the kinetic energy distribution of accelerated electrons and obtained power law size distributions with values in a range of $\alpha_E \approx 1.8 - 2.3$, but exponential distributions were found also, perhaps due to the neglect of Coulomb collisions in the model (Anastasiadis et al. 2004).

On a more generalized level, SOC processes can be interpreted in terms of scale-free networks, where avalanches act as network links, and loop footpoints as network nodes (Paczuski and Hughes 2004). SOC is then a mechanism for generating scale-free networks. Probability functions of network nodes, however, yield rather high size distribution slopes ($\alpha_n = 3.5 - 4.8$) for network node sizes (Gheibi et al. 2017; Daei et al. 2017), which is not consistent with any SOC model.

Solar flare models have been proposed in terms of cascades of reconnecting magnetic loops (Hughes et al. 2003), evolving in space and time. Misaligned magnetic field lines can break and reconnect, producing nanoflares in avalanche-like processes (Antolin et al. 2021). Loops injected at small scales are anchored in footpoints of opposite polarity, moving randomly on the solar surface. Nearby footpoints of the same magnetic polarity aggregate, and loops can reconnect when they collide, possibly triggering a cascade of further reconnection. Numerical simulations that mimic a power law distribution of flare energies emerges are associated with a scale-free network of loops, indicating self-organized criticality (Hughes et al. 2003). The simulations of Hughes et al. (2003), however, yield a steeper power law slope ($\alpha_E = 3.0$) than predicted by the standard fractal-diffusive SOC model ($\alpha_E = 1.44$). Similar MHD-driven simulations of magnetic reconnection events come closer ($\alpha_E = 1.76 \pm 0.02$) to the theoretically predicted value ($\alpha_F = 1.8$).

There is observational evidence that the mechanism of solar nanoflares may be close to the standard flare model, which are found to have typical values of $T_e = 3.1$ MK for the plasma temperature, $E_{th} = 6.2 \times 10^{25}$ *erg* for the thermal energy, and an electric-current distribution that peaks at a height of $h \approx 1.5$ Mm (Ulyanov et al. 2019).

The appearance of nonpotential magnetic fields could be a fundamental process leading to self-organized criticality in the Quiet-Sun-like supergranular network (Chesny et al. 2016). Comparisons of the scaling properties of EUV intensity fluctuations in coronal holes and quiet-Sun regions indicate ubiquitous magnetic reconnection processes (Cadavid et al. 2019a, 2019b).

6.7 Summary

A key result of solar flare statistics is the continuity of size distributions over nine orders of magnitude, consisting of nanoflares (detected in EUV), microflares (detected in soft X-rays), and large flares (detected in hard X-rays), covering a range of $E \approx 10^{24} - 10^{33}$ ergs in energy. The FD-SOC model predicts power law distribution functions with a slope of $\alpha_A = 2.33$ when the energy of flare events are derived from the flare event 2-D area A, but a flatter slope of $\alpha_F = \alpha_V = 1.80$, if the flare energies are derived from the volume-integrated total flux of the 3-D flare volume V. These predictions match the observations of EUV nanoflares and microflares. These scaling laws imply more energy is distributed at large flare sizes ($\alpha_E = 1.44 \le \alpha_{crit} = 2.0$) and thus makes nanoflares less important for coronal heating.

Theoretical predictions of the waiting time distribution cover a range of $\alpha_\tau = 2 + 1/p \approx 2.5 - 3.0$, This includes processes with a linear time evolution $(p = 1) \mapsto \alpha_\tau \approx 3$, and processes with a nonlinear time evolution $(p = 2) \mapsto \alpha_\tau \approx 2.5$, such as a quadratically (or exponentially growing) flare instability. Such scaling laws are numerically simulated with cellular automaton codes, and are applied to the time evolution of coronal loops, magnetic field line breading, and magnetic reconnection processes.

7

Solar Photospheric Events

We focus on phenomena of magnetic energy dissipation events in the solar photosphere. We apply the model of self-organized criticality (SOC) to the observed (occurrence frequency) size distributions and fractal dimensions. The solar photosphere is mostly observed by means of Stokes polarimetry in strong iron atomic lines. Although the avalanching events in the photosphere exhibit quite different morphologies, called active regions, sunspots, plages, magnetic elements, magnetic brightenings, magnetic bright points, and so on, they can all be quantified by standard SOC parameters (length scale, area, volume, flux, fluence, energy, and time duration). We will test whether the observed size distribution functions (e.g., Figures 7.1 and 7.2), the fractal dimensions (Table 7.1), and the underlying scaling laws are consistent with the standard fractal-diffusive self-organized criticality (FD-SOC) model (Section 2.7).

7.1 Solar Photospheric Events: Size Distributions

Examples of photospheric magnetic features are shown in Figure 7.3, ranging from the largest (sunspot) structure (Figure 7.3a) to the smallest "salt-and-pepper" texture (Figure 7.3b).

A dataset obtained from 12 different spatial locations and time intervals, observed with the Interface Region Imaging Spectrograph (IRIS) at a wavelength of 1,400 Å, is shown in Figure 7.1 (Nhalil et al. 2020; Aschwanden and Nhalil 2022a, 2022b). The corresponding size distributions of event areas A have been fitted with power law functions $N(A)dA \propto A^{-\alpha_A} dA$. Obviously, all 12 distributions are close to a power law function, and the power law slopes are all similar (Figure 7.2). The total number of automatically detected events amounts to 23,633, for all analyzed 12 datasets together, varying from 65 to 4,725 events per IRIS dataset (Figure 7.1). The power law slope fits vary from the lowest value $\alpha_A = 2.14$

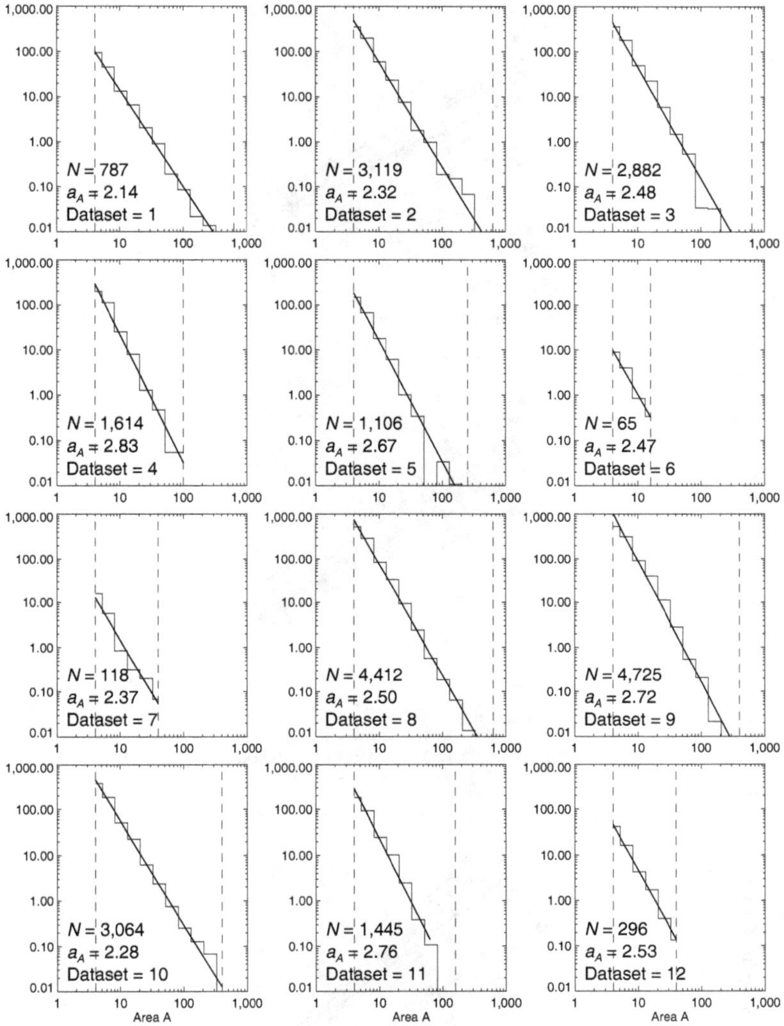

Figure 7.1 Size distributions of event areas A for 12 datasets observed with IRIS SJI 1,400Å in the transition region of different active regions (Aschwanden and Nhalil 2023).

(dataset 1) to the highest value $\alpha_A = 2.83$ (dataset 4), and have a mean slope α_A^{obs} and standard deviation of (Figure 7.2a),

$$a_A^{obs} = 2.51 \pm 0.21, \tag{7.1}$$

which is close to the theoretically predicted value α_A^{theo} of the standard FD-SOC model (Table 2.1),

$$a_A^{theo} = 1 + \frac{(d-1)}{D_A} = (7/3) \approx 2.33. \tag{7.2}$$

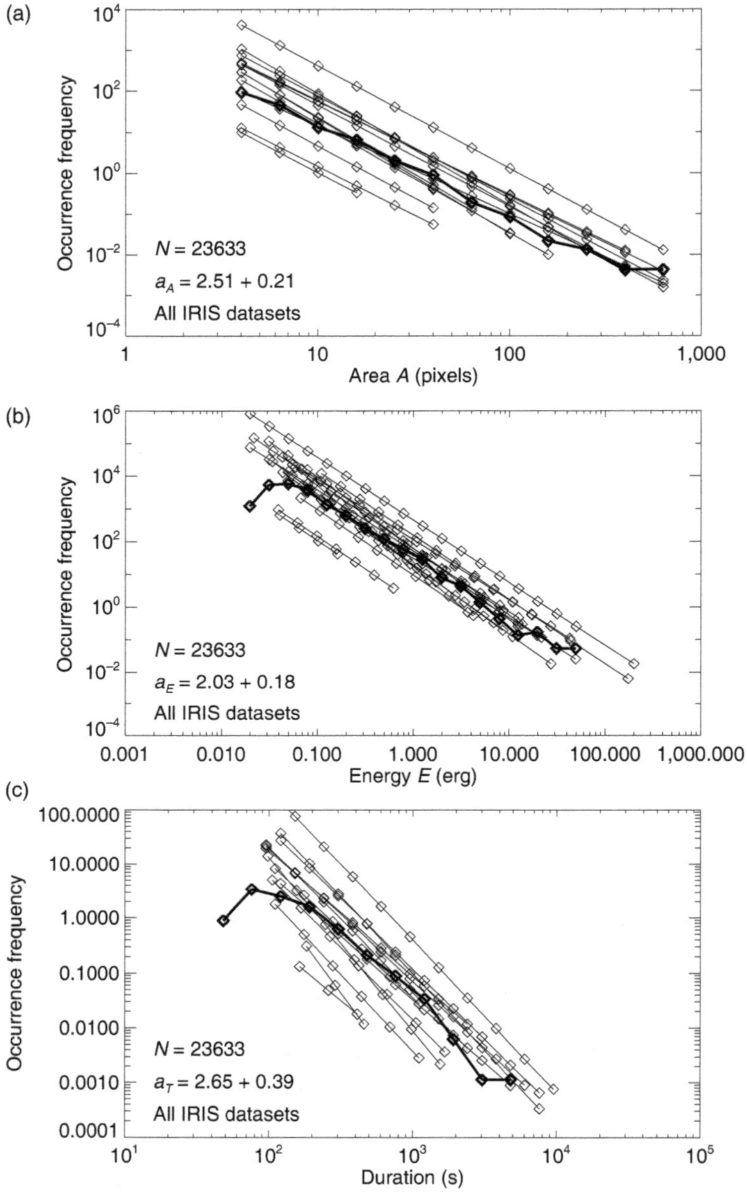

Figure 7.2 Power law fits to the size distributions of three SOC parameters: the event area A (a), the radiated energy flux F (b), and the time duration T (c). Individual fits to each of the 12 IRIS datasets are indicated with a thin line style, while the fit to all events combined is indicated with a thick line style, and the power law slopes are given in each panel (Aschwanden and Nhalil 2023).

Table 7.1 *The fractal (Hausdorff) dimensions D_A are listed for granules, plages, sunspots, and active regions. Statistical means and standard deviations of each of the four groups are indicated with numbers in boldface. The reference A+N (2022) stands for Aschwanden and Nhalil (2022).*

Phenomenon in photosphere	Data, method	Fractal dimension D_A	Reference
Granulation		**1.21 ± 0.07**	**Mean**
Granules	5,750 Å, PA	1.25	Roudier and Muller (1986)
Granules	5,257 Å, PA	1.30	Hirzberger et al. (1997)
Granular cells	5,257 Å, PA	1.16	Hirzberger et al. (1997)
Granules	WL, PA	1.09	Bovelet and Wiehr (2001)
Granules	MDI/SOHO, PA	1.25	Paniveni et al. (2005)
Supergranulation	3,934 Å, Ca II K, PA	1.23 ± 0.02	Paniveni et al. (2010)
Plages		**1.26 ± 0.13**	**Mean**
Plage, TR	IRIS 1,400 Å	1.09	Nhalil et al. (2020), A+N (2022)
Plage, TR	IRIS 1,400 Å	1.20	Nhalil et al. (2020), A+N (2022)
Plage, TR	IRIS 1,400 Å	1.36	Nhalil et al. (2020), A+N (2022)
Plage, TR	IRIS 1,400 Å	1.45	Nhalil et al. (2020), A+N (2022)
Plage, TR	IRIS 1,400 Å	1.33	Nhalil et al. (2020), A+N (2022)
Plage, TR	IRIS 1,400 Å	1.16	Nhalil et al. (2020), A+N (2022)
Plage, TR	IRIS 1,400 Å	1.14	Nhalil et al. (2020), A+N (2022)
Plage, TR	IRIS 1,400 Å	1.31	Nhalil et al. (2020), A+N (2022)
Sunspots		**1.54 ± 0.13**	**Mean**
Sunspots	IRIS 1,400 Å	1.75	Nhalil et al. (2020), A+N (2022)
Sunspot/plage	IRIS 1,400 Å	1.52	Nhalil et al. (2020), A+N (2022)
Sunspot/plage	IRIS 1,400 Å	1.35	Nhalil et al. (2020), A+N (2022)
Sunspot/plage	IRIS 1,400 Å	1.56	Nhalil et al. (2020), A+N (2022)
Active regions		**1.59 ± 0.20**	**Mean**
Active regions	BBSO, MG, LA	1.56 ± 0.08	Lawrence (1991)
	BBSO		Lawrence and Schrijver (1993)
Active region plages	6,302 Å, Fe I, LA	1.54 ± 0.05	Balke et al. (1993)
Active regions	7,929 Å, LA	1.86 ± 0.08	Meunier (1999)
Active regions	7,929 Å, PA	1.58 ± 0.18	Meunier (1999)
Small scales	6,302 Å, Fe I, PA	1.41 ± 0.05	Janssen et al. (2003)
Active regions	6,768 Å, Ni I	1.80 ± 0.09	Meunier (2004)
- Cycle minimum	6,768 Å, Ni I	1.31 ± 0.22	Meunier (2004)
- Cycle rise	6,768 Å, Ni I	1.80 ± 0.16	Meunier (2004)
- Cycle maximum	6,768 Å, Ni I	1.76 ± 0.04	Meunier (2004)
Active regions	6,768 Å, Ni I, BC	1.35 ± 0.30	McAteer et al. (2005)
	5,250 Å, Fe I	1.5	Ioshpa et al. (2008)

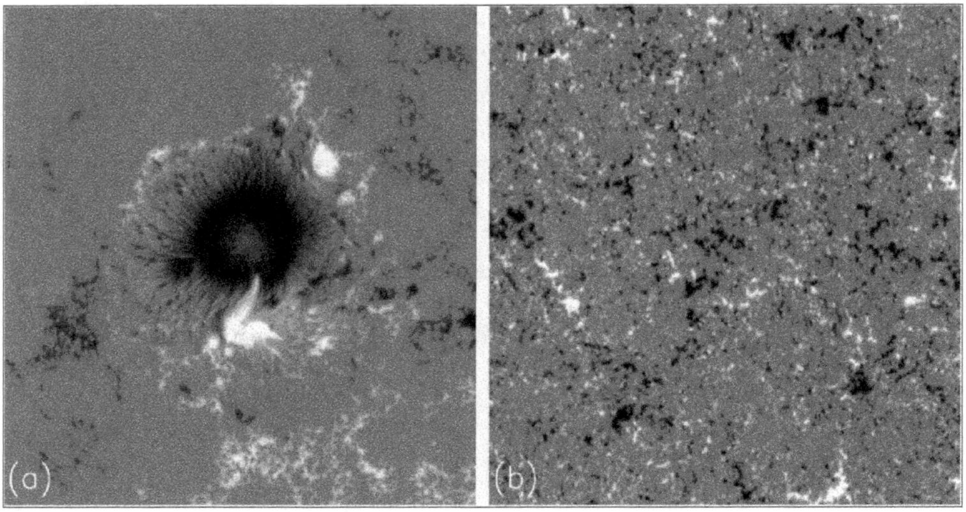

Figure 7.3 Examples of a large sunspot structure (a) and the salt-and-pepper structure in the Quiet Sun (b), observed with the Spectro-Polarimeter (SP) onboard *SOT/Hinode*, which takes line profiles of the Stokes V parameter in the blue wing of the 6320 Å spectral absorption line, recorded over a slit of 0.15″ (Parnell et al. 2009).

The theoretically predicted value for the fractal dimension is $D_A = D_2 = d - (1/2) = (3/2) = 1.50$, based on the dimensionality $d = 2$. Fitting the mean flux distributions, $N(F)dF \propto F^{-\alpha_F} dF$, yields the following mean for all 12 IRIS datasets (Figure 7.2b),

$$a_F^{obs} = 2.03 \pm 0.18, \tag{7.3}$$

which matches the theoretically predicted value α_F^{theo} of the 3-D mean flux model (Table 2.1),

$$a_F^{theo} = 1 + \frac{(d-1)}{\gamma D_V} = 1 + 2/D_V = (9/5) = 1.80, \tag{7.4}$$

based on the standard FD-SOC parameters $d = 3, \gamma = 1, D_V = D_3 = d - (1/2) = 2.5$.

Fitting the size distributions of event durations, $N(T)dT \propto T^{-\alpha_T} dT$, yields the following mean for all 12 IRIS datasets (Figure 7.2c),

$$a_T^{obs} = 2.65 \pm 0.39, \tag{7.5}$$

which shows a significant deviation from the standard FD-SOC model (Table 2.1), for the standard FD-SOC parameters $d = 3$ and $\beta = 1$,

$$a_T^{theo} = 1 + (d-1)\beta/2 = 2.00. \tag{7.6}$$

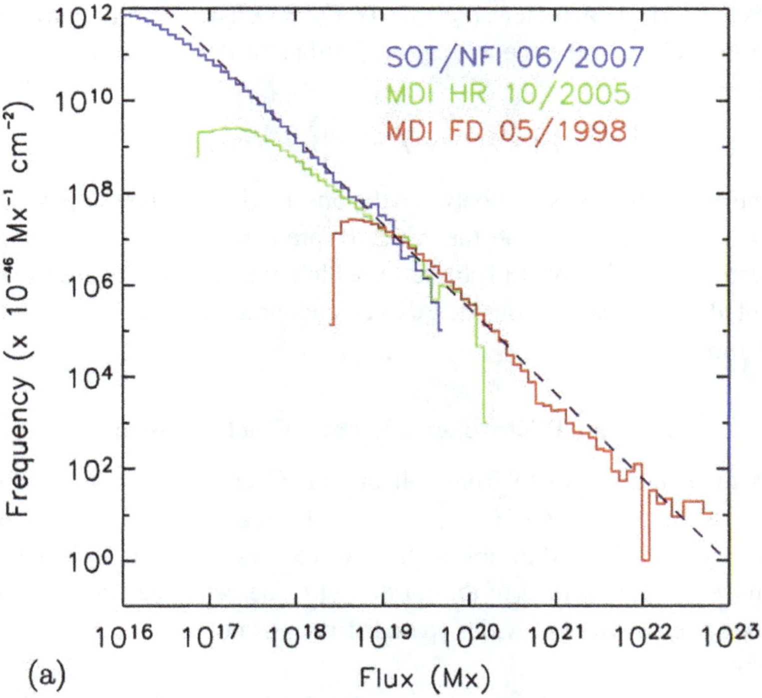

Figure 7.4 Histogram of magnetic flux features observed with SOT/Hinode and MDI/SOHO. The dashed line represents a power law fit with a slope of $\alpha = 1.85$. Note the range of five decades in magnetic flux (Parnell et al. 2009).

This discrepancy indicates a deficit of large time durations T in the automated detection algorithm scheme, which is likely to be caused by the imposed time duration limit (Nhalil et al. 2020). Hence, the standard FD-SOC model is a powerful tool to detect inconsistencies in the completeness, truncation, and sampling of SOC event parameters.

Another dataset containing probably the largest amount of data on fractal structures in magnetograms has been synthesized by Parnell et al. (2009). Particularly interesting is the comparison with magnetic flux events. It was discovered that the size distribution of solar magnetic fluxes F measured in solar magnetograms extends over more than five decades in flux, with a power law index of $\alpha_F = 1.85 \pm 0.14$ in the range of $F = 2 \times 10^{17} - 10^{23}$ Mx (Parnell et al. 2009; Figure 7.4),

$$N(F) \propto F^{-\alpha_F} \propto F^{-(1.85\pm0.14)} \qquad \text{(observations)}. \tag{7.7}$$

Since our standard FD-SOC model is based on the proportionality between flux and volume, $F \propto V^{\gamma}$ (Eq. 2.43), we expect that the power law slopes of these quantities to be identical, that is, $\alpha_F = \alpha_V$ (for $\gamma = 1$). These most accurate measurements agree well with the theoretical predictions of the standard FD-SOC model (Table 2.1),

$$N(F) \propto F^{-\alpha_F} \propto F^{-(1.80)} \qquad \text{(theory)}. \tag{7.8}$$

Hence, the excellent agreement between the observed power law slopes of magnetic flux $\alpha_F = 1.85 \pm 0.14$ with the theoretically predicted value $\alpha_F = 1.80$ based on the standard FD-SOC model (Table 2.1), which extends over five decades in flux (Parnell et al. 2009), can be considered as an encouraging validation of the standard FD-SOC model.

7.2 Solar Photospheric Events: Fractal Dimension

"Fractals in nature originate from self-organized critical dynamical processes," according to Bak and Chen (1989). In astrophysical systems, the fractal dimension can only be measured in objects that we can resolve spatially, such as on the Sun or in the magnetosphere of the Earth. Let us describe the fractal nature of the solar photosphere, which provides spatial structures not only in active regions, but also in Quiet-Sun regions.

To measure the fractality, the five most commonly used fractal analysis techniques are: (i) the perimeter area method, (ii) the linear-size area method, (iii) the box-counting (mono-fractal) method, (iv) the multifractal method (McAteer et al. 2005), and (v) the structure function (Abramenko 2005). Fractal (or multifractal) analysis has been performed in 23 studies (Table 7.1), using white-light data and magnetograms, most frequently measured from the Zeeman effect, such as from the Fe XIV 5303 Å line. Using the box-counting method, the fractality of a structure or a phenomenon is measured by covering an object with boxes of varying sizes L, and the fractal dimension D_A is defined by the scaling relationships (Eqs 2.37 and 2.38),

$$A \propto L^{D_A}, \qquad D_A = \frac{\ln(A)}{\ln(L)}, \tag{7.9}$$

where D_A is the fractal (Hausdorff) dimension, which corresponds to an *Euclidean dimension d* whenever the fractal dimension D_A matches an integer number, that is, $d = 1, 2, 3$. A structure with a 1-D geometry is represented by a 1-D curve or line L, a 2-D geometry with an area A, and a 3-D geometry with a volume V. Using solar imaging data, we have to be aware that the solar surface is oriented by an angle corresponding to the cosine of the line-of-sight angle. The curvature of the photosphere should not affect measurements of the 2-D fractal area in locations near the center of the solar disk, but is expected to converge to a 1-D structure near the solar limb due to projection effects, although a dependence of the fractal

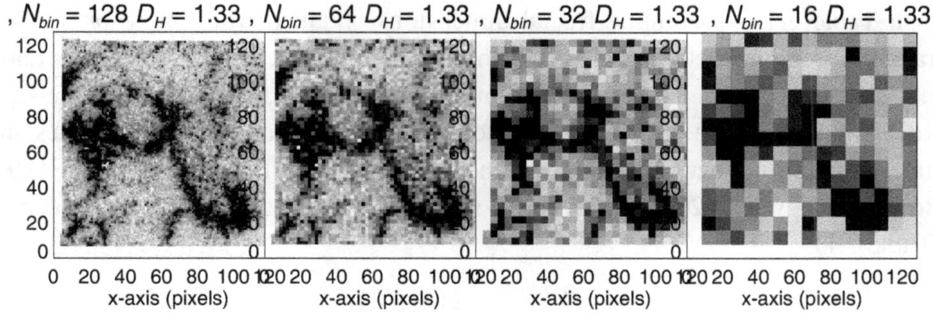

Figure 7.5 Intensity images of one dataset (#8), observed with IRIS SJI 1,400 Å in an active region. The four panels render varying spatial resolutions, with $N_{bin} = 128, 64, 32, 16$, while the fractal dimension stays constant ($D_A = 1.33$) (Aschwanden and Nhalil 2022).

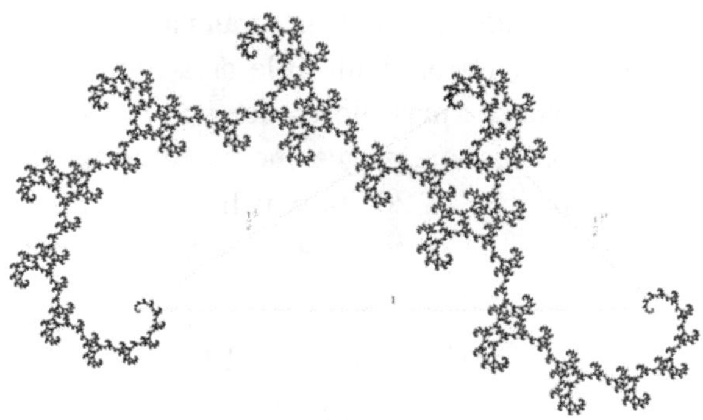

Figure 7.6 This numerically calculated fractal pattern is called a *golden dragon* and has a Hausdorff dimension of $D_A = 1.61803$ (https://en.wikipedia.org/wiki/List_of_fractals_by_Hausdorff_dimension). Note the similarity with dataset 8 in Figure 7.5 (Aschwanden and Nhalil 2022).

dimension on the center-limb distance has not been found yet (Janssen et al. 2003). Alternatively, the absence of a center-limb effect could be explained in terms of the isotropic nature of magnetic point charges buried below the photosphere (see nonlinear force-free magnetic field models defined in Aschwanden 2013).

The fractal nature of the 12 IRIS datasets is rendered in Figure 7.5, where the black areas correspond to zones with enhanced emission, and the white areas correspond to the background with weak emission. The successive reduction of spatial resolution is shown in Figure 7.5 for $N_{bin} = 128, 64, 32, 16$ pixels, which all converge to the same fractal dimension of $D_A \approx 1.33$, confirming the validity of a fractal dimension. An example of a theoretical fractal pattern with a close resemblance to the observed transition region patterns of dataset 8 (Figure 7.5) is shown in Figure 7.6, which is called the "golden dragon fractal" and has a Hausdorff dimension of $D_A = 1.61803$.

In Table 7.1, we compile fractal dimensions obtained from photospheric and transition region fractal features, which may be different from coronal and flare-like size distributions. The fractal dimension has been measured in white-light wavelengths with the perimeter-area method, containing dominantly granules and supergranulation features (Roudier and Muller 1986; Hirzberger et al. 1997; Bovelet and Wiehr 2001; Paniveni et al. 2010), which exhibit a mean value of (Table 7.1),

$$D_A^{gran} = 1.21 \pm 0.07. \tag{7.10}$$

We have to be aware that white-light emission originates in the solar photosphere, which has a lower altitude than any transition region or coronal feature. The relatively low value obtained for the fractal dimension of granulation features (compared with $D_A = 1.0$ for linear or curvilinear elements and $D_A = 2.0$ for area-filling elements) thus indicates that the granulation features seen in optical wavelengths are almost curvilinear (with little area-filling geometries), which is expected for sparse photospheric mass flows along curvilinear flow lines.

A second feature we consider are plages, measured in the transition region with IRIS at 1,400 Å data (Nhalil et al. 2020), which have formation temperatures of $T_e \approx 10^{3.7} - 10^{5.2}$ K in the lower transition region, exhibiting a mean value of (Table 7.1),

$$D_A^{plage} = 1.26 \pm 0.13. \tag{7.11}$$

These sets of IRIS measurements exhibit a relatively low value for the fractal dimension, similar to the photospheric granulation features. Based on this low fractal dimension, photospheric flows appear to be organized along curvilinear features, rather than solid-area geometries.

A third feature that we investigate are sunspots (Nhalil et al. 2020; Aschwanden and Nhalil 2022), which reveal higher values of fractal dimension, namely, (Table 7.1),

$$D_A^{sunspot} = 1.54 \pm 0.16. \tag{7.12}$$

Apparently, sunspots organize fractal features into space-filling geometries, where fragmentation into smaller and smaller fractal features is suppressed, because of the strong magnetic fields that control the penumbral flows of sunspots.

A fourth feature is the active region, observed in photospheric magnetograms and analyzed with the linear-area method (Lawrence 1991; Lawrence and Schrijver 1993; Meunier 1999, 2004; Balke et al. 1993; Janssen et al. 2003; Ioshpa et al. 2008), or with the box-counting method (McAteer et al. 2005). The mean value of fractal dimensions measured in active regions is found to be (Table 7.1),

$$D_A^{AR} = 1.59 \pm 0.20. \tag{7.13}$$

Apparently, active regions organize magnetic features into space-filling, area-like geometries, similar to sunspot features (Figure 7.3).

Comparing the measurements of the fractal dimensions from the photosphere, transition region, and corona, we arrive at four groups that have nonoverlapping means and standard deviations of their fractal dimensions (Table 7.1). None of the fractal structures reaches an Euclidean dimension, which underscores the ubiquity of fractal geometries that needs to be included in any physical scaling law.

The physical understanding of solar (or stellar) granulation has been advanced by numerical magneto-convection models and n-body dynamic simulations, which predict the evolution of small-scale (granules) into large-scale features (meso- or supergranulation), organized by surface flows that sweep up small-scale structures and form clusters of recurrent and stable granular features (Hathaway et al. 2000; Berrilli et al. 2005; Rieutord et al. 2008, 2010). Recent convection models and n-body dynamic simulations predict the evolution of small-scale features (i.e., granules) and establish that sites where recurrent and stable granular features are observed exhibit a clustering spatial scale of ≈ 8 Mm and time scales around 10 min (Berrilli et al. 2005). The fractal structure of the solar granulation is obviously not a self-organizing feature, but is created by a combination of subphotospheric magneto-convection and surface flows, which are turbulence-type phenomena.

The analysis presented here demonstrates that we can distinguish between (i) physical processes with sparse curvilinear flows, as they occur in granulation, mesogranulation, and supergranulation, and (ii) physical processes with near space-filling flows, as they occur in the chromospheric evaporation process in solar flares. IRIS data can therefore be used to diagnose the degree of mass flows in the transition region. Moreover, reliable measurements of the fractal dimension yields realistic plasma filling factors that are important in the estimation of radiative energies and hot plasma emission measures.

Testing SOC models includes also the validity of the power law assumption in the size distribution and waiting time distribution. Deviations from ideal power law slopes in the size and duration distributions have been found in a number of cases, which were quantified with log-normal fits (Abramenko and Yurchyshyn 2010c), bimodal distributions of magnetic fields and sunspot areas (Tlatov and Pevtsov 2014), broken power laws in the flux and magnetic field (Javaherian et al. 2017), but they do not adequately fit a Weibull distribution neither (Shapoval et al. 2018). Different power law slopes of the flux F, size S, and lifetime T were found for flaring and nonflaring active regions observed with HMI/SDO (Moradhaseli et al. 2021).

Deviations from the standard SOC model was also noticed in the degree of diffusivity, and the distribution of waiting times. The fractal dimension of the pattern of sites available to random walk is $\beta < 1$ for the subdiffusive range, and tends to be $\beta = 1$ in the normal diffusion range (Cadavid et al. 1999). The waiting times at the trap sites was found to be a truncated Lévy type (power law) (Cadavid et al. 1999).

While earlier studies mostly deal with mono-fractal models (e.g., Aschwanden 2022a, 2022b; Aschwanden and Nhalil 2022), more recent studies detect multifractal characteristics or log-normal distributions, which nevertheless are considered to be indicative of multifractality (Abramenko 2014). In the standard FD-SOC model, we quantify monofractal dimensions with mean values D_A and D_V (Eq. 2.39–2.40), while multifractlity is modeled with the ranges $[D_{A,min}, D_{A,max}]$, and $[D_{V,min}, D_{V,max}]$, as defined in Eqs (2.39) and (2.40). Both flaring and nonflaring active regions are found to exhibit significant fractality, multifractality, and non-Kolmogorov turbulence, but none of these three parameters could discriminate between active regions with major flares and minor (flare-quiet) ones (Georgoulis 2011). However, a discrimination between flaring and nonflaring active region was accomplished with multifractality measures using higher flux sensitivity and higher spatial and temporal resolution (Giorgi et al. 2015). A method of multifractal segmentation was also used to detect emergence of new magnetic field (Golovko and Salakhutdinova 2015, 2018).

7.3 Magnetic Field Diffusion

The transport of magnetic field observed on the solar surface defines the fundamentally important scaling law between spatial scales L and temporal scales T, which represents one of the key assumptions in the concept of nonlinear avalanches governed by SOC. In Section 1.8, we introduced the cases of classical diffusion ($\beta = 1$), subdiffusivity ($\beta < 1$), and superdiffusivity ($\beta > 1$). The fact that the standard SOC model matches solar observations well in most of the data, has the important implication that classical diffusion is a viable approximation of cellular automaton models as well as other SOC models. On the other hand, observations report also subdiffusive as well as superdiffusive transport, which may reflect different degrees of magnetic diffusivity in the photosphere and the corona.

Magnetic field diffusion is generated by convection on the solar surface and below and is manifested in the flow of ubiquitous magnetic structures (Leighton 1964), such as granules, mesogranules, supergranules, and minigranular structures below spatial scales of $L \lesssim 600$ km (Abramenko et al. 2012). The transport of magnetic elements (magnetic bright points) was found to be subdiffusive for times less than 20 minutes (which is about the lifetime of a granule), but consistent with classical diffusion for times larger than 25 min (Cadavid et al. 1999). The subdiffusive transport is explained by random walkers that are being trapped at stagnation points in the intercellular pattern (Cadavid et al. 1999). Giannattasio et al. (2013) report a double-regime also, superdiffusivity ($\beta = 1.34 \pm 0.02$) up to granular spatial scales ($L \lesssim 1500$ km) and slight superdiffusivity ($\beta \approx 1.20 \pm 0.05$) up to supergranular scales. A high number of magnetic elements were detected in a single supergranular cell, yielding exponents of $\beta = 1.27 \pm 0.05$ and $\beta = 1.08 \pm 0.11$ in the (supergranular) network at smaller and larger scales, and $\beta = 1.44 \pm 0.08$ in the internetwork regions (Giannattasio et al. 2014a), which is of the Lévy type transport, that is, $L \propto T^{\beta/2}$ with $\beta > 1$. Pair separation of magnetic elements yields an even more superdiffusive regime of $\beta = 1.55 \pm 0.05$ (Giannattasio et al. 2014b). Using a method called *diffusion entropy analysis*, the Lévy-type transport is confirmed further (Giannattasio et al. 2019, 2020; Giannattasio and Consolini 2021).

7.4 Photosphere–Corona Coupling

Sandpiles, the paradigm of SOC systems, may have multiple zones with avalanching behavior, some near the top of the sandpile cone where randomly dropped sand grains cause instabilities and exponential growth behavior, and other zones further down the inclines, where exponential growth stopped and the motion of avalanching slows down. We could characterize the two phases with an exponential growth phase and a saturation phase (Figure 1.2). It is widely believed that the initial exponential growth phase occurs where the magnetic field is most non-potential, which is most likely in the lower corona and beneath (in the transition region, chromosphere, and photosphere). Consequently one expects a coupling and correlation between the lower and the upper corona. Since the magnetic topology in active regions as well as in Quiet-Sun regions reveals a high degree of geometric and magnetic complexity, the photosphere-corona coupling is not linearly correlated, but rather exhibits a stochastic coupling (Uritsky et al. 2013). This was proven by comparing the size distributions of length scales L, areas A, volumes V, sizes S, time durations T, and energies E between (coronal) EUVI/STEREO and (photospheric) MDI/SOHO data (Uritsky et al. 2013). By contrast, while energy release in the Quiet-Sun was found to be close to random, following meso- and supergranulation patterns in the underlying photosphere, flaring active regions exhibit clustering in their waiting times (Uritsky and Davila 2014).

Turbulence parameters of the photospheric magnetic field, observed in active regions, reveal a turbulence phase prior to a solar flare, and coincide with the maximum of the correlation length, which was interpreted to be a trace of an avalanche of coronal reconnection events (Abramenko et al. 2013). The structure function, intermittency, and multifractality were found to be correlated with the flaring rate and thus allows flare predictivity to some extent (Abramenko and Yurchyshyn 2010c; Abramenko 2015).

7.5 Solar Photospheric Events: Turbulence

In contrast to the concept of SOC, which is produced by nonlinear avalanche processes and exhibits power law-like size distributions, a competing interpretation that is raised in a number of publications is turbulence, which displays a power spectrum that has the shape of a power law function also. This power spectrum is generally shown as a function of the wave number. Some of the publications that suggest the turbulence interpretation in photospheric events, instead of SOC interpretation, are given in Abramenko et al. (2003, 2010), Georgoulis et al. (2005), Abramenko and Yurchyshyn (2010a, 2010b, 2010c, 2020), Pavlos et al. (2012), Abramenko (2014), Vlahos and Isliker 2016, Rincon et al. (2017), and Dudok de Wit (2020).

Observations revealed that the magnetic field had been increasing for 6–33 min before each flare and reached a maximum value 3–14 min before the peak of the hard X-ray emission, which suggests the existence of a turbulent phase prior to a solar flare (Abramenko et al. 2003). Since the temporal evolution of the SOC scaling laws shows a distinct behavior starting a few hours prior to a flare, this temporal behavior can be used in flare prediction (Georgoulis et al. 2005). Nonlinear processes are multiplicative in nature and involve inter-scale energy exchange, which is the method of evolution of any turbulent medium (Abramenko and Yurchyshyn 2010a). Active regions with higher flare productivity exhibit steeper power spectra, and the regime of the fully developed turbulence occurs in decaying active regions and in emerging active regions, while well-developed active regions display underdeveloped turbulence with strong magnetic dissipation at all scales (Abramenko and Yurchyshyn 2010b). Turbulent magnetized plasma in a nonperturbed solar atmosphere exhibit multifractality and log-normal distributions (Abramenko 2014). Other researchers found Tsallis nonextensive statistics, intermittent turbulence, SOC, and chaos dynamics in solar plasmas (Pavlos et al. 2012). It was proposed that the scale-free energy distribution of solar flares is largely due to the twist exerted by the vorticity of the turbulent photosphere (Mendoza et al. 2014). SOC systems can be driven by small ambient magnetic fields $(\delta B/B) \ll B$ (weak turbulence) or by large amplitude magnetic fluctuations $(\delta B/B) \gg B$ (strong turbulence or fully developed turbulence) (Vlahos and Isliker 2016). Multiscale photospheric flows (granulation, mesogranulation, and supergranulation) was also described in terms of an anisotropic extension of the Bolgiano–Obukhov theory of hydrodynamic stratified turbulence (Rincon et al. 2017). The Parker Solar Probe (PSP) reveals switchbacks in the near-Sun magnetic field, where the waiting time statistics shows evidence of aggregation, and the spectral properties of the inertial range turbulence differ inside and outside of switch-back structures (Dudok de Wit 2020).

7.6 Summary

The fractal nature in avalanching systems with SOC is investigated here for phenomena in the solar photosphere and transition region. In the standard SOC model, the fractal Hausdorff dimension is expected to cover the range of [1, 2], with a mean of $D_A = 1.5$ for 2-D observations projected in the plane-of-sky, and the range of [2, 3], with a mean of $D_V = 2.5$ for real-world 3-D structures. Observations of magnetograms and with IRIS (1,400 Å) reveal four groups: (i) photospheric granulation with a low fractal dimension of $D_A \approx 1.2$; (ii) transition region plages with a low fractal dimension of $D_A \approx 1.3$; (iii) sunspots at transition region heights with an average fractal dimension of $D_A \approx 1.5$; and (iv) active regions at photospheric heights with an average fractal dimension of $D_A \approx 1.6$. Phenomena with a low fractal dimension indicate sparse curvilinear flows (as they occur in granulation), while high fractal dimensions indicate near space-filling flows (as they occur in the chromospheric evaporation process in solar flares). Investigating the SOC parameters, we find a good agreement for the event areas A and mean radiated fluxes F in events in transition region plages. Another deviation from the standard SOC model (with classical diffusion, $\beta \approx 1.0$) is the anomalous subdiffusive behavior with $\beta \approx 1.1 - 1.5$, which is explained by random walkers that are being trapped at stagnation points in the intercellular pattern.

8

Solar Radio Bursts

Solar radio bursts are usually subdivided into incoherent (gyroemission, gyroresonance, gyrosynchrotron emission, and free–free emission) and coherent emission mechanisms (electron beam instability, loss-cone instability, and microwave amplification by stimulated emission of radiation (MASER)). Since incoherent emission mechanisms scale with the volume of the emitting source, which could be a solar flare region or a magnetic reconnection region, we expect some proportionality between the flare volume and the associated radio burst flux, such as for microwave (MW) bursts and type IV bursts (produced by gyrosynchrotron emission). Consequently we expect power law slopes of their size distributions that are similar to other incoherent emission mechanisms of flares (e.g., bremsstrahlung in hard X-rays or free-free emission in soft X-rays, which exhibit a typically observed and also a theoretically predicted value of $\alpha_F = 1.80$). On the other hand, since coherent emission mechanisms produce a highly nonlinear response to some wave-particle instability, their emitted intensity flux is expected to scale nonlinearly with the flare volume, and thus may produce quite different size distributions. We present a compilation of size distributions gathered from solar radio bursts in Table 8.1, which we subdivide into three groups: incoherent emission (MW), coherent emission (III, DCIM-P), and random noise (I, DCIM-S, MW-S). The first two groups exhibit power law-like size distributions, while the third group exhibits a size distribution with an exponential cutoff.

8.1 Incoherent Emission Solar Radio Bursts: Size Distributions

Microwave bursts, which are typically observed in frequencies of $\nu \approx 1 - 15$ GHz, have been found to exhibit size distributions with power law slopes within a range of $\alpha_F \approx 1.7 - 1.9$ (Akabane 1956; Kundu 1965; Kakinuma et al. 1969; Das et al. 1997; Nita et al. 2002; Kou et al. 2020), similar to the size distributions observed

Table 8.1 *Frequency distributions measured from solar radio bursts, classified into three groups: (i) incoherent emission (MW=microwave); (ii) coherent emission (type III bursts, decimetric pulsations DCIM-P), and iii) random noise (type I storms, decimetric millisecond spikes DCIM-S), and microwave spikes (MW-S). Coherent and incoherent emission mechanisms have power law-like distribution functions, while random noise processes have a distribution function with an exponential-like cutoff.*

Power law slope of mean flux α_F	Log range	Waveband frequency f	Radio burst type	References
Incoherent				
1.8	2	3 GHz	MW	Akabane (1956)
1.5	2	3, 10 GHz	MW	Kundu (1965)
1.8	2	1, 2, 3.75, 9.4 GHz	MW	Kakinuma et al. (1969)
1.9–2.5	2	3.75, 9.4 GHz	MW	Kakinuma et al. (1969)
1.65	2	2.8 GHz	MW	Das et al. (1997)
1.81 – 1.82	4	0.100–2 GHz	MW	Nita et al. (2002)
1.85 ± 0.16	1.5	0.1–14 MHz	MW	Kou et al. (2020)
Coherent				
1.26–1.69	3	110 kHz-4.9 MHz	III	Fitzenreiter et al. (1976)
1.28	2	100 MHz-3 GHz	III	Aschwanden et al. (1998)
1.45 ± 0.31	3	100 MHz-3 GHz	III	Aschwanden et al. (1998)
1.33 ± 0.11	3	100 MHz-3 GHz	DCIM-P	Aschwanden et al. (1998)
1.22–1.25	2.5	650–950 MHz	III	Das et al. (1997)
1.16	2.5	3–10 MHz	III	Eastwood et al. (2010)
Random				
2.9–3.6	1	164, 237 MHz	I	Mercier and Trottet (1997)
2.99 ± 0.63	1	100 MHz-3 GHz	DCIM-S	Aschwanden et al. (1998)
7.4 ± 0.4	0.5	4.5–7.5 GHz	MW-S	Ning et al. (2007)
4.8 ± 0.1	0.5	185–198 MHz	I	Iwai et al. (2013)

in solar hard X-ray and soft X-ray bursts, which implies a near-proportionality between the flare volume and the radio flux.

Akabane (1956) sampled 587 MW bursts observed at the Ottawa station, 148 bursts at the Toyokawa station, and 386 bursts at the Tokyo Astronomical Observatory at a frequency of $\nu \approx 3$ GHz, and found for each of the three datasets a power law slope of $\alpha_F = 1.8$ that extends over two decades. It is amazing that these pioneering observations already perfectly agree with the value of the peak flux $\alpha_F = 1.8$ theoretically predicted by the fractal-diffusive self-organized criticality (FD-SOC) model (Section 2.7).

Kundu (1965) observed MW bursts at $\nu = 3$ GHz and $\nu = 10$ GHz, and obtained a size distribution with a slope of approximately $\alpha_F = 1.5$.

Kakinuma et al. (1969) observed MW bursts at $\nu = 1, 2, 3.75, 9.44$ GHz, which show power law distributions with a slope of $\alpha_F = 1.8$ and match the theoretical prediction of the standard FD-SOC model. A second dataset at $\nu = 3.75$ and $\nu = 9.4$ GHz yielded a slope in the range of $\alpha_F = 1.9 - 2.5$, which is close to the standard FD-SOC model.

Das et al. (1997) observed microwave bursts at $\nu = 2.8$ GHz and derived a power law slope of $\alpha_F = 1.65$, covering two decades of inertial range.

Nita et al. (2002) analyzed a statistical sample of 518,060 microwave bursts observed during 40 years (1960–1999) by the National Geophysical Data Center (NGDC) of the National Oceanic and Atmospheric Administration (NOAA) at frequencies around $\nu \approx 2$ GHz. A power law distribution function was fitted over an inertial range of ≈ 4 decades (Figure 8.1), and a slope of $\alpha_F = 1.82 \pm 0.01$ was found during the solar maximum, and a slope of $\alpha_F = 1.81 \pm 0.02$ during the solar minimum, which both are consistent with the theoretical prediction $\alpha_F = 1.80$ of the standard FD-SOC model. Self-consistent values in the range of $\alpha_F \approx 1.6 - 1.8$ were obtained in each investigated frequency band, $\nu =$ (0.1–0.9 GHz, 1.0–1.7 GHz,

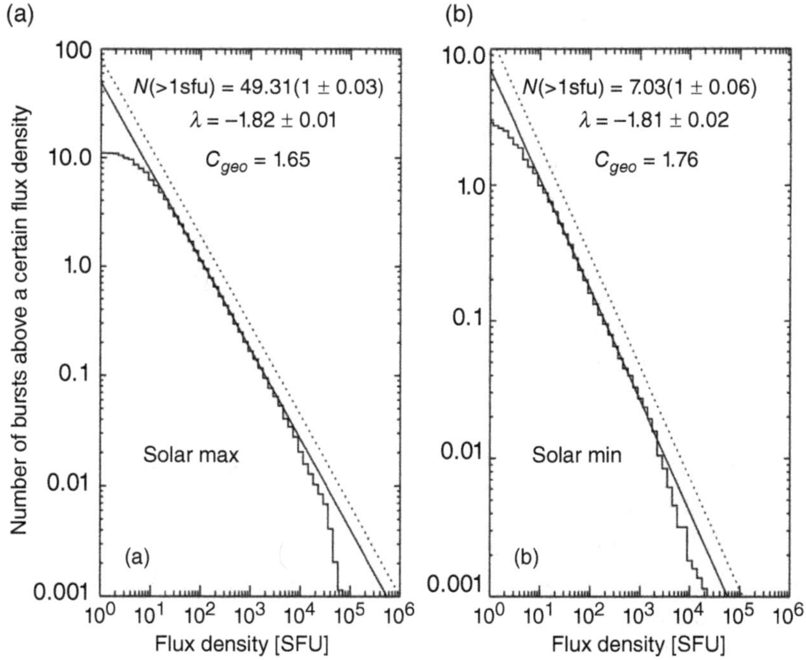

Figure 8.1 Cumulative number of solar radio bursts at frequencies above 2 GHz (a) during solar maximum years and (b) during solar minimum. The histogram shows the best power law fit to the observed distribution, and the dotted line shows the fit after correction with a geographical factor (Nita et al. 2002).

2.0–3.8 GHz, 4.9–7.0 GHz, 8.4–11.8 GHz, 15.0–37.0 GHz, <2 GHz, >2 GHz). The sampled radio flux is interpreted in terms of gyrosynchrotron emission at high frequencies ($\nu > 2$) GHz to predominantly plasma emission below ($\nu < 2$) GHz. Apparently the power law slope $\alpha_F = 1.8$ of the size distributions of mean fluxes F is an universal constant that does not depend on the frequency, the observed wavelength, or (solar) activity level, as expected from the standard FD-SOC model. Of course, the universality in the standard FD-SOC model rests on the assumption that the observed radio flux F is proportional to the fractal volume V, namely, $F \propto V^\gamma$ (Eq. 2.43), where the proportionality is rendered with the linear power law index $\gamma = 1$.

Kou et al. (2020) investigated *interplanetary type III radio bursts (IT3)* in the decameter-kilometric range and found that the GOES flux of the IT3-associated and CME-associated events exhibit a power law with a slope of $\alpha_F = 1.85 \pm 0.16$, which is consistent with the theoretical standard FD-SOC model. Selecting events that are not IT3-associated or CME-associated events yields steeper slopes ($\alpha_F >$ 2.0), which may indicate that CMEs play a role in producing IT3s (Kou et al. 2020).

8.2 Coherent Emission Solar Radio Bursts: Size Distributions

Type III bursts are produced by plasma emission excited by an electron beam-driven instability, which is a coherent emission process, detectable in the solar corona and interplanetary space. They exhibit flatter size distributions than MW bursts, in the order of $\alpha_F \approx 1.2 - 1.5$ (Fitzenreiter et al. 1976; Das et al. 1997; Aschwanden et al. 1998), as listed in Table (8.1).

Fitzenreiter et al. (1976) observed interplanetary type III bursts with the IMP-6 satellite during May–July 1971 at frequencies from 110 kHz to 4.9 MHz and found power law distributions of their fluxes F with slopes in the range of $\alpha_F = 1.26 - 1.69$, where flatter power law slopes ($\alpha_F = 1.33$) occur at low frequencies (110 kHz), and steeper power law slopes ($\alpha_F = 1.69$) are found at high frequencies (4,900 kHz). This directivity of low-frequency solar type III radio bursts, which is correlated with the direction of the mean interplanetary magnetic field, is interpreted as beaming of the observed burst radiation along the magnetic field direction. The flux size distribution has therefore a frequency dependence $F(\nu)$ and thus size distributions of coherent radio bursts are not universal.

Aschwanden et al. (1998) identified 882 type III bursts in the metric and decimetric frequency range of $\nu = 100$ MHz–3 GHz, using the Zurich radio spectrometer *Phoenix*, which yield a power law slope of $\alpha_F = 1.28$, similar to the results of Fitzenreiter et al. (1976). Analyzing flare substructures with the same instrument (PHOENIX), in the same frequency range of $\nu = 100$ MHz$-$3 GHz, Aschwanden et al. (1998) detected 4000 solar radio type III bursts, which are mostly found to be correlated with hard X-ray peaks, and yield a power law of $\alpha_F = 1.45 \pm 0.31$, covering three decades in radio flux.

Other datasets containing decimetric pulsation (DCIM-P) type bursts yield $\alpha_F = 1.33 \pm 0.11$. The datasets in the range of $\alpha_F \approx 1.22 - 1.45$ (type III, DCIM-P) indicate coherent plasma emission (with $\gamma \gtrsim 2$ in the flux–volume scaling law, $F \propto V^\gamma$ (Eq. 2.43). This can be explained by a nonlinear scaling $F \propto V^\gamma$ between radio flux F and the flare volume V. The standard SOC model predicts a power law distribution for the radio flux, $N(F) \propto F^{-\alpha_F}$, with a slope of $\alpha_F = 1 + (d - 1)/(D_V\gamma) = 1.8$ (Table 2.1: for Euclidean dimension $d = 3$, mean fractal dimension $D_V = 2.5$, and linear flux–volume scaling $\gamma = 1$.) For the case of nonlinear flux–volume scaling, say for quadratic flux–volume scaling ($\gamma \approx 2$), the power law slope changes to $\alpha_F = 1 + (d - 1)/(D_V\gamma) = 1.4$, which explains the trend that coherent radio emission ($\alpha_F \approx 1.4$) have flatter power law slopes than incoherent radio bursts ($\alpha_F \approx 1.8$).

Eastwood et al. (2010) sampled radio bursts from a type III noise storm and finds a power law slope of $\alpha_F^{cum} = 2.16$ for the cumulative flux size distribution, which corresponds to a slope of $\alpha_F = \alpha_F^{cum} - 1 = 1.16$ and fits a coherent radio emission mechanisms, as it is expected for type III-producing electron beams. Furthermore, Eastwood et al. (2010) modeled the time-dependent variation of the flare rate with piece-wise Bayesian blocks and demonstrated this way that significant deviations from a pure power law in the flux size distribution exist.

The statistics of radio bursts thus tell us two things: (i) coherent radio bursts have size distributions that are not consistent with the standard FD-SOC model as expected from the nonlinear scaling, $F \propto V^\gamma$, with $\gamma \gtrsim 2$, and (ii) incoherent radio bursts are consistent with the standard FD-SOC model as expected from the linear scaling, $F \propto V^1$. Thus the standard FD-SOC model provides a reliable diagnostic which radio bursts are produced by a coherent radiation mechanism, and which ones are produced by an incoherent radiation mechanism.

8.3 Random Noise Solar Radio Bursts: Size Distributions

Mercier and Trottet (1997) sampled solar radio bursts with the Nançay Radioheliograph in the frequencies of 164, 237, 327, 410, and 435 MHz during 11 solar radio type I bursts and found size distributions with slopes of $\alpha_F \approx 2.9 - 3.6$, which were interpreted as signatures of nanoflares. However, the inertial range (or fitted power law range) of the observed flux densities does not extend over more than one decade, which is more typical for a random noise distribution (with exponential cut-off) rather than for a power law extending over multiple decades, as expected for a SOC model.

Ning et al. (2007) observed high-frequency drift radio structures with the Purple Mountain Observatory in the frequency range of 4.5–7.5 GHz, and found power laws for the peak flux ($\alpha_F = 7.38 \pm 0.40$). Such high values for the peak flux are typical for a random noise process (with exponential cutoff), rather than for a power law SOC process.

Iwai at al. (2013) observed solar radio bursts with the radio telescope AMATERAS and found size distributions with slopes of $\alpha_F = 4.8 \pm 0.1$, covering a relatively small range of about one decade, which is typical for a random noise process, rather than a SOC process. The burst types were interpreted as type I noise storms.

In contrast to coherent and incoherent solar radio bursts, some very steep size distributions have been observed, such as $\alpha_F \approx 3 - 5$ for type I bursts (Mercier and Trottet 1997; Iwai et al. 2013), or $\alpha_F \approx 3 - 7$ for decimetric and microwave millisecond spike bursts (DCIM-S) (Aschwanden et al. 1998), and microwave spikes (MW-S) (Ning et al. 2007). Size distributions generally show an exponential cutoff at the upper end with the largest and rarest events, which produce steep slopes when fitted with a power law function ($\alpha_F \approx 3 - 7$). Stochastic models of clustered solar type III bursts produce power law-like size distributions with an exponential cutoff (Isliker et al. 1998).

8.4 Summary

From the statistics of solar radio bursts, we learn that we can discriminate between three diagnostic regimes (as listed in Table 8.1): (i) the incoherent regime where the radio burst flux is essentially proportional to the flare volume (with a power law slope of $\alpha_F \approx 1.8$), as it occurs for gyroemission, gyroresonance emission, gyrosynchrotron emission; (ii) the coherent regime that implies a nonlinear scaling between the radio flux and the flare volume $F \propto V^\gamma$ with $\gamma \approx 2$ and $\alpha_F \approx 1.4$; as it occurs for the electron beam instability, the loss-cone instability, or maser emission; and (iii) the exponential regime that does not display a power law function, but rather an exponential cutoff as expected for random noise distributions. Thus, the power law slopes offer a useful diagnostic to verify the flux–volume scaling law, and to discriminate between coherent and incoherent radio emission processes, as well as to distinguish between SOC processes (if power law-like) and non-SOC processes (if exponential-like size distributions). An additional diagnostic comes from the inertial range of power law fits: SOC-related power law size distributions should extend over multiple decades, while power law ranges of less than one decade are most likely not related to SOC processes.

9

Coronal Mass Ejections

Every main sequence star is losing mass, caused by dynamic phenomena in its atmosphere that accelerate plasma or particles beyond the escape speed. Inspecting the Sun, our nearest star, we observe two forms of mass loss: the steady solar wind outflow and the sporadic ejection of large plasma structures, termed coronal mass ejection (CME). The solar wind outflow amounts to $\approx 2 \times 10^{-10}$ (g cm^{-2} s^{-1}) in coronal holes, and $\lesssim 4 \times 10^{-11}$ (g cm^{-2} s^{-1}) in active regions. The phenomenon of a CME occurs with a frequency of a few events per day, carrying a mass in the range of $m_{\rm CME} \approx 10^{14} - 10^{16}$ g, which corresponds to an average mass loss rate of $m_{\rm CME}/(\Delta t \cdot 4\pi R_\odot^2) \approx 2 \times 10^{-14} - 2 \times 10^{-12}$ (g cm^{-2} s^{-1}), which is $\lesssim 1\%$ of the solar wind mass loss in coronal holes, or $\lesssim 10\%$ of the solar wind mass in active regions. The transverse size of CMEs can cover from a fraction, up to more than a solar radius, and the ejection speed is in the range of $v_{\rm CME} \approx 10^2 - 2 \times 10^3$ (km s^{-1}). Ambiguities from line-of-sight projection effects make it difficult to infer the geometric shape of CMEs. Possible interpretations include flux ropes, semi-shells, or bubbles (Figure 9.1).

There is a general consensus that a CME is associated with a release of magnetic energy in the solar corona, but its relation to the flare phenomenon is controversial. Even big flares (at least GOES M-class) have no associated CMEs in 40% of the cases (Andrews 2003). A long-standing debate focused on the question of whether a CME is a by-product of the flare process or vice versa. This question has been settled in the view that both CMEs and flares are quite distinctly different plasma processes, but related to each other by a common magnetic instability that is controlled on a larger global scale. A CME is a dynamically evolving plasma structure, propagating outward from the Sun into interplanetary space, carrying a frozen-in magnetic flux and expanding in size. If a CME structure travels from a sub-solar point radially toward the Earth, it is called a *halo-CME* or an *Earth-directed* event.

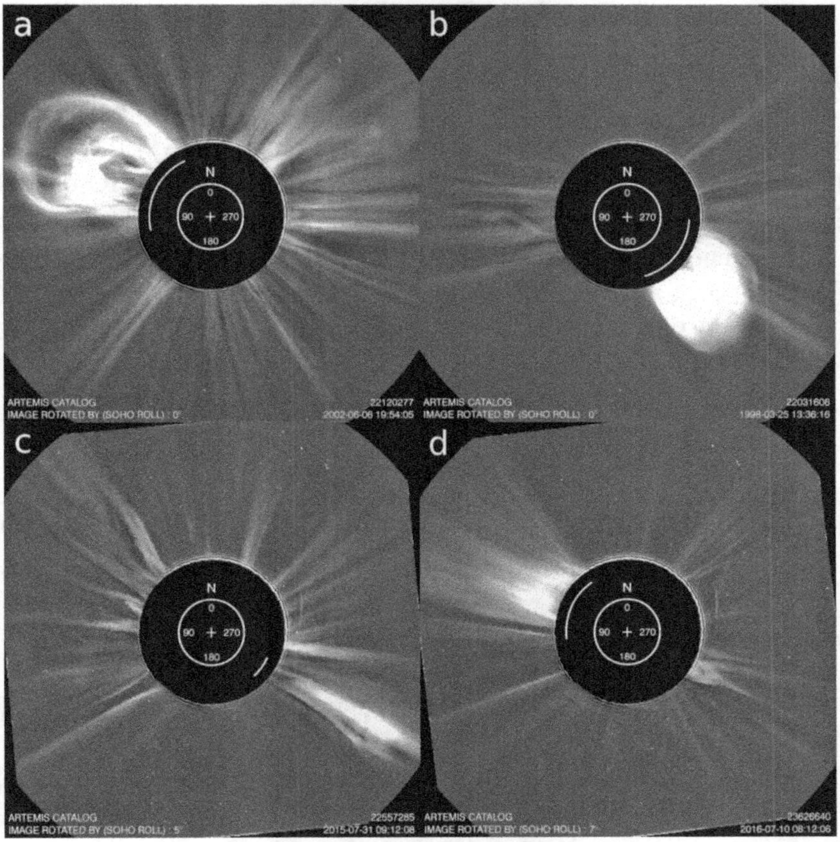

Figure 9.1 Typical examples of well-defined morphologies of CMEs, obtained from observations with the white-light coronagraph LASCO onboard SOHO: flux rope (a), loop (b), jet (c), and other (d) (Lamy et al. 2019).

Coronal mass ejections are launched as a consequence of a magnetic instability in the solar corona, in most cases accompanied by a solar flare, which most likely is driven by a magnetic reconnection process (Lippiello et al. 2008), as the high correlation between the relative helicity magnitude and the free magnetic energy demonstrates (Georgoulis et al. 2019). The two phenomena (flares and CMEs) are therefore strongly connected and a similar statistics of size distributions and waiting times is expected for both. However, the energy parameter in CME statistics requires a specific definition, such as the kinetic energy of CMEs, $E_{kin} = (1/2) m_{\rm CME} v_{\rm CME}^2$ (Li et al. 2016; Aschwanden 2016), the angular width $w_{\rm CME}$ (D'Huys et al. 2016), or the gravitational energy E_{grav} (Aschwanden 2016). In addition to the size distributions of CME energies, the size distributions of CME waiting times ($\tau_{\rm CME}$) has also been characterized. Typical examples of CMEs observed with the LASCO/SOHO white-light coronagraph are shown in Figure 9.1.

9.1 Coronal Mass Ejections: Size Distributions

Yashiro et al. (2006) gathered a catalog of the flux F, the fluence E, and the duration T of solar flares, listed in the solar geophysical data (SGD), in the GOES soft X-ray flare catalog, and in the online solar event reports provided by NOAA. In addition, the CME flare catalog was obtained from LASCO/SOHO data, during the epoch of 1996–2005. Based on these two types of flare catalogs (Table 9.2), it was determined that the size distribution exhibits power law functions, but have different slopes: $\alpha_F = 2.16 \pm 0.03$, $\alpha_E = 2.01 \pm 0.03$, and $\alpha_T = 2.87 \pm 0.09$ for all flares, $\alpha_F = 1.98 \pm 0.05$, $\alpha_E = 1.79 \pm 0.05$, and $\alpha_T = 2.49 \pm 0.09$ with CMEs; and $\alpha_F = 2.52 \pm 0.03$, $\alpha_E = 2.47 \pm 0.11$, and $\alpha_T = 3.22 \pm 0.15$ without CMEs (Figure 9.2). These values can be compared with the predictions of the FD-SOC

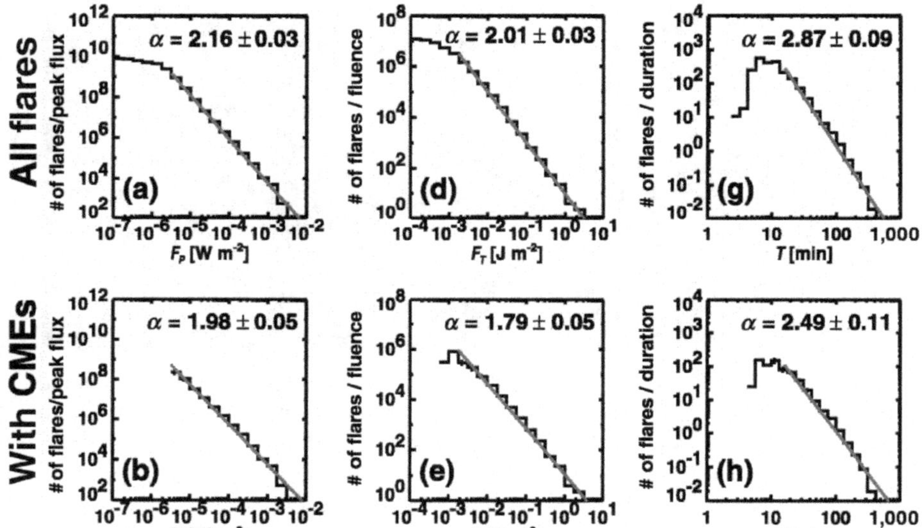

Figure 9.2 Flare frequency distributions as a function of flux (a–c), fluence (d–f), and duration (g–i) for all flares (a,d,g), flares with CMEs (b,e,h), and flares without CMEs (c,f,i), respectively. The power law index of each distribution is shown in each panel. Flares without CMEs have steeper power law slopes than with CMEs (Yashiro et al. 2006).

Table 9.1 *Size distributions of CME parameters (flux, fluence, and duration) from SOHO/LASCO data.*

Power law slope of flux	log range	Power law slope of fluence	log range	Power law slope of duration	log range	References
2.16 ± 0.03	3	2.01 ± 0.03[a]	3	2.87 ± 0.09	2	Yashiro et al. (2006)
1.98 ± 0.05	2.5	1.79 ± 0.05[b]	3	2.49 ± 0.11	1.5	Yashiro et al. (2006)
2.52 ± 0.03	2	2.47 ± 0.11[c]	1	3.22 ± 0.15	1	Yashiro et al. (2006)
1.735 ± 0.116[d]						D'Huys et al. (2016)
1.657 ± 0.017[e]						D'Huys et al. (2016)
		2.0 ± 0.4[f]				Aschwanden (2016)
		1.4 ± 0.1[g]				Aschwanden (2016)
		2.0 ± 0.3[h]				Aschwanden (2016)
		1.4 ± 0.3[i]				Aschwanden (2016)
1.98		**2.00**		**2.87**	Median	All observations
1.80		**1.44**		**2.00**	FD-SOC	Prediction

[a]All flares; [b]with CMEs; [c]without CMEs; [d]stealth CMEs; [e]classical CMEs; [f]gravitational CME energy; [g]kinetic CME energy; [h]gravitational+kinetic CME energy; [i]thermal CME energy.

model, $\alpha_F = 1.80$, $\alpha_E = 1.44$, and $\alpha_T = 2.00$ (Table 2.1). The observed power law slopes for flares, for both with and without CMEs, are steeper than the critical value of $\alpha_E > 2.0$, which was interpreted as dominant nanoflare heating in the coronal heating scenario Yashiro et al. (2006). This interpretation, however, appears to be incorrect, because no preflare background was subtracted in the GOES data, as noted earlier (Table 5.1) in the same soft X-ray flare data set analyzed by (Yashiro et al. 2006). Neglecting flare-unrelated background causes a bias that results into steeper power law slopes. Consequently, we observe discrepancies between the no-background subtracted power law slopes calculated by Yashiro et al. (2006) and the standard FD-SOC model (Table 9.1).

D'Huys et al. (2016) categorized CME events into classical flare-associated CME events, and into CME events without low coronal signatures, so-called *stealth CMEs* (Robbrecht et al. 2009a, 2009b), using the CACTUS/LASCO CME catalog. They found notable differences between these two classes of angular widths, $\alpha_w = 0.97 \pm 0.41$ for stealth CMEs, versus $\alpha_w = 1.50 \pm 0.34$ for classical CMEs. These values were improved by using the alternative Maximum Likelihood Estimator (MLE), to $\alpha_w = 1.735 \pm 0.116$ for stealth CMEs, versus $\alpha_w = 1.657 \pm 0.017$ for classical CMEs, which is close to the theoretical prediction of the flux, $\alpha_F = 1 + (d-1)/(D_V\gamma) = 9/5 = 1.80$ (Table 2.1). Interestingly, the power law slope does not depend on the stealth criterion, and the angular width $w_{\rm CME}$ of a CME appears to be a good proxy for the length scale of a CME volume.

Table 9.2 *Waiting time distribution of CMEs from SOHO/LASCO data.*

Power law slope of waiting times α_τ	log range	Instrument, detection algorithm	Observations times [years]	References
2.36 ± 0.11	2	LASCO/SOHO	1996–2001	Wheatland (2003)
1.36 ± 0.11	2	LASCO/SOHO	1996–1998	Wheatland (2003)
2.98 ± 0.20	1.5	LASCO/SOHO	1999–2001	Wheatland (2003)
2.4	1	LASCO/SOHO	1996–2003	Yeh et al. (2005)
$1.85 - 1.90$	1	LASCO/SOHO	1996–2003	Yeh et al. (2005)
2.00 ± 0.05	3	LASCO/SOHO	1996–2008	Li et al. (2016)
2.2	1.5	LASCO/ARTEMIS	1996–2019	Lamy et al. (2019)
2.3	1.5	LASCO/SEEDS	1996–2019	Lamy et al. (2019)
2.5	1	LASCO/CDAW	1996–2019	Lamy et al. (2019)

Aschwanden (2016) analyzed 399 GOES M- and X-class events, using LASCO/SOHO and AIA/SDO data. Size distributions of various CME energies were determined, for gravitational CME energies ($\alpha_{grav} = 2.0 \pm 0.4$), for kinetic CME energies ($\alpha_{kin} = 1.4 \pm 0.1$), and the sum of gravitational and kinetic energies ($\alpha_{tot} = 2.0 \pm 0.3$). These values are commensurable with the theoretical predictions of the FD-SOC model, for fluences ($\alpha_E = 1.44$), fluxes ($\alpha_F = 1.80$), and CME volumes ($\alpha_V = 1.80$). Refined CME energy models with a larger statistics (860 events) yield similar values, that is, $\alpha_E = 1.4 \pm 0.3$ (Aschwanden 2017).

9.2 Coronal Mass Ejections: Waiting Time Distributions

Wheatland (2003) examined 4645 CME events observed with SOHO/ LASCO during 1996–2001, as well as soft X-ray flares observed by GOES during the same period. He found similar waiting time distributions for both phenomena, as shown in Figure 9.3. The waiting-time distribution is flat for short waiting times and turns into a power law-like distribution for long waiting times, with slopes of $\alpha_\tau = 2.36 \pm 0.11$ (all LASCO CMEs), $\alpha_\tau = 1.86 \pm 0.14$ (during 1996–1998 near solar minimum), and $\alpha_\tau = 2.98 \pm 0.20$ (during 1999–2001 near solar maximum), compared with $\alpha_\tau = 2.26 \pm 0.11$ (all GOES flares), $\alpha_\tau = 1.75 \pm 0.08$ (during 1996–1998), and $\alpha_\tau = 3.04 \pm 0.19$ (during 1999–2001) (Figure 9.3a–c). Wheatland (2003) subdivided the time series into Bayesian blocks with constant rates and was able to fit the distributions with a nonstationary Poisson process for both flares and CMEs (Figure 9.3d, e). Thus, the fits are consistent with theoretical models of nonstationary Poisson processes, for which a power law tail results

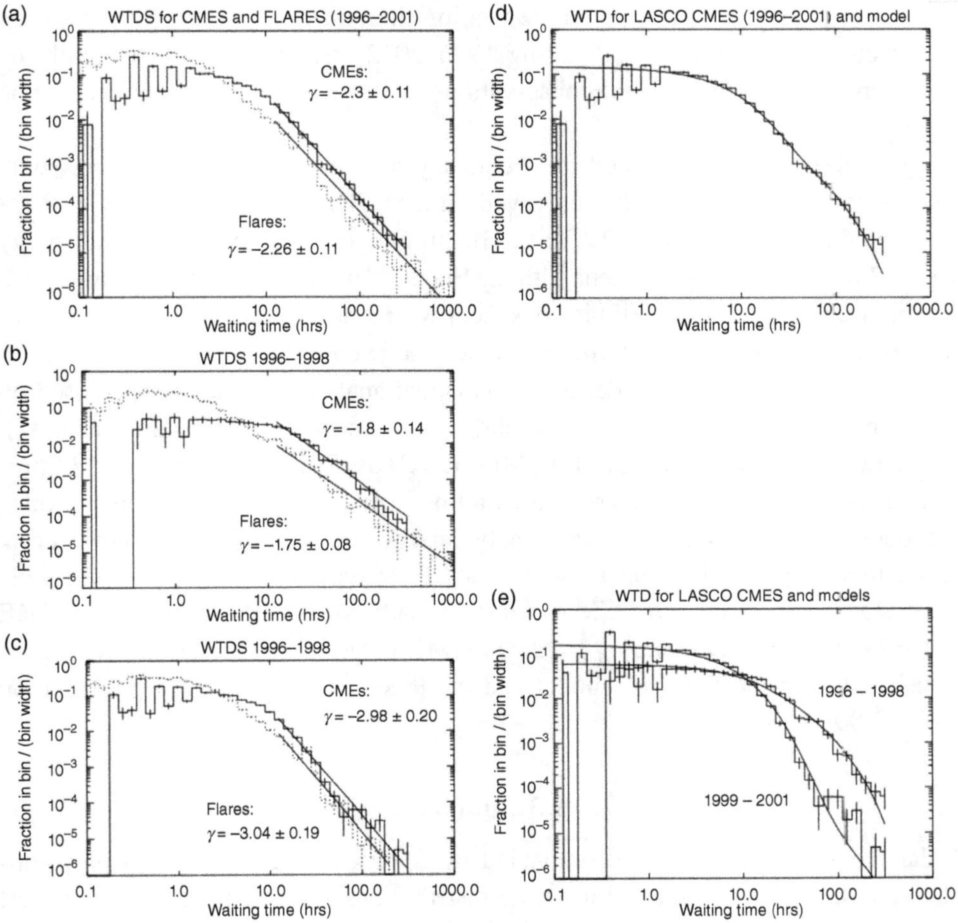

Figure 9.3 Waiting time distributions of CMEs observed with LASCO/SOHO and flares observed with GOES, for the whole time interval 1996–2001 (a), and the two epochs near solar minimum 1996–1998 (b) and near solar maximum 1999–2001 (c) with fitted power law tail. (d, e) The same three distributions fitted with a nonstationary Poisson process model with a time-varying flare rate (Wheatland 2003).

with a maximum slope of $\alpha_\tau \lesssim 3$ (Figure 9.3). It is also satisfactory to see an identical variation of the power law slope for CMEs and flares during the solar cycle, which implies a synchronous variation of the flaring rate $\lambda(t)$. This supports the notion that CMEs and flares are driven by the same magnetic instability, which was also concluded by Yeh et al. (2005) with similar data. The same data were also analyzed by Moon et al. (2003) and fitted with the same model of a nonstationary Poisson process developed by Wheatland (2000a), without finding an excess of short waiting times that could possibly indicate sympathetic CMEs.

Telloni et al. (2014) compared the waiting time distribution function of CMEs observed with LASCO/SOHO during 1996–2012 and found that a Weibull distribution function did fit the observations better than a Poisson distribution function.

Li et al. (2016) investigated the frequency distributions of 23,728 flares with GOES data and 14,017 CMEs observed with SOHO/LASCO during 1996–2008 and found that the flux ($\alpha_F = 2.13 \pm 0.03$), duration ($\alpha_T = 2.55 \pm 0.07$), and energy ($\alpha_E = 2.08 \pm 0.08$) are consistent with the FD-SOC model and that the waiting time distribution ($\alpha_\tau = 2.00 \pm 0.05$) is consistent with a nonstationary Poisson process. The energy of CMEs was inferred from the kinetic energy.

Lamy et al. (2019) did undertake a statistical analysis of solar CMEs based on 23 years of LASCO/SOHO observations, using a manual event catalog (CDAW), and four automated catalogs (ARTEMIS, CACTus, SEEDS, and CORIMP), providing size distributions of mass rates, waiting times, periodicities, angular widths, latitudes, speeds, acceleration, and kinetic energy. The size distribution of waiting times reveals power law functions, Poisson distributions, and Weibull functions, with slopes of $\alpha_\tau = 2.2$ for CME detection with ARTEMIS, $\alpha_\tau = 2.3$ for CME detection with SEEDS, and $\alpha_\tau = 2.5$ for CME detection with CDAW, which are consistent with the FD-SOC model and the theoretical waiting time distribution ($\alpha_\tau = 2.00$); see Chapter 3.

9.3 Summary

Solar flares are commonly accompanied by CMEs, and thus CMEs display similar size distributions and waiting time distributions as solar flares do. However, some studies report relatively steep power law slopes with values of $\alpha_F > 2$, which most likely are caused by a bias due to neglecting background subtraction in GOES data. The datasets from LASCO/SOHO are not affected by this background bias, because the white light background from CMEs appears to be sufficiently faint or nonexisting. Waiting time distributions are sampled from a variety of CME and flare catalogs, such as CDAW, LASCO/SOHO, ARTEMIS, CACTus SEEDS, and CORIMP. These waiting time distributions are found to be consistent with the theoretical prediction of the standard FD-SOC model, that is, $\alpha_\tau = 1 + 2/p$ (with p the nonlinearity index).

10

Solar Energetic Particle Events

Solar energetic particle (SEP) events represent a subset of large solar flares that produce protons, electrons, and helium ions, with energies of $\gtrsim 25$ keV to 1 GeV. Evidence for two different physical mechanisms for acceleration of SEPs arose 50 years ago, with radio observations of type III bursts (producing outward streaming electron beams), and type II bursts (from coronal and interplanetary shock waves), which translate into *impulsive SEPs* and *gradual SEPs* (Reames 2013). It was noted early on that the size distribution of peak counts of SEP events is flatter ($\alpha_E \approx 1.2 - 1.4$) than those of flare electromagnetic emission (Hudson 1978). Size distributions of the fluences of SEP events were found in a typical range of $\alpha_E \approx 1.2 - 1.4$ (Van Hollebeke et al. 1975; Cliver et al. 1991; Gabriel and Feynman 1996; Mendoza et al. 1997; Perez Enriquez and Miroshnichenko 1999; Miroshnichenko et al. 2001; Gerontidou et al. 2002; Gabriel and Patrick 2003; Xapsos et al. 2006; Belov et al. 2007; Cliver et al. 2012; Kahler 2013; Cliver and D'Huys 2018; Peng et al. 2023). The power law slopes of the size distributions of fluxes and fluences are listed in Table 10.1, which clearly exhibit a much flatter range ($\alpha_F \approx \alpha_E \approx 1.2 - 1.4$) than those measured in hard X-rays (Table 4.1) and soft X-rays (Table 5.1).

10.1 SEP: Size Distributions

We review the observations of size distributions (or occurrence frequency distributions) of solar energetic particles (SEP), in chronological order.

Van Hollebeke et al. (1975) performed a statistical study of 185 SEP events, using the Goddard cosmic ray experiments on the satellites IMP IV and V, consisting of a low-energy detector that detects protons and alpha particles in the range of 4–20 MeV nucleon^{-1} and a medium energy detector, which registers alpha particles in the range of 20–80 MeV nucleon^{-1}. A concomitant solar flare was identified in 68% of the observed SEP events. The size distribution of the SEP events was

Table 10.1 *Frequency distributions of solar energetic particle (SEP) events.*

Power Law slope of flux α_F	Power Law slope of fluence α_E	Spacecraft	Energy range E_{min}	Reference
1.15 ± 0.05		IMP4, 5	20–80 MeV protons	Van Hollebeke et al. (1975)
1.13 ± 0.04		IMP8	24–43 MeV protons	Cliver et al. (1991)
1.30 ± 0.07		IMP8	3.6–18 MeV electrons	Cliver et al. (1991)
	1.32 ± 0.05	IMP, OGO	>10 MeV protons	Gabriel and Feynman (1996)
	1.24 ± 0.04	IMP, OGO	>30 MeV protons	Gabriel and Feynman (1996)
	1.32 ± 0.07	IMP, OGO	>60 MeV protons	Gabriel and Feynman (1996)
1.32 ± 0.05			>10 MeV protons	Mendoza et al. (1997)
1.37 ± 0.05			>10 MeV protons	Miroshnichenko et al. (2001)
1.35 ± 0.2			>10 MeV protons	Gerontidou et al. (2002)
	1.34 ± 0.02		>10 MeV protons	Belov et al. (2007)
	1.46 ± 0.03		>100 MeV protons	Belov et al. (2007)
	1.22 ± 0.05		>10 MeV protons	Belov et al. (2007)
	1.26 ± 0.03		>100 MeV protons	Belov et al. (2007)
	1.28 ± 0.08^a		>25 MeV protons	Cliver and D'Huys (2018)
	1.51 ± 0.05^b		>25 MeV protons	Cliver and D'Huys (2018)
	1.16 ± 0.12^c		>25 MeV protons	Cliver and D'Huys (2018)
	1.31 ± 0.24^d		>25 MeV protons	Cliver and D'Huys (2018)
1.91 ± 0.30	1.69 ± 0.30		>100 MeV protons	Peng et al. (2023)
1.32	**1.32**			Median
1.80	**1.44**			FD-SOC prediction

[a]Western hemisphere SEP events, with maximum likelihood estimator, [b]SEP-associated events, with standard power law fitting, [c]Western hemisphere SEP events, with maximum likelihood estimator, and [d]SEP-associated events, with standard power law fitting.

found to be $\alpha_E = 1.15 \pm 0.05$, which is significantly flatter than what was reported earlier from hard X-rays, soft X-rays, and radio fluxes ($\alpha_E \approx 1.4$), as pointed out by Hudson (1978).

Cliver et al. (1991) obtained size distributions of fluxes of solar energetic proton (24–43 MeV) and electron (3.6–18.5 MeV) events observed with the NASA/GSFC particle detectors on IMP-8. The differential fluxes of the 92 proton events in the sample have a power law size distribution with a slope of $\alpha_F = 1.13 \pm 0.04$ and a slope of $\alpha_F = 1.30 \pm 0.07$ for the 67 electron events, respectively, which also exhibits a significantly flatter slope than the hard X-ray distributions of flares, which prompted Hudson to suggest that proton flares behave somehow differently from "ordinary" flares. However, Kurt (1990) compiled observations of electromagnetic fluences and particle fluxes and finds that both types of emission can be described in general by a differential power law function with a slope of

$\alpha_E = 1.45 \pm 0.15$, and concludes (Kurt 1990), contrary to Hudson (1978), that protons account for a proportional fraction of the total flare energy budget for proton energies < 20 MeV (Kuznetsov and Kurt 1991), that is, all flares are proton flares.

Gabriel and Feynman (1996) collected spacecraft data from the IMP 1, 2, 3, OGO 1, and IMP 5, 6, 7, 8, which recorded daily average fluxes integrated over the time of the event, consisting of proton fluxes above the energy thresholds of 10, 30, and 60 MeV, during the years 1965–1989. For these three thresholds, power law size distributions with slopes of $\alpha_E = 1.32 \pm 0.05$ ($E > 10$ MeV), $\alpha_E = 1.24 \pm 0.04$ ($E > 30$ MeV), and $\alpha_E = 1.32 \pm 0.07$ ($E > 60$ MeV) were found. Note that this power law slope range of $\alpha_E \approx 1.2 - 1.4$ exhibits the same trend of significantly flatter slopes than found for hard X-ray flares ($\alpha_E \approx 1.48$) as commented by Hudson (1978).

Mendoza et al. (1997) compiled data from the solar cycles 19–22 (covering the years 1955–1993), and found power law size distributions for protons above an energy threshold of $E > 10$ MeV, with a slope of $\alpha_E = 1.32 \pm 0.05$, when averaging two different methods. Some difference is noted between the ascending and descending phase of the solar cycle.

Miroshnichenko et al. (2001) sampled 320 flare-associated SEP events and found power law distributions extending over four orders of magnitude, and slopes of $\alpha_E = 1.37 \pm 0.05$. SEP events with associated sudden storm commencements (SSC) yield deviations from straight power laws, suggesting two different particle acceleration mechanisms (e.g., flares, shock waves).

Gerontidou et al. (2002) combined 147 solar proton events with proton energies >10 MeV and peak intensities of >10 protons cm^{-2} s^{-1} sr^{-1} (particle flux unit), measured at Earth during 1976–1999. The corresponding size distribution was found to be a power law with a slope of $\alpha_E = 1.35 \pm 0.2$.

Gabriel and Patrick (2003) evaluate the predictive power of SEP events. Because of the high energies of SEP events, which can harm astronauts or electronic equipment in space, statistical information that improves their predictability is highly desirable. The occurrence of SEP events has been assumed to be random so far, but SOC models and related observations suggest that power law distribution functions are more realistic, especially due to the clustering of large SEP events near the maximum of the solar cycle.

Xapsos et al. (2006) demonstrate with statistical studies, using daily and monthly fluences of solar protons, measurements by IMP-8 and GOES over 28 years, long-term correlations, fractal analysis, power law fitting, and waiting time analysis, that it is not possible to predict the time of occurrence of SEP events within narrow limits.

Belov et al. (2007) combined data of solar proton enhancements (SPE) with soft X-ray flares from GOES and confirm the finding of power law functions

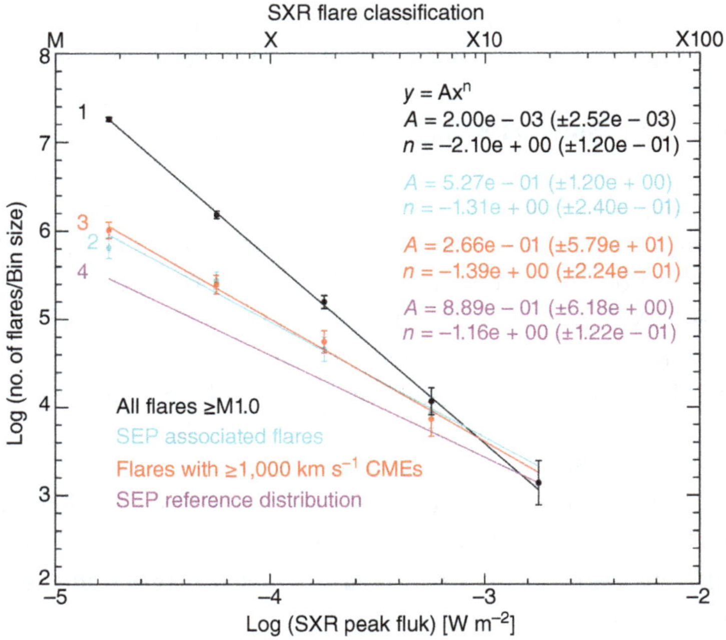

Figure 10.1 Size distributions for: (1) peak 1–8 Å fluxes of all ≥M1.0 soft X-ray flares; (2) subset of flares associated with > 10 MeV proton events; (3) subset of flares with ≥ 1,000 km s^{-1} CMEs; and (4) peak proton fluxes of > 10 MeV SEP events. All SEP events were observed during 1996–2005 (adapted from Cliver et al. 2012).

that extend over 3 to 5 orders of magnitude, a range that agrees with the sensitivity above the smallest registered peak values. They find power law slopes of $\alpha_E = 1.34 \pm 0.02$ and $\alpha_E = 1.46 \pm 0.04$ for the two thresholds $E_P > 10$ MeV and $E_P > 100$ MeV. For SPEs associated with flares located to the West of 20°, the values are $\alpha_E = 1.22 \pm 0.05$ and $\alpha_E = 1.26 \pm 0.03$).

Cliver et al. (2012) interpreted the discrepancy between flares and SEP events as a selection effect of SEP events that are preferentially associated with larger flares, and thus the SEP events are drawn from a subset of flares that do not form a statistically representative sample. This interpretation has been demonstrated by sampling subsets of flares that are associated with > 10 MeV proton events, or with ≥1,000 km s^{-1} CMEs, which exhibit a similar flat power law slope as the SEP events themselves (Figure 10.1). Note that a power law function fits the size distribution of SEP fluences only in the low-fluence part, while the high-fluence part is better fitted by an exponential cutoff function, that is, $N(E) \propto E^{-\alpha_E} \times \exp -(E/E_0)$, based on SEP data from 41 solar cycles from the year 1561 to today (Miroshnichenko and Nymmik 2014).

Kahler (2013) challenges the interpretation of a selection bias and suggests that the difference can be explained by the dimensionality of the SOC system. If we take the general expression of the power law slope α_E for energy or fluences, we expect in the standard FD-SOC model, for classical diffusion ($\beta = 1$) and incoherent processes ($\gamma = 1$), and inserting the mean value of the fractal dimension $D_d = d - 1/2$, a power law slope of

$$\alpha_E = 1 + \frac{(d-1)}{(D_V\gamma + 2/\beta)}. \tag{10.1}$$

This yields $\alpha_E = 9/7 \approx 1.44$ in 3-D space, but a flatter slope of $\alpha_E = 9/7 \approx 1.29$ in 2-D space. This is conceivable if the avalanche spreads over a 2-D surface only, such as a reconnection current sheet, or the surface of a shock wave. This would invalidate a close physical connection between flares and SEP events and provide this way a diagnostics of the dimensionality of the particle acceleration process in flares and SEP events.

Cliver and D'Huys (2018) used the maximum likelihood estimator to determine power law size distributions of peak proton fluxes of SEP events observed during 1997–2016. They find a slope of $\alpha_E = 1.28 \pm 0.08$ for western hemisphere SEP events, and $\alpha_E = 1.51 \pm 0.05$ for SEP-associated events. These values are somewhat steeper than the slopes determined previously with power law fitting (Cliver et al. 2012), $\alpha_E = 1.16 \pm 0.12$ for western hemisphere SEP events, and $\alpha_E = 1.31 \pm 0.24$ for SEP-associated events. The flatter slopes are likely due to the fact that CME-selected SEP events are subject to a threshold speed of $v_{\rm CME} \geq 400$ km s^{-1}, which represents the slow solar wind component.

Peng et al. 2023 inquires the size distribution of solar GeV flares and finds $\alpha_F = 1.91 \pm 0.33$ for the >100 MeV peak flux, $\alpha_E = 1.69 \pm 0.19$ for the fluence, and $\alpha_T = 2.10 \pm 0.35$, which roughly agree with the theoretical prediction ($\alpha_F = 1.80$, $\alpha_E = 1.44$, $\alpha_T = 2.00$). This is the only SEP study that agrees with the FD-SOC model, which indicates that the SEP particle fluxes are driven by a different physical mechanism than flare-related hard X-ray fluxes.

10.2 Summary

The size distribution of SEP events, which represent a more energetic subset than flare events, is mostly found to follow power law distribution functions, rather than Poissonian random distribution functions. However, the numerical value of the power law slope is generally flatter ($\alpha_E \approx 1.2 - 1.4$) than the slopes of the flare size distributions in hard X-rays, soft X-rays, and EUV ($\alpha_E \approx 1.44$) (Hudson 1978), which can be explained in at least four different ways: (i) normal flares and proton flares are produced by two fundamentally different acceleration mechanisms (magnetic reconnection-induced acceleration versus shock acceleration); (ii) proton flares behave differently than normal flares; (iii) the fractal dimensionality of SEP events is different from normal flares; and (iv) proton flares are subject to a selection bias toward the most energetic events, and thus are not a representative sample of large flares. Nevertheless, the standard fractal-diffusive SOC model can explain the observed slopes of SEP size distributions, but observations reveal deviations from straight power law functions, or broken power law slopes, and thus are not unique and need to be modeled in more detail.

11

Solar Wind

The solar wind is a turbulent magneto-fluid, consisting of charged particles (electrons, protons, alpha particles, and heavy ions) with typical energies of 1–10 keV, which escape the Sun's gravity field because of their high kinetic (suprathermal) energy and the high temperature ($T \gtrsim 1.0$ MK) of the solar corona. The solar wind has two different sources, depending on its origin, namely, a fast solar wind with a speed of $v \lesssim 800$ km s^{-1}, originating from open-field regions in coronal holes, and a slow solar wind with a speed of $v \lesssim 400$ km s^{-1}, originating from low latitudes in the surroundings of coronal streamers. The dynamics of the solar wind was originally explained by Parker (1958) as a supersonic outflow that can be derived from a steady-state solution of the hydrodynamic momentum equation. Later refinements take the super-radial expansion of the coronal magnetic field into account, as well as the average macroscale and fluctuating mesoscale electromagnetic field in interplanetary space, and the manifold microscale kinetic processes (such as Coulomb collisions and collective wave–particle interactions). The properties of the solar wind that can be measured from the solar corona throughout the heliosphere are plasma flow speeds, densities, temperatures, magnetic fields, wave spectra, and particle composition, which all exhibit complex spatiotemporal fluctuations. Most of the observations of the solar wind were made in-situ (with the Mariner, Pioneer, Helios, ISEE-3, IMP, Voyager, ACE, WIND, Cluster, Ulysses, or STEREO spacecraft), complemented by remote-sensing imaging (with STEREO, the Parker Solar Probe [PSP], and Solar Orbiter), as well as radio scintillation measurements.

The dynamics of the solar wind is often characterized by the MHD turbulent cascade model. The solar wind power spectrum exhibits fully developed turbulence of the Kolmogorov type, with a power spectrum of $P(f) \propto f^{-5/3}$, in interplanetary space and near Earth (Figure 11.1), while the input spectrum in the lower corona is of the 1/f-noise type, $P(f) \propto f^{-1}$ (Matthaeus and Goldstein 1986; Nicol et al. 2009). The MHD turbulent cascade starts at the largest scales fed by MHD waves with a f^{-1} noise spectrum in the lower corona, while turbulent interactions produce

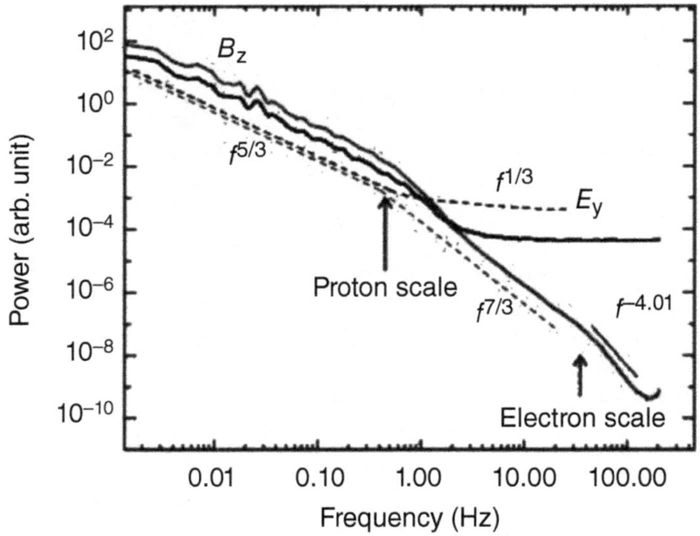

Figure 11.1 A spectrum of the solar wind is shown, based on *CLUSTER* observations from large scales ($\approx$ 105 km) down to small scales ($\approx$ 3 km) is shown, with the proton and electron gyroradius scale indicated. The solar wind spectrum is interpreted in terms of a turbulent MHD cascade, with the theoretically predicted slopes of $f^{-5/3}$ and $f^{-7/3}$ from gyrokinetic theory. The plot proves that the energy continues cascading below the proton scale down to the electron scale, where it is converted to heat (via electron Landau damping resonance) causing the steepening of the B_z spectrum to f^{-4} (Howes et al. 2008; Sahraoui et al. 2009; credit: ESA, Cluster mission).

a cascade of energy through vortices and eddies to progressively smaller sizes with a spectrum of $f^{-5/3}$, and final energy dissipation at the smallest scales by heating of electrons, with a spectrum of $f^{-11/3}$ (Meyrand and Galtier 2010). The analysis of MHD turbulence in solar wind data includes determining power spectra and structure functions, waiting time distributions of solar-wind bursts, identifying the phenomenology or MHD turbulence (Kolmogorov 1941; Kraichnan 1974), characterizing self-similarity and intermittency, and identifying the most intermittent structures, such as shock waves, small random events, current cores, and 1-D current sheets (e.g., Horbury et al. 1997; Veltri et al. 1999).

11.1 Solar Wind: Size Distributions

Recent interpretations of the dynamics of the solar wind include self-organized criticality (SOC) models. Essentially, exponentially growing instabilities of waves and particles produce nonlinear avalanches after they exceed a threshold in excess of the noise fluctuations in the solar wind. Power law-like size distribution functions are the hallmarks of SOC processes. In the following, we review relevant observations of size distributions that give evidence for SOC systems in the solar wind, and discuss them in chronological order.

Freeman et al. (2000) calculate the probability density function of energies E, durations T, and waiting times τ, from bursts in the Earth–Sun component of the energy Poynting flux, $F_P \approx \mathbf{E} \times \mathbf{H} \approx v(B_y^2 + B_z^2)/\mu_0$, assuming a radial solar wind. The measurements are carried out in situ at 1 AU in the solar wind, using data from the MFI and SWE experiments onboard the WIND spacecraft. They find that these probability density functions are approximate power law functions, that is, $N(x) \approx x^{-\alpha_x}$, from which (their Fig. 1) we measure slopes of $\alpha_E \approx 1.8$, $\alpha_T \approx 2.2$, and $\alpha_\tau \approx 1.67$. These values roughly match the theoretical predictions of the FD-SOC model, $\alpha_E \approx 1.44$, $\alpha_T \approx 2.0$, and $\alpha_\tau \approx 2.0$, which yield evidence for a SOC interpretation of the solar wind (Freeman et al. 2000; Moloney and Davidsen 2011). Although the power law shape of the waiting time distribution of the solar wind bursts is not exponential, as expected for the original stationary BTW model (Boffetta et al. 1999; Freeman et al. 2000), it can be reproduced with a nonstationary Poisson process (Wheatland et al. 1998; Aschwanden and McTiernan 2010). This was also demonstrated with MHD simulations (Watkins et al. 2001; Greco et al. 2009), and by cellular automaton simulations with correlations in the driver that produces a "colored" power spectrum (Sanchez et al. 2002), that is, f^{-1} (pink noise); f^{-2} (red noise); or f^{-3} (black noise).

Moloney and Davidsen (2011) used magnetic field measurements from the SWE PAM and MAG instruments onboard the ACE spacecraft and found size distributions with power law slopes of $\alpha_E \approx 1.60$ for the energy E; $\alpha_T \approx 2.46$ for the event durations, and $\alpha_\tau \approx 1.53$ for the waiting times, which are similar to the slopes derived from WIND data (Freeman et al. 2000). For the definition of the energy of a solar wind fluctuation event, the Akasofu parameter is defined (in SI units) as,

$$E_B = v \frac{B^2}{\mu_0} l_0^2 \sin^4(\theta/2), \tag{11.1}$$

where v is the solar wind velocity, B the magnetic field, $\mu_0 = 4\pi \times 10^{-7}$ the permeability of free space, $l_0 \approx 7R_E$, and $\theta = \arctan(|B_y|/B_z)$. Measurements were mostly available at a time resolution of 64 s and 16 s for the solar wind velocity and the 3-D magnetic field, accumulating up to $\approx 2 \times 10^7$ datapoints during the analyzed four years (2000–2003).

Moloney and Davidsen (2014) investigated the statistics of extreme bursts in the solar wind and found no significant variation during the solar cycle, which is manifested in the stationary slope of the size distributions of duration times T and energies E of extreme bursts in the solar wind. This provides a solar cycle invariant constraint between possible sizes and durations of solar wind bursts

(Tindale et al. 2018). Measurements of kinetic energy fluctuations and the magnetic field constrain then also the velocity distributions of particles in the solar wind that can be tested with SOC models (Pavan and Vinas 2019).

Dudok de Wit et al. (2020) discovered so-called switch-back structures with PSP, using the MAG magnetometer from the FIELDS consortium, which is able to run cadences down to 292.97 samples per second, during the first perihelion pass on November 6, 2018. This dataset produced waiting time distributions (with a slope of $\alpha_T = 1.4 - 1.6$), residence time distributions, and power spectra, which all revealed power law distribution functions (the hallmarks of SOC systems), extending over $\approx$5 orders of magnitude (Dudok de Wit et al. 2020).

Aschwanden and Dudok de Wit (2021) analyzed the waiting time distributions of GOES data (1974–2012) during observations with PSP and found the following empirical scaling law,

$$\alpha_T = 1.38 + 0.01 \times SN, \tag{11.2}$$

where SN is the sunspot number, which is a convenient indicator of the solar cycle activity level (in the range of $\alpha_T \approx 1.4 - 3.2$ during the analyzed solar cycles). This property of the time-dependent power law slope $\alpha_T(t)$ that depends only on the global magnetic flux (quantified by the sunspot number) is not predicted by the FD-SOC model (which does not have any magnetic variable). The novel phenomena of the so-called switchback events were sampled in situ with PSP in the solar wind at a distance of $\approx 36 R_{\odot} = 0.166$ *astronomical units (au)*. Switchback events have durations of less than 1 s to more than an hour. Hallmarks of switchback events are reversals in the radial field component B_r (with respect to the Parker spiral geometry, which can produce deflection angles from a few degrees to nearly 180° in the fully anti-sunward direction. A switchback can be quantified by the magnetic potential energy,

$$E_P = -\mathbf{B} \cdot < \mathbf{B} >, \tag{11.3}$$

or by the normalized deflection angle μ,

$$z = \frac{1}{2}(1 - \cos \mu) \quad \text{for } 0 \le z \le 1. \tag{11.4}$$

Power law slopes of $\alpha_T \approx 1.2 - 1.6$ were also found in solar switchback events, as observed with PSP during the solar minimum, while steeper slopes are expected during the solar maximum. The annual variability of switchback events in the heliospheric solar wind and solar flare and CME rates (originating in the photosphere and lower corona) are highly correlated.

Chen & Hu (2022) used data from the first six perihelion encounters of the PSP spacecraft and determined the size distributions of event durations ($T \approx 10 - 10^3$;

slope $\alpha_T = 1.76$), and small scale sizes of events ($L \approx 10^{-5} - 10^{-3}$ AU; slope $\alpha_L = 1.87$). The events detected here are called *small-scale flux ropes*, which could possibly be modeled with a SOC system, or alternatively, in terms of turbulence due to the process of local Alfvénic alignment.

11.2 Solar Wind: Waiting Time Distributions

The observed waiting time distributions yield approximate power laws, obtained either by direct log-log fitting or by fitting Pareto-type II size distributions, which cover a relatively wide range of power law slopes (Table 11.1), $\alpha_\tau = 1.67$ (Freeman et al. 2000), $\alpha_\tau = 1.6$ (Moloney and Davidsen 2011), $\alpha_\tau = 1.4 - 1.6$ (Dudok de Wit et al. 2020), and $\alpha_\tau = 1.2 - 1.6$ (Aschwanden and Dudok de Wit 2021). Since our FD-SOC model does not contain magnetic field parameters, a wide range of power law slopes is expected, depending on the variability range of solar activity. If we include the sunspot number as a proxy for the magnetic field activity, one finds an excellent correlation between the observed sunspot number $SN(t)$ and the waiting time power law slopes $\alpha_\tau(t)$ (Eq. 11.2; Figure 11.2), which has predictive power. If one selects an arbitrary time instant or associated sunspot number in the range of $0 < SN < 200$, we can predict the power law slope of waiting times $1.4 \lesssim \alpha_\tau \lesssim 3.4$, according to the relationship of Eq. (11.2) (Aschwanden and Dudok de Wit 2021). The corresponding prediction explains the role of the solar minimum in the waiting time distribution throughout the heliosphere (Nurhan et al. 2021).

Table 11.1 *Power law slopes measured in the size distributions of magnetic field fluctuations in the solar wind. The burst energy E is defined as the area-integrated and time-integrated Poynting flux, derived from the Akasofu parameter, or from the geometry of magnetic switchback events (in the case of PSP data).*

Instrument	Power law slope of energy α_E	Power law slope of duration α_T	Power law slope of waiting time α_τ	Reference
WIND	1.8	2.2	1.67	Freeman et al. (2000)
ACE	1.53	2.46	1.60	Moloney and Davidsen (2011)
PSP			1.4–1.6	Dudok de Wit et al. (2020)
PSP			1.2–1.6	Aschwanden and Dudok de Wit (2021)
PSP		1.76		Chen and Hu (2022)
	1.67	**2.20**	**1.55**	Median
	1.44	**2.00**	**2.00**	FD-SOC prediction

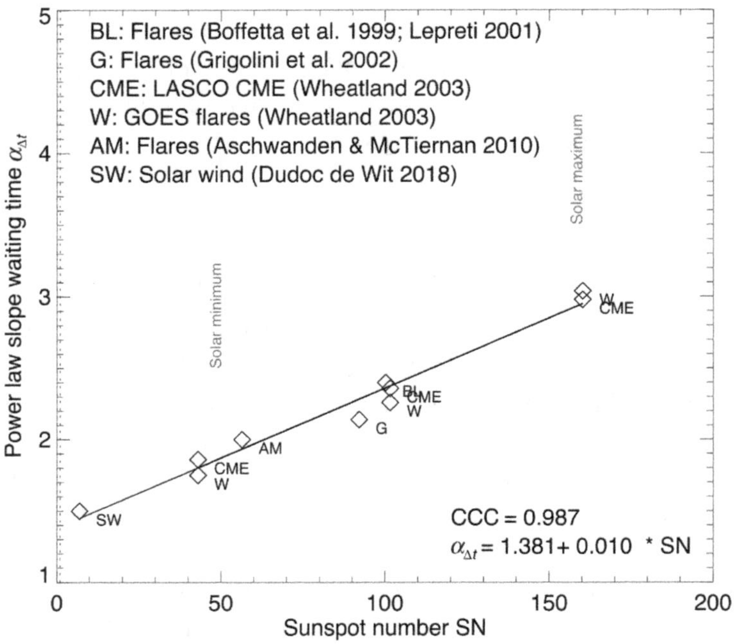

Figure 11.2 Scatterplot of power law slope α_τ of waiting time distributions vs. the sunspot number SN, with linear regression fit (solid line) and cross-correlation coefficient CCC (Aschwanden and Dudok de Wit 2021).

11.3 Solar Wind: Fractal Geometry

Self-organized criticality systems produce fractal spatiotemporal structures. The fractal nature of magnetic energy density fluctuations in the solar wind has been verified observationally (Hnat et al. 2007; Rypdal and Rypdal 2011). Moreover, solar wind turbulence is found to be multifractal, requiring a generalized model with multiple scaling parameters to analyze intermittent turbulence (Macek and Szczepaniak 2008; Macek and Wawrzasek 2009; Macek 2010), although a single generalized scaling function is sometimes sufficient too (Chapman and Nicol 2009; Rypdal and Rypdal 2011). However, the fractal geometry of solar wind bursts seems not to be self-similar, since the ratio of kinetic (E_k) to magnetic energy ($E_B \propto B^2$) is frequency-dependent, with a magnetic energy spectrum of $E_B^{-5/3}$ and a kinetic energy spectrum of $E_k^{-3/2}$ (Podesta et al. 2006a, 2006b, 2007a, 2007b). It was suggested that the interplanetary magnetic field (IMF) is clustered (self-organized) by low-frequency magnetosonic waves, leading to a fractal structure with a Hausdorff dimension of $D = 4/3$ and a turbulent power spectrum with $\nu^{-5/3}$ (Milovanov and Zelenyi 1999). Multifractal analysis of time series of solar wind speeds were validated by Garcia-Marin et al. (2013).

11.4 Solar Wind: SOC versus MHD Turbulence

Boffetta et al. (1999) argue that the statistics of solar flare (laminar or quiescent) waiting times indicate a physical process with complex dynamics with long correlation times, such as in chaotic models, in contradiction to stationary SOC models that predict Poisson-like statistics. They consider chaotic models that include the destabilization of the laminar phases and subsequent restabilization due to nonlinear dynamics, as invoked in their shell model of MHD turbulence.

Similarly, Lepreti et al. (2001) attribute the origin of the observed waiting time distribution to the fact that the physical process underlying solar flares is statistically self-similar in time and is characterized by a certain amount of "memory." They find that the power law distribution can be modeled by a Lévy function which can explain a power law exponent of $\alpha_\tau = 3$ in the waiting time distribution.

Grigolini et al. (2002) develop a technique called diffusion entropy method to reproduce the observed waiting time distribution function, which evaluates the entropy of the diffusion process generated by the time series. Note that classical diffusion has been employed in SOC models (Aschwanden 2012a; 2012b), which may be related to the diffusion entropy method, since both models produce a similar scaling of short waiting times, with power law slopes in the range of $\alpha_\tau \approx 1.4 - 1.6$ (Table 11.1). The change of the power law index from $\alpha_\tau > 3$ to $\alpha_\tau < 3$ has been interpreted in terms of a phase transition from the Gaussian to the Lévy basin of attraction (Grigolini et al. 2002).

It is understood that a large part of the energy dissipation must occur at scales that are smaller than the resolved observations (≈ 100 km on the solar surface), perhaps as small as 10–100 m. The scale of Kolmogorov turbulent cascade dissipation determines the rate of coronal heating (Buchlin et al. 2006). The interplanetary space is subject to very high Reynolds numbers, which produces time scales in a wide range, from sub-second scales up to the solar rotation period of 26 days (Bruno et al. 2007). It appears that the solar memory lasts from hours to decades (Aschwanden and Johnson 2021). The lack of strict self-similarity and the lack of equi-partition between magnetic and kinetic fluctuations suggests that propagating MHD waves and static structures are convected by the wind (Bruno et al. 2007). SOC and intermittent turbulence appears to coexist (Uritsky et al. 2007). The avalanching systems may be understood in terms of intermediate driving, while the ratio of driving rate to dissipation rate is taken to zero in pure SOC systems (Chapman et al. 2009). Besides the two regimes of the Kolmogorov spectrum, the proton inertial range and the electron gyroradius, a third regime of the magnetic fluctuation index of $-11/3 \approx 3.7$ has been predicted for the power spectrum at the smallest spatial scales (Meyrand and Galtier 2010). More recent measurements yield a fluctuation index of ≈ 4.24 (Lion et al. 2016).

11.5 Summary

In the end, can we claim that the dynamics of the solar wind is consistent with a SOC system? Observationally we find that magnetic field and kinetic energy fluctuations measured in the solar wind exhibit power law distributions, which is consistent with a SOC system. One argument against the SOC interpretation is the observed power law distribution of waiting times (Boffetta et al. 1999), but this argument applies only with respect to the original BTW model, while it presents no obstacle for nonstationary Poisson processes. Another (Occam's razor) argument was that a SOC interpretation is not needed when turbulence can already explain solar wind spectra (Watkins et al. 2001). Considering the spatial structure of the solar wind, a fractal (or multifractal) property has been identified, which is another hallmark of SOC models. What about the driver, instability, and avalanches expected in a SOC system? The driver mechanism is the acceleration of the solar wind in the solar corona itself, a process that basically follows the hydrodynamic model of Parker (1958), and may be additionally complicated by the presence of nonlinear wave–particle interactions, such as ion-cyclotron resonance (e.g., for a recent review, see Ofman 2010). Then, the instability threshold, triggering extreme bursts of magnetic field fluctuations, and the avalanches of solar wind SOC events can be caused by dissipation of Alvén waves, onset of turbulence, or by the ion-cyclotron instability. Thus, in principle, the generalized SOC concept can be applied to the solar wind, if there is a system-wide threshold for an instability that causes extreme magnetic field fluctuations. On the other side, the MHD turbulent cascade model explains naturally two particular spatial scales with enhanced energy dissipation (i.e., the proton and the electron gyroradii), which is in contrast to the scale-freeness of energy dissipation in classical SOC models. Nevertheless, the solar wind dynamics can be described by multiple models that do not exclude each other: (1) the MHD turbulent cascade model describes the power spectrum of the solar wind, (2) the kinetic theory captures the microscopic physics of wave–particle interactions and the evolution of particle velocity distributions in the solar wind, and (3) SOC models quantify the statistics and macroscopic size distributions of extreme events and scaling laws of solar wind parameters.

12
Magnetospheric Phenomena

In the Earth's magnetosphere, a number of phenomena have been interpreted as features of SOC systems, such as geomagnetic storms, substorms, current disruptions, magneto-tail current disruptions and associated magnetic field fluctuations, bursty bulk flow events, and auroras visible in UV and optical wavelengths. Recent reviews on magnetospheric phenomena with focus on SOC aspects are given in Sharma et al. (2016) and Aschwanden et al. (2016).

Most magnetospheric phenomena result from the interaction of the Earth's (or some other planet's) magnetic field with the ambient solar wind in the heliosphere, at the magnetopause, in the magnetotail, and in the polar regions of the planet. The solar wind transports disturbances associated with solar flares in form of coronal mass ejections, shock waves, and solar energetic particles, causing geomagnetic storms, substorms, and auroral activities. The solar wind is thus the driver of the geomagnetic processes in space weather. Substorms involve many processes, including magnetic reconnection, ballooning-mirror modes, and current disruption, which cause a fast unloading of the highly stressed geotail system (Papadopoulos et al. 1993; Baker et al. 1996; Horton and Doxas 1996). In addition, multiscale intermittent turbulence of overlapping plasma resonances play an important role in substorms (Chang 1999a). Brief magnetospheric disturbances occur when the interplanetary magnetic field (IMF) flips southward, which triggers magnetic reconnection at the dayside magnetopause and transfers momentum and energy from the solar wind to the magnetosphere. Part of the transferred energy is stored in the magnetotail, where also magnetic reconnection and field relaxation events can occur during *magnetospheric substorms*. A magnetospheric substorm has three phases (Figure 12.1): (i) the growth phase (when energy from the solar wind is transferred to the dayside magnetosphere), (ii) the substorm expansion phase (when the energy stored in the magnetotail is released, the magnetosphere relaxes from the stretched tail, and the tail snaps into a more dipolar configuration and energizes particles in the plasma sheet), and (iii) the recovery phase (during which the magnetosphere

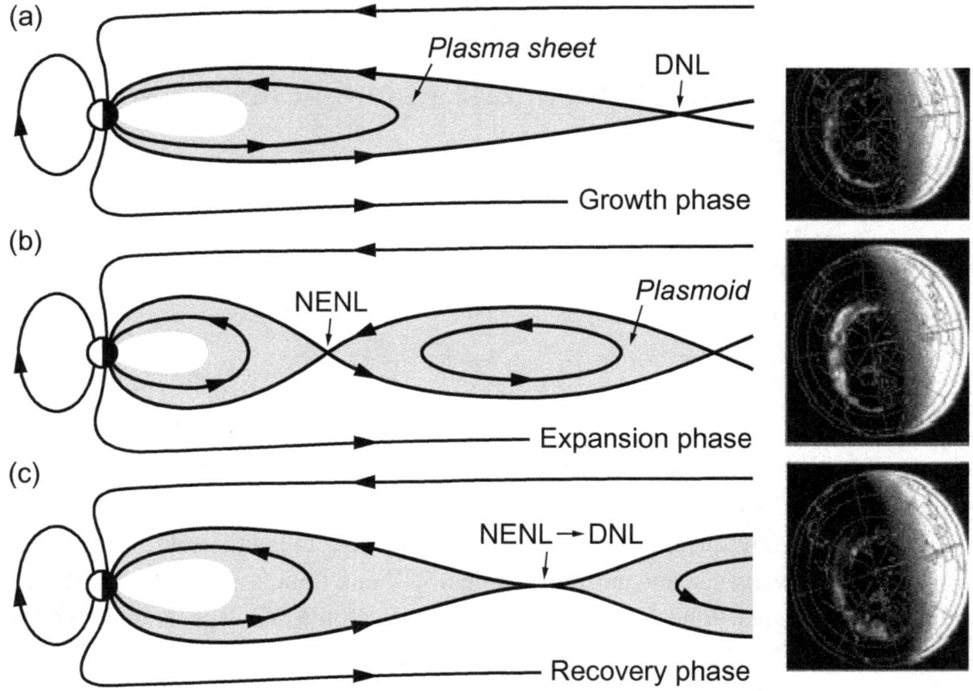

Figure 12.1 The three phases of a geomagnetic substorm are shown: the growth phase (a), the expansion phase (b), and the recovery phase (c) (Baumjohann and Treuman 1996). The accompanying three auroral images were obtained with the IMAGE WIC instrument (credit: NASA).

returns to its quiet state). The whole process causes changes in the auroral morphology (Figure 12.1) and induces currents in the polar ionosphere, where the resulting heating leads to auroral displays. The frequency of substorms is about six per day on average, but larger during geomagnetic storms.

12.1 Magnetospheric Phenomena: Size Distributions

How did the SOC concept came into play for magnetospheric processes? The bursty nature of magnetospheric phenomena, such as localized current disruptions in auroral blobs (Lui et al. 1988), bursty bulk flow events in the geotail (Angelopoulos et al. 1996, 1999), and the power law magnetic field spectra in the magneto-tail (Hoshino et al. 1994), have been interpreted in terms of an open, dissipative nonlinear system near a forced or self-organized critical state (Chang 1998a; Chang 1998b; 1999a, 1999b; Klimas et al. 2000; Chang et al. 2003; Chapman and Watkins 2001; Consolini and Chang 2001; Consolini 2002). Probability or size distributions with a power law shape (Table 12.1), the hallmark of SOC systems, have been measured from auroral blobs in UV (Lui et al. 2000; Uritsky et al. 2002, 2003,

Table 12.1 *Frequency distributions measured from magnetospheric phenomena. Values determined with nonstandard methods are marked with parentheses. The datasets of Uritsky et al. (2002) refer to different observing periods, the datasets of Kozelov et al. (2004) to different luminosity threshold levels, and the datasets of Uritsky et al. (2008) to different latitude zones (HL = high latitude events, LLs = low-latitude small-scale events, and LLl = low-latitude large-scale events). The medians of observed power law distributions and the theoretical predictions based on the FD-SOC model are marked in boldface (Aschwanden 2012a).*

Phenomenon	Power law slope of area α_A	Power law slope of flux α_F	Power law slope of fluence α_E	Power law slope of durations α_T	References
Geotail flow bursts				1.59 ± 0.07	Angelopoulos et al. (1999)
AE index				1.24	Takalo et al. (1993, 1999a)
AU index				1.3	Freeman et al. (2000)
Aurora UV (substorms)	(1.21 ± 0.08)	(1.05 ± 0.08)			Lui et al. 2000
Aurora UV (quiet)	(1.16 ± 0.03)	(1.00 ± 0.02)			Lui et al. 2000
Aurora UV Jan 1997	1.73 ± 0.03	1.66 ± 0.03	1.46 ± 0.04	2.08 ± 0.12	Uritsky et al. (2002)
Aurora UV Feb 1997	1.74 ± 0.03	1.68 ± 0.03	1.39 ± 0.02	2.21 ± 0.11	Uritsky et al. (2002)
Aurora UV Jan 1998	1.81 ± 0.04	1.73 ± 0.02	1.62 ± 0.03	2.24 ± 0.11	Uritsky et al. (2002)
Aurora UV Feb 1998	1.92 ± 0.04	1.82 ± 0.03	1.61 ± 0.04	2.39 ± 0.11	Uritsky et al. (2002)
Aurora UV	1.85 ± 0.03	1.71 ± 0.02	1.50 ± 0.02	2.25 ± 0.06	Kozelov et al. (2004)
Aurora TV 2.0 kR	1.98 ± 0.04	2.02 ± 0.02	1.74 ± 0.03	2.53 ± 0.07	Kozelov et al. (2004)
Aurora TV 2.5 kR	1.85 ± 0.04	1.92 ± 0.02	1.66 ± 0.04	2.38 ± 0.05	Kozelov et al. (2004)
Aurora TV 2.R kR	1.86 ± 0.05	1.84 ± 0.03	1.60 ± 0.02	2.33 ± 0.06	Kozelov et al. (2004)
Aurora UV HL	1.87 ± 0.05	1.81 ± 0.02	1.57 ± 0.02	2.30 ± 0.11	Uritsky et al. (2008)
Aurora UV LLs	2.11 ± 0.16	2.16 ± 0.09	1.83 ± 0.04	3.21 ± 0.33	Uritsky et al. (2008)
Aurora UV LLl	1.09 ± 0.14	1.32 ± 0.14	1.04 ± 0.12	1.26 ± 0.44	Uritsky et al. (2008)
Outer radiation belt		1.5 – 2.1	1.5 – 2.7		Crosby et al. (2005)
Ionospheric disturbances				1.8-2.5	Bristow (2008)
	1.85	**1.80**	**1.62**	**2.27**	Median
	2.33	**1.80**	**1.44**	**2.00**	FD-SOC prediction

2006) and optical light (Kozelov et al. 2004), from the auroral electron jet index (AE) (Takalo et al. 1993; Consolini 1997, 2002), from magnetospheric substorm-related tail current disruptions (Consolini and Lui 1999), from geotail flow bursts (Angelopoulos et al. 1999), from ionospheric velocity fluctuations driven by IMF (Bristow 2008), and from electron bursts in the outer radiation belt (Crosby et al. 2005, 2011). A power law slope of $\alpha_T \approx 1.3$ was also determined for waiting times in an AE index time series (Lepreti et al. 2004). Critical finite-size scaling and a fractal dimension of $D_A = 1.54 \pm 0.02$ was found for auroral blobs (Uritsky et al. 2006), which agrees with the mean-value estimate $D_A = 2 - 1/2 = 1.5$ in the FD-SOC model (Section 2.7). The power law behavior, as observed in many magnetospheric phenomena (Table 12.1), provides the main basis for interpretation as SOC processes. The measurements listed in Table (12.1) were obtained from UVI onboard the POLAR spacecraft (Lui et al. 2000; Uritsky et al. 2002), from all-sky TV cameras at the Barentsburg Observatory (Kozelov et al. 2004), the GEOTAIL spacecraft (Angelopoulos et al. 1999), from the WIND spacecraft (Freeman et al. 2000b), with the SuperDARN radar network (Bristow 2008), and the STRV microsatellites (Crosby et al. 2005).

The agreement between the statistics of large auroral events observed with POLAR /UVI (Uritsky et al. 2002) and small auroral events observed with a TV camera (Kozelov et al. 2004) is accurately matching (Figure 12.2), and covers a combined (but not overlapping) range of 10 orders of magnitude in energy. However, we can see in Table 12.1 a glaring discrepancy in some measurements made by Lui et al. (2000), while those made by Uritsky et al. (2002) agree surprisingly well with the predictions of the standard fractal-diffusive SOC model (Aschwanden 2012a). The measurements by Lui et al. (2000) yield much flatter power law slopes, close to unity. This discrepancy has been convincingly explained in terms of a different methodology and event definition: The statistics carried out by Lui et al. (2000) refers to equidistant time snapshots that count large avalanche events multiple times, while the analysis of Uritsky et al. (2002) determines the time-integrated avalanche areas and energies, consistent with standard definitions of SOC parameters (also used in the FD-SOC model: see Section 2.7). Another anomaly that was found is the latitude dependence of the size distribution of auroral events (see Uritsky et al. 2008 in Table 12.1), which indicates substantially different scaling regimes of bursty energy dissipation in the inner and outer portions of the geotail plasma sheet (Uritsky et al. 2008, 2009).

12.2 Magnetospheric Phenomena: Extreme Storms

The largest geomagnetic storms are of highest interest, since they cause the relatively largest damage in space-weather scenarios, and in upper limits of SOC

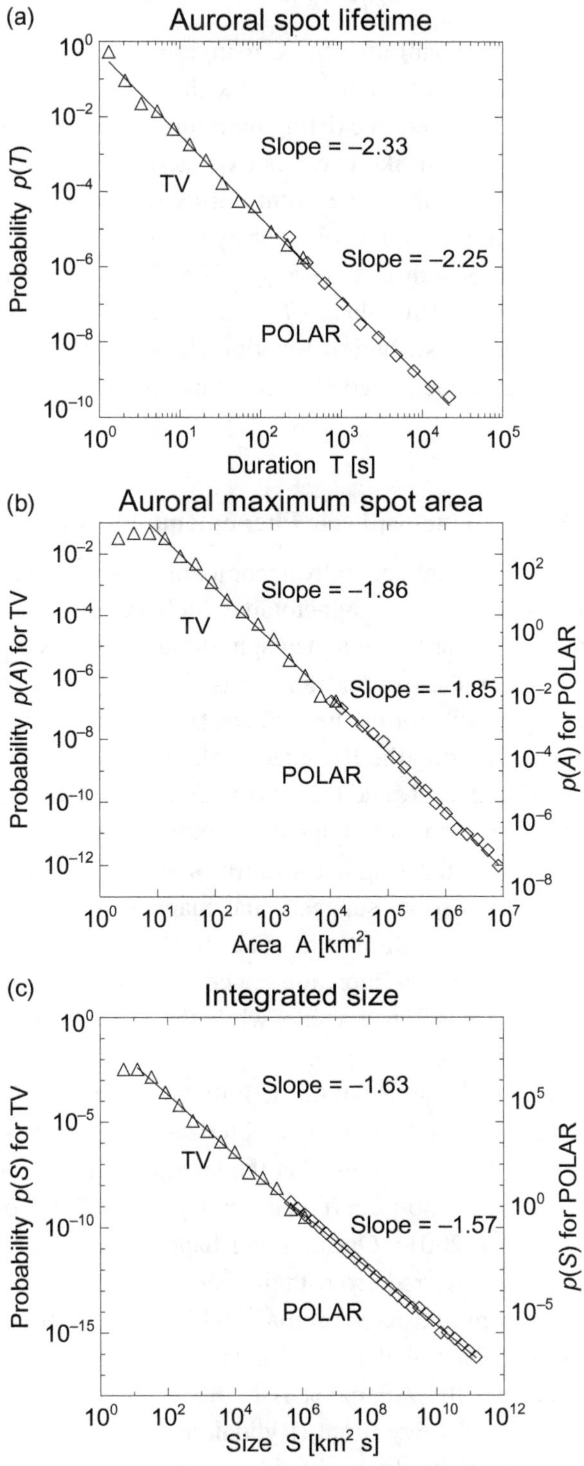

Figure 12.2 Combined probability distributions of auroral blob parameters (a: event duration T; b: maximum area A; c: time-integrated size S, approximately proportional to the or total energy E) measured with ground-based TV cameras (Kozelov et al. 2004) and with the UVI/ POLAR spacecraft (Uritsky et al. 2002).

size distributions. The probability of a Carrington-like geomagnetic storm was estimated to be between 0.46% and 1.88% (with a 95% confidence), which is a much lower value than reported in existing literature (Morina et al. 2019). Another study noted that a Carrington-like event is expected to be observed on the Earth with a frequency not higher than one event during ≈ 500 years (Yermolaev et al. 2013). Based on geomagnetic data for solar cycles 14-24, the average return rate for a storm as intense as March 1989 ($-F_{stm} = 565$ nT) is less than 4.1 solar cycles (Love 2021). Maximum storm values $-D_{stm}$, as ranked per solar cycle, are well represented by a Weibull distribution function (Love 2021) but can also be represented by log-normal, generalized Pareto, or maximum-likelihood distributions (Love 2020).

12.3 Magnetospheric Phenomena: Substorms

Magnetospheric substorms are generally accompanied by sporadic and intermittent large magnetic fluctuations in the magnetotail, which are related to the disruption of the cross-tail current. The onset of magnetospheric substorms is interpreted in terms of SOC and second-order phase order transitions (Consolini et al. 2001; Sitnov et al. 2002). The behavior of substorms are sudden transitions of the magnetosphere, which have been studied using the Bargatze et al. (1985) dataset of the solar wind, induced by the electric field vB_S and the auroral electron jet index AL (Sitnov et al. 2000). The hysteresis phenomenon typical of first-order phase transitions, and the power law dependence in the singular spectra of vB_s and AL, that is typical for a second-order phase transition, suggests that magnetospheric substorms exhibit signatures of both self-organization (without criticality) and SOC (with criticality) (Sitnov et al. 2000). The self-organization concept reflects the nonlinear organized behavior of the magnetosphere as a whole, while the SOC approach emphasizes its multiscale aspects (Sharma et al. 2001).

The ensemble average dynamics of the activity bursts in the AE index was found to be essentially scale-free and can be characterized in terms of spreading critical exponents, that is, $S \approx T^{1+\beta+\gamma}$, where S is the size parameter, T is the time duration, β is the diffusive transport coefficient, and γ is the flux–volume correlation coefficient (Uritsky et al. 2001). On the other hand, the size distributions of AE-index burst parameters, inferred from their thermodynamics, is found not to be scale-invariant (Consolini and Kretzschmar 2007). From statistical tests in terms of self-similar motion, fractional Lévy flights, and fractional Brownian motion, it was demonstrated that the AE index is better approximated by a multifractal model than by a fractional Lévy flight (Rypdal and Rypdal 2010). Re-examining the statistical properties of bright, night-side auroral emission events as observed with UVI onboard POLAR, the conclusion was drawn that the auroral dynamics is a reflection of a critical state in the magneto-tail that is based on the dynamics of turbulent reconnection in the mid-tail plasma sheet (Klimas et al. 2010).

12.4 Magnetospheric Phenomena: Auroras

The first results from spatiotemporal statistical analysis of ionospheric emission as observed by the Ultraviolet Imager (UVI) onboard the POLAR spacecraft during four months of 1997–1998, resulting into 12,300 images with auroral displays has been presented in Uritsky et al. (2002). The size (probability p) distributions of these events over the lifetime T, maximum area A, integrated area S, maximum power W, and integrated energy output yield the following power law relationships: $p \approx T^{-2.2}$, $p \propto A^{-1.8}$, $p \propto S^{-1.6}$, $p \propto W^{-1.7}$, $p \propto E^{-1.5}$, over a wide range of about 10 decades, which clearly reproduce the avalanche behavior of a SOC system in a self-organized critical state, as the evaluation of spreading critical exponents show (Uritsky et al. 2003). The energy and duration probability distributions of particle precipitation events (visible in the auroral displays) obey finite-size scaling relations and are therefore indicative of a SOC dynamical state (Uritsky et al. 2006a, 2010). However, the fractal structure can be time-dependent during the substorm evolution. It was then shown that under nonstationary driving conditions, the magnetosphere can deviate from its steady critical state (Uritsky et al. 2006b).

12.5 Magnetospheric Phenomena: AE Index

The auroral electron jet index (AE) is a time-dependent variable that provides a suitable measure to monitor solar activity. Low frequency 1/f-like fluctuations of the AE-index thus was considered as a manifestation of SOC in the magnetosphere (Uritsky and Pudovkin 1998). It was suggested that the solar wind acts as a SOC state, and the AE index time series indicates a nearly identical SOC state (Price and Newman 2001). The AE index is derived from geomagnetic variations in the horizontal component, while four auroral electron jet definitions (AE, AL, AO, and AU) are used to define a global measure of the auroral zone magnetic activity. A time series analysis of the AL index revealed that characteristics of self-similarity and SOC prevails also in (AL data Choe et al. 2002). *High-intensity, long-duration, continuous AE activity (HILDCAA)* are magnetic/ionospheric events that occur during high-speed solar wind streams and were considered as substorm expansion events with SOC characteristics (Tsurutani et al. 2004), or Alfvén waves with typical periods of $\approx 1 - 10$ h (Souza et al. 2016). Evidence for self-organization was also found in the G index of interplanetary scintillation (IPS) data, which suggests that the IPS medium is close to a critical state, explaining the noncyclic components of solar activity (Perez-Enriquez et al. 2006). The burst lifetime distribution functions of the solar wind have been compared with the contemporaneous SYM-H index, which confirmed that the SYM-H scaling is robust (Wanliss and Weygand 2007). A fractal analysis of ultra-low-frequency (ULF) emission observed from ground revealed SOC characteristics (Smirnova and Hayakawa 2007). Other characteristics are the D time series that obey a nonextensive energy distribution function with the Tsallis entropic index q (Balasis et al. 2011).

12.6 Magnetospheric Phenomena: Radio Bursts

Radio bursts in the magnetosphere have been found in the Earth's bow shock, as well as near the Earth's poles. Bursty Langmuir-like waves driven by electron beams in the Earth's foreshock have properties that are inconsistent with the standard paradigm of uniform exponential growth saturated by nonlinear processes or with standard SOC models. Instead, the stochastic growth theory has been employed to explain the foreshock waves and type III solar bursts Cairns and Robinson (1999).

Another type of radio bursts in the magnetosphere is the so-called Auroral Kilometric Radiation (AKR), which were interpreted in terms of electron cyclotron maser emission in the case of AKR data from the planet Jupiter. Data analyzed from terrestrial AKR sources were observed from the POLRAD spacecraft, which revealed SOC scaling parameters of $\alpha_F \approx 2.5$ (Marek and Schreiber 2017). Statistical data were obtained from 240 AKR dynamic spectra, containing 3,750 short-time AKR bursts each, and recorded onboard the Interball-2 mission (POLRAD and MEMO experiments), confirming a marginal power law-like size distribution (Marek and Schreiber 2022). However, the power law range in the size distribution is found to cover a relatively small inertial range (of less than one order of magnitude), which does not support an interpretation in terms of SOC models.

12.7 Magnetospheric Phenomena: Cellular Automata

While cellular automaton algorithms are designed to mimic the nonlinear dynamics of sandpiles as pioneered by Bak et al. (1987), they are "physics-free." In order to give them a physical meaning, a transformation of the cellular automaton framework into discretized magneto-hydrodynamic equations (MHD), based on Ohm's law, has been attempted (Vassiliadis et al. 1998). However, the magnetic field cannot be represented properly with this approach, because a magnetized plasma would produce a highly anisotropic model (of next-neighbor interactions). In a numerical simulation of a 1-D cellular automaton, it was found that system-wide avalanches produce a single-peaked component with a well-defined mean and standard deviation (which is also called a finite system-size effect), on top of the nominal power law size distribution produced by all smaller avalanches (Chapman et al. 1998, 1999; 2001).

The SOC interpretation of magnetospheric phenomena has stimulated cellular automaton simulations and alternative aspects of SOC modeling, such as the power law robustness under varying loading (Watkins et al. 1999), the discretization in terms of MHD equations (Takalo et al. 1999a, 1999b), renormalization group analysis (Tam et al. 2000; Chang et al. 2004), the scaling of the critical spreading exponents (Uritsky et al. 2001), phase transition-like behavior (Sitnov et al. 2000, 2002; Sharma et al. 2001), aspects of percolation and branching theory (Milovanov et al. 2001; Zelenyi and Milovanov 2004), chaotic turbulence models (Kovacs et al. 2001), a coupled-map lattice model of magneto-tail current sheet (Takalo et al. 2001), forced SOC models (Consolini 2001; Chang et al. 2003), modeling of energetic particle spectra in magneto-tail (Milovanov and Zelenyi 2002), MHD modeling of the plasma sheet dynamics near a SOC state (Klimas et al. 2004), aspects of complexity systems (Dendy et al. 2007), the framework of thermodynamics of rare events (Consolini and Kretzschmar 2007), kinetic theory of linear fractional stable motion (Watkins et al. 2009b), avalanching with an intermediate driving rate (Chapman and Watkins 2009; Chapman et al. 2009), and multifractal and fractional Lévy flight models (Zaslavskii et al. 2007, 2008; Rypdal and Rypdal 2010), modeling of substorm dynamics with SOC (Bolzan 2018).

12.8 Magnetospheric Phenomena: MHD Turbulence

The observation of power law size distributions in solar flares (Boffetta et al. 1999) raised the question whether these power laws are evidence for avalanching SOC systems, or for turbulent cascades. The same question has been asked for the solar wind: Are power law and interburst interval distributions in the solar wind evidence for turbulence of dissipative SOC (Freeman et al. 2000). Based on a wavelet analysis of intermittent events in geomagnetic time series, it was argued that *near-SOC* or chaotic turbulence can explain the observed features without ambiguity (Kovacs et al. 2001). Multi-spacecraft tracing of the high-latitude magnetopause indicate that the turbulent boundary layer is a persistent feature of the cusp (Savin et al. 2002). Observations of the plasma pressure distribution in the magnetosphere with data from the Interball/Tail probe reveals transverse and field-aligned currents, which can be interpreted in terms of turbulent plasma transport as well as SOC (Antonova 2004). Sporadic perturbations in the magnetosphere are believed to be associated with multiscale turbulent activity in the tail (Uritsky et al. 2006a). A numerical simulation study of turbulent current and vorticity structures yields SOC features that are consistent with MHD and an interpretation of intermittency in terms of propagation of local instabilities (Uritsky et al. 2010). It has been suggested that the auroral dynamics is a reflection of a critical state in the magneto-tail that is based on the dynamics of turbulent reconnection in the mid-tail plasma sheet (Klimas et al. 2010). MHD simulations of a 2-D plasma current-sheet produces multiscale correlations in the velocity field that are indistinguishable from certain intermittent turbulence theories (Klimas et al. 2017). A study of SOC systems under the influence of turbulent motion in the environment investigates the role of isotropy and anisotropy, applied to Kolmogorov turbulence and renormalization group theory (Antonov et al. 2020). A complementary study focuses on the effects of turbulent environment and random noise on self-organized critical behavior of universality versus nonuniversality (Antonov et al. 2021).

12.9 Magnetospheric Phenomena: Forced SOC

Extending the original definition of SOC of Bak et al. (1987), Chang (1999a, 1999b, 2000) introduced the term *forced criticality*, which differs from Bak's SOC model in the sense that it is critical without self-organization, under the influence of external forcing. Coherent magnetic structures approach each other and merge or scatter under the influence of external forcing, turbulence, or self-organization, until a power law-like spectrum of size scales occurs. It is now argued that the magnetospheric system is driven to a critical or near-critical state as a result of the continuous loading and subsequent unloading above a critical current (Chang 1992, 1999a, 1999b; Consolini and de Michelis 2001; Horton and Doxas 1996). Examples of forced SOC include intermittent turbulent behavior of the sporadic and localized reconnection processes in the Earth's magneto-tail, ion hole/double layers, fast electron holes, kinetic and inertial Alfvén structures, and coherent structures associated with whistler resonances (Chang 2000; Chang et al. 2001). The technique of dynamical renormalization group is introduced and applied to the study of 2-D intermittent MHD fluctuations and/or forced SOC and other types of topological plasma transitions (Chang and Wu 2002; Chang et al. 2003, 2004). The analysis of the near-Earth magnetosphere and auroral phenomena with multiscale, cross-scale coupling, and reorganization suggest that the magnetosphere is in a forced and/or SOC state (Lui 2002).

12.10 Magnetospheric Phenomena: SOC Forecasting

The Earth's magnetosphere is a large-scale natural system driven by the turbulent solar wind and exhibits nonequilibrium phenomena (Sharma and Kaw 2005), including SOC discussed here. In general, the properties of such systems are characterized as a combination of global and multiscale features and have been studied extensively using the techniques of nonlinear dynamics and complexity science (Sharma 1995; Klimas et al. 1996; Vassiliadis 2006). The first evidence of global coherence of the magnetosphere was obtained from time series data of the AE index in the form of low-dimensional dynamics (Vassiliadis et al. 1990; Sharma et al. 1993). This result is consistent with the morphology of the magnetosphere derived from observations and theoretical understanding (Siscoe 1991), and simulations using global MHD models (Lyon 2000; Shao et al. 2003). The recognition of the low-dimensional dynamics of the magnetosphere has stimulated a new direction in the studies of the solar wind–magnetosphere coupling and such systems in nature. Among these is the forecasting of the global conditions of space weather, viz. the AL and AE indices for substorms (Vassiliadis et al. 1995; Ukhorskiy et al. 2002, 2003, 2004; Chen and Sharma; 2006) and the disturbance time index Dst for magnetic storms (Valdivia et al. 1996; Boynton et al. 2011). The forecasting of regional space weather requires data from the spatially distributed stations around the globe (Valdivia et al. 1999a, 1999b; Chen et al. 2008), and the predictability is largely determined by the availability of long time series data from the network of observing stations. The spatiotemporal dynamics of many systems are studied using such data, including the images obtained from satellite-borne imagers, by defining new variables computed from the data. For example, the fragmentation parameter (Rosa et al. 1998, 1999) represents the complexity of the spatial structure and has been used to model the dynamics of the solar atmosphere using the hard X-ray images from the SOHO spacecraft. Further, the low dimensionality of the magnetosphere has stimulated the development of models with a small number of equations (Vassiliadis et al. 1993; Horton and Doxas 1996).

The multiscale nature of the magnetosphere, expressed in many ways including the power law dependence of the scales, is a reflection of turbulence and plays an essential role in the accuracy of the forecasts. An early recognition of this was in the analogy of the dynamics of the magnetosphere to turbulence generated by a fluid flow past an obstacle (Rostoker 1984). The power law dependence of the AE index and of the solar wind provided quantitative measures of the power law indices and also the differences (Tsurutani et al. 1990). The scaling laws, which have been studied in detail using techniques such as the structure functions (Takalo et al. 1993), have many implications. The first is the characterization in terms of SOC, as discussed earlier in this section. The second is that the predictability of a multiscale system could not be quantified in terms of the characteristic quantities such as the Lyapunov exponents (Vassiliadis et al. 1991) in a low-dimensional dynamical system. The presence of many scales as well as the nonequilibrium nature imply that predictions should be based on the statistical properties of the dynamical trajectories, for example, using a mean-field approach (Ukhorskiy et al. 2004). Further, this approach is suitable for analyzing the predictability of extreme events

(Sharma et al. 2012). Although the power law size distributions can be reproduced with single power law SOC models, recent observations of all-sky auroras have suggested a dual regime, where the smaller events have a steeper slope (VallieRes-Nollet et al. 2010). Phase coherence analysis shows the presence of large-amplitude, quasi-perpendicular Alfvén vortex-like structures in the fast solar wind turbulence (Lion et al. 2016). Solar wind–magnetosphere interaction and the injection of large quantity of plasma particles into the Earth's magnetosphere are the primary reasons behind geomagnetic storm, aurora effects, and fluctuations observed in the terrestrial magnetic field, as sandpile-like cellular automatons simulations demonstrate (Banerjee et al. 2019).

12.11 Summary

In summary, let us ask: What is the merit of the SOC concept in the context of magnetospheric phenomena? The generalized fractal-diffusive SOC model (Sections 2.6–2.11) predicts the probability distribution functions for each parameter as a function of the dimensionality (d), diffusive spreading exponent (β), fractal dimension (D_d), and type of (coherent/incoherent) radiation process (γ). In addition, the special case of $d = 3$, $\beta = 1$, and $\gamma = 1$ is called the standard FD-SOC model. The waiting time distributions are predicted by the FD-SOC model to follow a power law with a slope of $\alpha_T \approx 2$ during active and contiguously flaring episodes, while an exponential cutoff is predicted for the time intervals of quiescent periods. This dual regime of the waiting time distribution predict both persistence and memory during the active periods, and stochasticity during the quiescent periods. All these predictions of the FD-SOC model provide useful constraints of the physical parameters and underlying scaling laws. Significant deviations from the size distributions predicted by the FD-SOC model could indicate problems with the measurements or data analysis, such as indicated by the contradicting results of Lui et al. (2000) versus Uritsky et al. (2002) in the case of auroral size distributions.

Let us emphasize again that the generic FD-SOC model is considered to have universal validity and explains the statistics and scaling between SOC parameters, but does not reveal the detailed physical mechanism that governs the instabilities and energy dissipation in a particular SOC process. The physical process may be well described by a number of established models, such as turbulence theory, kinetic theory, wave–particle interactions, and other branches of plasma physics. There was also a debate whether magnetospheric substorms are SOC or forced-SOC (e.g., Chang et al. 2003), an issue that largely disappears in our generalized FD-SOC concept, where a slow driver is required to bring the SOC system continuously near to the instability limit, but does not matter whether the driver is internally, externally, or globally organized.

13
Planetary Systems

13.1 Lunar Craters and Meteorites

An amazingly straight power law size distribution has been found for the sizes of lunar craters (Figure 13.1), with a cumulative power law slope of $\alpha_L^{cum} = 2.0$ over a size range of $L = 0.65 - 69,000$ m, which covers five orders of magnitude (Cross 1966), derived from crater statistics measured in pictures of the lunar probes *Ranger 7, 8, 9* combined with a lunar map of Wilkins (1946). Since a cumulative size distribution is flater than a differential size distribution (by a value of 1), this corresponds to a power law slope of $\alpha_L = \alpha_L^{cum} + 1 = 3.0$. A similar power law index of $\alpha_L = 2.75$ was found for the size distribution of meteorites and space debris from man-made rockets and satellites in the range of L = [10 μm, 10 cm] (Fig. 3.11 in Sornette 2004).

Given the ubiquitous power law shape of these size distributions, it is not far-fetched to consider the possibility of an interpretation in terms of a SOC model. Since lunar craters are believed to be produced by meteorite impacts, the directly observed meteorites and the lunar impact craters have the same origin in the solar system, although they cover different length scale ranges. We find that these observed power law slopes of $\alpha_L \approx 2.75 - 3.0$ agree remarkably well with the scale-free probability conjecture, which predicts a universal scaling exponent of $\alpha_L = 3$ (in 3-D space) (Eq. 1.1). Reservoirs of meteorites are slow SOC drivers, as well as small bodies with orbits in the solar system are slow SOC drivers also, providing projectiles for lunar or planetary impacts. The dissipated energy is essentially the kinetic energy of the projectiles, given by the relative velocity of the projectile (v_{proj}) and the target (v_{target}),

$$E_{kin} = \frac{1}{2} m_{proj} (v_{proj} - v_{target})^2 \geq \frac{1}{2} m_{proj} v_{inel}^2, \tag{13.1}$$

which has to exceed the critical threshold v_{inel} that is given by the limit between elastic and inelastic collisions. If the projectile hits the target below this threshold,

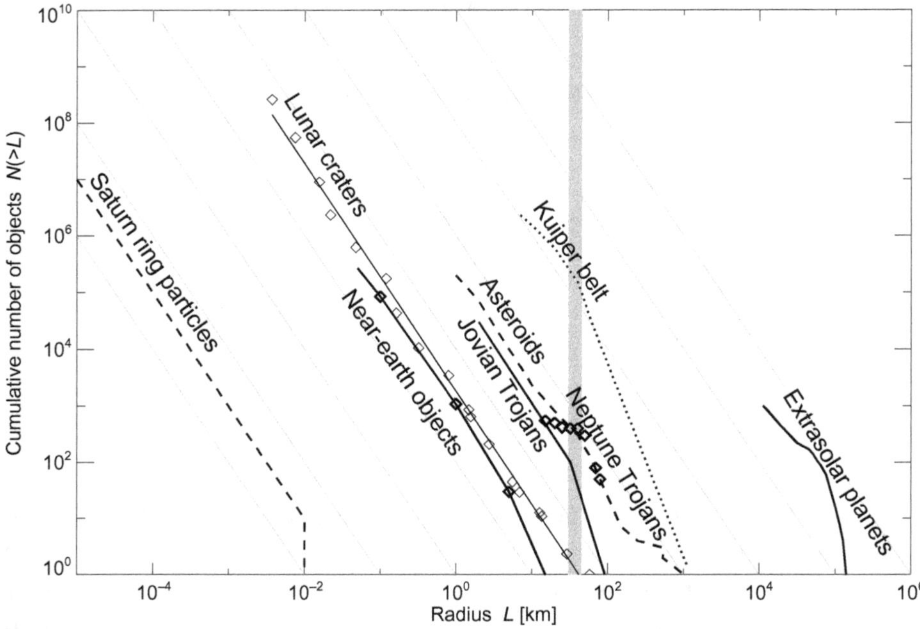

Figure 13.1 Cumulative size distribution of Saturn ring particles (Zebker et al. 1985), near-Earth objects (McFadden and Binzel 2007), Jovian Trojans (Jewitt et al. 2000), asteroids (Jedicke et al. 2002), Neptunian Trojans (Sheppard and Trujillo 2010), lunar craters (Cross 1966), Kuijper belt objects (Fraser and Kavelaars 2008; Fuentes and Holman 2008), and Earth-sized extra-solar planets (Catanzarite and Shao 2011). The gray diagonal lines indicate the prediction of the FD-SOC model, with a power law slope of $\alpha_L^{cum} = 2$ for the cumulative size distribution, corresponding to a power law slope of $\alpha_L = \alpha_L^{cum} + 1 = 3$ for the differential occurrence frequency distribution. A zone of paucity is indicated at a size range of $L = 30 - 45$ km identified from Neptunian Trojans (Sheppard and Trujillo 2010). A compilation is given in Aschwanden et al. (2016).

it will just bounce back by conservation of momentum, without producing an impact crater. If it hits the target with a larger velocity, the impact will produce a fractal-diffusive pattern of cracks on the projectile and target, similar to the rupture area during the energy release of an earthquake (see preface by Paul Charbonneau). In this analogy, lunar impact craters have much in common with earthquake "damage areas," which is considered as a SOC process.

The size distribution of meteorites and planetesimals may also be generated by a SOC process in the first place. The slow driver that provides the trickling of sand grains is the gravity-driven formation process of the solar system itself, which clumps the local molecular cloud into meteorites and planets. The aspect of self-organized criticality, which is a balance between the gravity and the frictional force that controls the critical angle of repose in Bak's sandpile, can be understood as a critical point between the condensation rate of planetesimals or meteorites (by self-gravity) and the diffusion rate (driven by thermal pressure and external gravitational disturbances). This critical threshold given by the balance of the condensation rate and the diffusion rate has to be exceeded in order to initiate the gravitational collapse that forms a solar system body. The gravitational collapse is the underlying instability in a physical SOC concept (Figure 1.2, right frame).

Hence, from such a generalized point of view, we may consider the meteorite formation as a SOC process and the resulting lunar cratering as the imprint of this process. The main benefit of the FD-SOC framework is the direct prediction of the scale-free size distribution of crater sizes, that is, $N(L) \propto L^{-3}$ (Eq. 1.1), which can also be used as a prediction for any other target in the solar system, such as cratering on Earth, Mars, or Mercury. This allows us, for instance, to predict the collisional probability of an asteroid hitting our Earth, although we have to take into account the variability of the impact rate, which varied drastically during the lifetime of our solar system. Both the Moon and the Earth were subject to intense bombardment between 4.0 and 3.7 billion years ago, which was the final stage of the sweep-up of debris left over from the formation of the solar system (Bottke et al. 2012). The impact rate at that time was thousands of times higher than it is today.

13.2 The Asteroid Belt

The asteroid belt is a large accumulation of irregular small solar system bodies orbiting the Sun between the orbits of Mars and Jupiter. The largest of these small bodies is Ceres, with a diameter of 1,020 km, followed by Pallas (538 km), Vesta (549 km), Juno (248 km), and extends down to the size of dust particles. While most planetesimals from the primordial solar nebula formed larger planets under the influence of self-gravity, the gravitational perturbations from the giant planets Jupiter and Saturn prevented a stable conglomeration of planetesimals in the zone between Mars and Jupiter. This fragmented soup of primordial planetesimals makes up the asteroid belt. The larger asteroids ($\geq$ 120 km) are believed to be primordial, while the smaller ones are likely to be a byproduct of fragmentation events (Bottke et al. 2005).

Statistics of the sizes of asteroids has been carried out in the *Palomar Leiden Survey* (Van Houten et al. 1970), in the *Spacewatch Surveys* (Jedicke and Metcalfe 1998), in a *Sloan Sky Survey* (Ivezic et al. 2001), and in the *Subaru Main-Belt Asteroid Survey* (Yoshida et al. 2003; Yoshida and Nakamura 2007). Most of these statistics yield a power law-like function for the cumulative size distribution (Figure 13.1). From these values α_L^{cum}, we can estimate the power law slopes of the differential size distributions $\alpha_L = \alpha_L^{cum} + 1$, which yield $\alpha_L = 2.8$ (Jedicke and Metcalfe 1998; Jedicke et al. 2002), a double power law of $\alpha_L = 2.3 - 4.0$

Table 13.1 *Frequency distributions measured from planetary phenomena.*

Phenomenon	Instrument	Power law slope of length α_L	Power law slope of fluence α_E	References
Terrestrial γ-ray flashes			1.3 – 1.7	Ostgaard et al. (2012)
Lunar craters	Ranger 7,8,9	3.0		Cross (1966)
Meteorites, space debris		2.75		Sornette (2004)
Asteroid belt	Spacewatch Surveys	2.8		Jedicke and Metcalfe (1998)
Asteroid belt (< 5 km)	Sloan Survey	2.3		Ivezic et al. (2001)
Asteroid belt (> 5 km)	Sloan Survey	4.0		Ivezic et al. (2001)
Asteroid belt	Subaru Survey	2.3		Yoshida et al. (2003) Yoshida and Nakamura (2007)
Jovian Trojans (< 40 km)	Hawaii 2.2 m	3.0 ± 0.3		Jewitt and Trujillo (2000)
Jovian Trojans (> 40 km)	Hawaii 2.2 m	5.5 ± 0.9		Jewitt and Trujillo (2000)
Neptune Trojans	Subaru Survey	5 ± 1		Sheppard and Trujillo (2010)
Kuijper belt objects		4.3		Fraser et al. (2008)
Saturn ring	Voyager 1	2.74 – 3.11		Zebker et al. (1985)
Saturn ring	Voyager 1	2.74 – 3.11		French and Nicholson (2000)
Extra-solar planets	Kepler	2.48		Catanzarite and Shao (2011)
		2.96	**1.50**	Median
		3.00	**1.44**	FD-SOC prediction

(Ivezic et al. 2001), and $\alpha_L = 2.3$ (Yoshida et al. 2003; Yoshida and Nakamura 2007); see compilation in Table 13.1. Observational selection effects in asteroid surveys, of course, affect the reported power law slopes, as discussed in Jedicke et al. (2002).

If the small bodies in the asteroid belt are formed by a SOC process, the scale-free probability conjecture predicts a size distribution of $N(L) \propto L^{-3}$, which is indeed close to what is observed (Figure 13.1). However, there are slight deviations from a single power law distribution for small and large bodies, which indicate some additional effects. Nevertheless, an almost scale-free behavior is observed for a range of $L \approx 0.4 - 50$ km, which makes it appropriate to consider the formation process in terms of a SOC system. As we discussed for the formation of meteorites earlier, the aspect of self-organized criticality can be understood as a critical point between the condensation rate of planetesimals or meteorites by self-gravity, and the diffusion rate driven by external gravitational disturbances, mostly from the giant planets Jupiter and Saturn. If this critical threshold of the ratio of the condensation rate to the diffusion rate exceeds the value of unity, the self-gravity force takes over and forms a small solar system body, which represents an avalanche process with a well-defined instability threshold.

13.3 Mars

It has also been suggested to apply SOC dynamics to Martian fluvial systems (Rosenshein 2003). The motivation was that complexity theory provides powerful methods to analyze, interpret, and model terrestrial fluvial systems, including the fractal structure of meandering, sediment dynamics, bedrock incision, and braiding.

Another application of SOC systems to Mars is the statistics of dust storms, especially the interannual variability of Mars global dust storms (Pankine and Ingersoll 2004a, 2004b). Previously it was thought that the threshold for wind speed for starting saltation and lifting dust from the Martian surface was a finely tuned process. In the study of Pankine and Ingersoll (2004a, 2004b), however, it was shown that the fine-tuning of this parameter could be the result of a negative feedback mechanism that lowers the threshold of the wind speed. In this way, the Martian atmosphere/dust system could organize itself as a SOC system, and no fine tuning of a critical threshold is required.

13.4 Saturnian Ring System

Saturn and Jupiter are the most massive planets in our solar system with a gravity that is sufficiently strong to keep numerous moons, rings, and ringlets in their strong gravitational field. The Saturn ring extends from 7,000 km to 80,000 km above Saturn's equator, consisting of particles ranging from 1 cm to 10 m, with a total mass of 3×10^{19} kg, which is comparable with the mass of its moon Mimas. Theories about the origin of Saturn's ring range from nebular material left over from the formation of Saturn itself, collisional fragmentation, to the tidal disruption of a former moon.

When *Voyager I* passed the orbit of Saturn, it carried out radio occultation observations, which were analyzed with a scattering model and yielded the size distribution of ring particles in the range of $L = 0.01 - 10$ m, being a power law distribution of $N(L) \propto L^{-3}$ (Figure 13.1). The results of eight size distributions

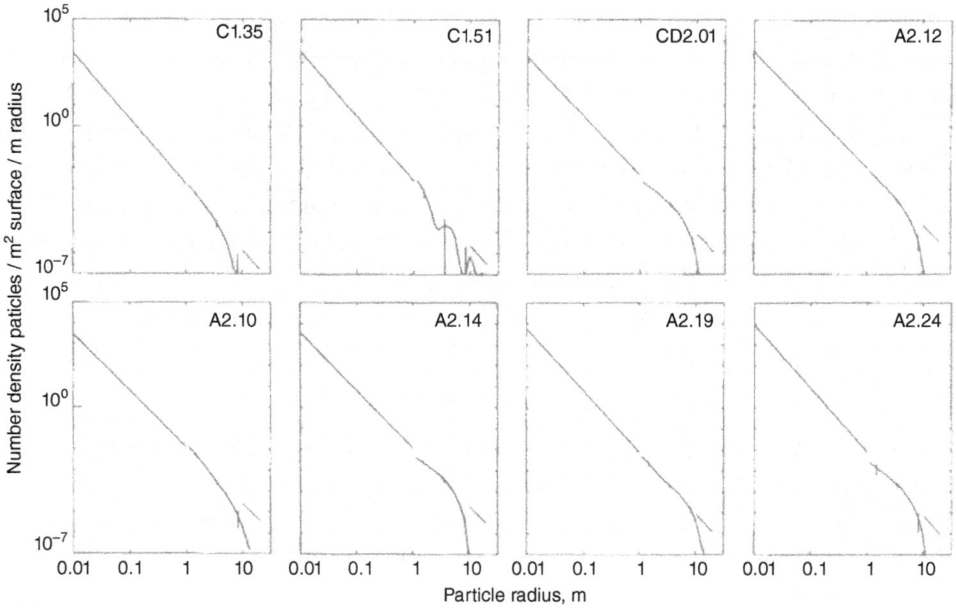

Figure 13.2 Measurements of the particle size distribution functions for eight ring regions with *Voyager 1* radio occultation measurements (Ring C: C1.35, C1.51; Cassini division: CD2.01; Ring A: A2.12, A2.10, A2.14, A2.19, A2.24). The slopes of the fitted power law functions in these 8 regions are: $\alpha_L = 3.11, 3.05, 2.79, 2.74, 2.70, 2.75, 2.93, 3.03$. The range of particle sizes is $L = 0.01 - 10$ m (Zebker et al. 1985).

obtained from eight different locations in Ring A, C, and the Cassini division are shown in Figure 13.2, which all are found to exhibit power law indices in the range of $\alpha_L = 2.74 - 3.11$ (Zebker et al. 1985). Related to this is a wavelet transform analysis of the Encke gap ringlets in Saturn's ring system (Bendjoya et al. 1993).

The coincidence of predicted and observed power law distributions for meteorites, lunar craters, asteroids, and Saturn ring particles may have all the same explanation, namely, SOC systems, although operating in different locations in our solar system, and in different ranges of length scales (Figure 13.1), which supports the universality of SOC systems. The critical threshold in all these systems is apparently given by the balance between the local self-gravity force and external gravity disturbances. All the gaps between the Saturn rings have been explained by mechanical resonances of Saturn's moons, which orbit outside the ring and amplify gravitational disturbances whenever two moons have an integer ratio of their orbital periods. Thus, we have all parts of a physical SOC system: the driver, the instability, and the avalanches. Saturn's moons are the driver of the system, because they provide random/periodic disturbances that lead to chaotic orbits of the ring particles. The instability is given by amplification of resonant orbits that leads to avalanches of particles, which clump in zones of nonresonant orbits. The appeal of SOC models is the simple way to predict the final size distribution of ringlets that result in the end, which cannot easily be predicted by celestial mechanics or chaos theory.

13.5 Jovian and Neptunian Trojans

The Jovian Trojans are two swarms of asteroids, which lead or trail Jupiter by ±60° on its orbit, known as the Lagrangian L4 and L5 points. The Jovian Trojans contain some 250 members. Their origin has been interpreted in terms of trapping of asteroidal fragments. A statistical analysis yielded a differential size distribution of $N(L) \propto L^{-3.0 \pm 0.3}$ in the size range of $L = 2 - 30$ km, and $N(L) \propto L^{-5.5 \pm 0.9}$ in the size range of $L = 50 - 84$ km (Jewitt et al. 2000).

Similarly, Trojans have been detected in the L4 and L5 regions of the planet Neptune, with a size distribution that approaches a power law slope of $\alpha_L = 5 \pm 1$ at the upper end (Sheppard and Trujillo 2010), while a flater slope is found at the lower end. The scarcity of intermediate- and smaller-sized Neptune Trojans (≲45 km), which is also found for other objects in the Kuijper belt, Jovian Trojans, and main belt asteroids, was interpreted in terms of a primordial origin, rather than a collisional or fragmentational origin, for which a size distribution of $N(L) \propto L^{-3}$ is expected in the SOC model. However, the smaller bodies of the Neptunian Trojans in the range of $L = 2 - 30$ km could still be consistent with a SOC origin, if they have the same distribution as Jovian Trojans (with $N(L) \propto L^{-3.0 \pm 0.3}$; Jewitt and Trujillo 2000). Their size range and distribution is close to that of asteroids (Figure 13.1).

13.6 Kuiper Belt Objects

The Kuiper belt is a region of our solar system beyond the orbit of Neptune (at 30 AU) out to ≈ 50 AU, consisting of many small bodies. A size distribution of $N(L) \propto L^{-4.3}$ was found for objects with $L \gtrsim 100$ km (Fraser et al. 2008; Fraser and Kavelaars 2008; Fuentes and Holman 2008). A comparison of the cumulative size distributions of Kuijper belt objects, Neptunian Trojans, Jovian Trojans, and asteroids is shown in Figure 13.1. Obviously, there is a paucity of objects in the zone of $L \approx 30 - 45$ km that shows up in the Neptunian Trojans and in the Kuijper belt objects (Figure 13.1). The data seem to be consistent with the predicted power law slope of $\alpha_L \approx 3$ only for small length scales of $L \approx 1 - 30$ km.

13.7 Extra-Solar Planets

The oligarchic growth of protoplanets has been brought into the context of a self-organized protoplanet–planetesimal system (Kokubo and Ida 1998). The growth and orbital evolution of protoplanets embedded in a swarm of planetesimals has been simulated with a 3-D n-body code, which shows the relative distribution of large planets that grow oligarchically, while most of the planetesimals remain small (Kokubo and Ida 1998).

Using the *Kepler* space telescope for search of Sun-like stars and (extra-solar) planets, a sample of over 150,000 stars was measured during the first four months of the mission. The Kepler science team determined sizes, surface temperatures, orbit sizes, and periods for over a thousand new planet candidates. From a size distribution of 1176 Earth-sized planet candidates within a range of $L = 2, \ldots, 20$ Earth radii, a power law distribution was found in the range of $L \approx 2 - 10$ Earth radii (Figure 13.1), with a power law slope of $\alpha_L = \alpha_L^{cum} + 1 = 1.48 + 1 = 2.48$, while the relatively narrow distribution falls of steeply between $L \approx 10 - 20$ Earth radii (Catanzarite and Shao 2011).

This sample from 1,176 different stars can be considered as a galactic SOC system, in which case a size distribution of $N(L) \approx L^{-3}$ is predicted by the FD-SOC model, which is close to the observed value of $N(L) \approx L^{-2.5}$ for a subset of Earth-like planets. The accretion of an Earth-like planet represents then an avalanche event, triggered by a gravitational instability in each stellar system. While the size distribution of planets may show a power law function, the spatial spacing between the planets is most likely not to be a power law (as expected for SOC systems), but rather governed by harmonic orbit resonances (Aschwanden and Scholkmann 2017; Scholkmann 2017). An alternative method of auto-regressive planet search has been applied to Kepler data (Casares et al. 2019a, 2019b).

13.8 Terrestrial Gamma-Ray Flashes

Terrestrial gamma-ray flashes (TGF) are gamma-ray bursts of terrestrial origin that have been discovered with the Burst and Transient Experiment (BATSE) onboard the Compton Gamma Ray Observatory (CGRO) and have been studied with RHESSI, Fermi, and AGILE since. These TGF bursts are produced by high-energy photons of energy >100 keV and last up to a few milliseconds. They have been associated with strong thunderstorms mostly concentrated in the Earth's equatorial and tropical regions, at a typical height of 15–20 km (Fishman et al. 1994; Dwyer and Smith 2005; Smith et al. 2005). The physical interpretation is that the TGF bursts are produced by bremsstrahlung of high-energetic electrons that were accelerated in large electric potential drops within thunderstorms. However the gamma-rays produced in thunderstorms (at 5 km) cannot readily propagate to higher altitudes due to atmospheric absorption. A mechanism for the generation of gamma rays that can reach the satellite-borne instruments is through the excitation of whistler waves by the relativistic electrons generated in the thunderstorms

(Kaw et al. 2001; Milikh et al. 2005). The whistler waves form a channel by nonlinear self-focusing, and the relativistic electrons propagate in this channel to higher altitudes (30 km). The gamma-rays generated at this altitude can escape the atmosphere and thus account for the BATSE/CGRO results.

A size distribution of the gamma-ray emission from TGF events needs to be corrected for the distance from the TGF-producing thunderstorm to the detecting spacecraft (in Earth orbit). In a combined analysis of TGF data from the RHESSI and Fermi satellites, corrected for their different orbits, different detection rates, and relative sensitivities, a true fluence distribution was derived, which was found to have a power law shape of $\alpha_E = 2.3 \pm 0.2$ if a sharp cutoff was assumed, or a slope of $\alpha_E \leq 1.3 - 1.7$ when a more realistic roll-over of the RHESSI lower detection threshold is assumed (Ostgaard et al. 2012).

We can consider a part of the Earth's atmosphere that contains a thunderstorm as a SOC system of finite size, where the electrostatic charging process represents the driver, the critical condition for electric discharging is given by an electric conductivity threshold, and the spontaneously triggered gamma-ray flashes or lightning represent the avalanches. The probability distribution function is then given by the scale-free probability conjecture (Eq. 1.1), which together with the fractal-diffusive transport predicts an energy or fluence distribution with a power law slope of $\alpha_E = 1.5$ in 3-D space, which matches the observed and corrected fluence distribution with a slope of $\alpha_E \approx 1.3 - 1.7$. The agreement with the standard FD-SOC model is consistent with an incoherent process for gamma-ray production, where the gamma-ray flux is proportional to the emitting volume of a TGF.

Another related phenomenon is that of atmospheric cascades, at least with respect to their origin in similar altitudes of the Earth's atmosphere, which has been considered as a SOC system and been studied by using Lévy flights and a nonextensive approach (Rybczynski et al. 2001).

13.9 Summary

Empirically we find that many phenomena in planetary science reveal power law-like size distributions with a power law slope of $N(L) \propto L^{-3}$ for differential size distributions, or $N^{cum}(L) \propto L^{-2}$ for cumulative size distributions. These observational results fully agree with theoretical predictions of the FD-SOC model. These observations include lunar craters, Saturn ring particles, near-Earth objects, Jovian Trojans, asteroids, Neptune Trojans, the Kuijper belt, and extra-solar planets. Mars fluvial systems and dust storms reveal fractal structures. Terrestrial gamma-ray flashes indicate also scale-free power law slopes. It appears that the scale-free behavior of planetary phenomena could result from both accumulation and/or fragmentation processes.

14
Stellar Systems

We can obtain information on spatial scales and spatiotemporal scaling laws from self-organized criticality (SOC) phenomena only from spatially resolved observations, which restricts us essentially to our solar system (i.e., from the Sun, the planets, and the magnetosphere), while such information from the rest of the universe is concealed by distance and cosmological time scale. Nevertheless, a large number of stellar phenomena has been attributed to SOC phenomena. SOC observables measured in stellar data include the time duration T, the peak flux F, and the fluence or the energy E of electromagnetic emission in some wavelength range, measured from a flux time profile of a stellar object with some automated event detection algorithm. In the following we will compile such observations from stellar flares, pulsar glitches, soft gamma-ray repeaters, magnetars, blazars, black-hole objects, and compare their size distributions with the predictions of the standard SOC model.

14.1 Stellar Flares: UBV and XUV Observations

We compile a dozen studies that were made in the pre-Kepler (spacecraft) era, which contain statistics of radiated stellar flux F, in form of occurrence frequency distributions (also called *size distributions*), which may be approximated with a power law function with slope α_F, that is, $N(F)dF \propto F^{-\alpha_F} dF$ (see Table 14.1). Stellar flares are generally defined in terms of temporary increases of the observed flux above the preflare and postflare quiescent level. From soft X-ray and EUV spectroscopy, flare temperatures of $T_e \approx 10 - 100$ MK have been determined in some of the stellar flares, exceeding solar flare temperatures ($T_e \approx 5 - 35$ MK). Consequently, the same physical interpretation in terms of SOC avalanches, driven by magnetic reconnection with subsequent heating of chromospheric plasma, has been proposed for stellar flares, in analogy to their solar analogs, although their total emission measure is a few orders of magnitude larger than for solar flares (Aschwanden and Güdel 2008).

Table 14.1 *The power law slopes α_F of the occurrence frequency size distributions observed with UBV, EXOSAT, EUVE, HSP/HST, EUVE, and XMM-Newton in fluxes of stellar flares.*

Star	Spectral type	Instrument	Number of flares or events	Differential slope α_F	Reference
CN Leo	dM6.5 Ve	UBV	111	1.50	Lacy et al. (1976)
UV Cet	M6.0e	UBV	107	1.51	Lacy et al. (1976)
Wolf 424 AB	M5.5	UBV	11	1.55	Lacy et al. (1976)
YZ CMi	M5 V	UBV	62	1.58	Lacy et al. (1976)
EQ Peg	M4 Ve	UBV	58	1.50	Lacy et al. (1976)
EV Lac	M3.5	UBV	22	1.59	Lacy et al. (1976)
AD Leo	M3.5eV	UBV	9	1.54	Lacy et al. (1976)
YY Gem	dM1e	UBV	6	1.17	Lacy et al. (1976)
M dwarfs	dM	EXOSAT	13	1.52 ± 0.42	Collura et al. (1988)
M dwarfs	dM	EXOSAT	20	1.70 ± 0.10	Pallavicini et al. (1990)
RS CVns	G	EUVE	30	1.60 ± 0.32	Osten and Brown (1999)
YZ CMi	M5 V	HSP/HST	54	2.25 ± 0.31	Robinson et al. (1999)
G dwarfs	G	EUVE	25	2.15 ± 0.15	Audard et al. (1999)
HD 2726	F2 V	EUVE	15	2.43 ± 1.07	Audard et al. (2000)
47 Cas	G0-5 V	EUVE	12	2.62 ± 1.85	Audard et al. (2000)
EK Dra	G1.5 V	EUVE	15	1.78 ± 0.62	Audard et al. (2000)
κ Cet 1994	G5 V	EUVE	5	2.55 ± 0.31	Audard et al. (2000)
κ Cet 1995	G5 V	EUVE	10	2.45 ± 1.08	Audard et al. (2000)
AB Dor	K1 V	EUVE	16	1.76 ± 0.74	Audard et al. (2000)
ϵ Eri	K2 V	EUVE	12	2.38 ± 1.50	Audard et al. (2000)
GJ 411	M2 V	EUVE	15	1.57 ± 0.57	Audard et al. (2000)
AD Leo	M3.5 V	EUVE	12	1.65 ± 1.17	Audard et al. (2000)
EV Lac	M4.5 V	EUVE	12	1.75 ± 1.18	Audard et al. (2000)
CN Leo 1994	M6.5 Ve	EUVE	13	2.24 ± 0.63	Audard et al. (2000)
CN Leo 1995	M6.5 Ve	EUVE	15	1.59 ± 0.86	Audard et al. (2000)
FK Aqr	dM2e	EUVE	65	2.60 ± 0.34	Kashyap et al. (2002)
V1054 Oph	M3 Ve	EUVE	85	2.74 ± 0.35	Kashyap et al. (2002)
AD Leo	M3.5 V	EUVE	80	2.17 ± 0.03	Kashyap et al. (2002)
AD Leo	M3.5 V	EUVE	75	2.32 ± 0.11	Kashyap et al. (2002)
AD Leo	M3.5 V	EUVE	...	2.25 ± 0.25	Güdel et al. (2003)
AD Leo	M3.5 V	EUVE	...	2.30 ± 0.10	Arzner and Güdel 2004
HD 31305	M, K	XMM-Newton	...	2.20 ± 0.30	Arzner et al. (2007)
TMC	M, K	XMM-Newton	...	2.40 ± 0.50	Stelzer et al. (2007)

The flare energy E is often estimated from the fluence or total counts (in a given wavelength), which is defined as the time integral of the flux (intensity or luminosity) over the flare duration. Some datasets contain flares from a single star only, other datasets combine flare events from different stars of the same stellar spectral type, while older studies are based on cooling times in hydrodynamic flare models (Haisch et al. 1983; Mullan et al. 2006; Cassak et al. 2008). Newer studies define the flux of a stellar flare in terms of the flux radiated in a given wavelength range. In cases where the cumulative size distribution is given (with slope α^{cum}), we convert it to the slope $\alpha = (\alpha^{cum} + 1)$ of the corresponding differential size distribution.

For the uncertainty σ_α of the power law slope we use the generic relationship $\sigma_\alpha = \alpha / \sqrt{n_{ev}}$ (Clauset et al. 2009), where n_{ev} represents the number of flare events that contribute to the linear regression fit of the power law slope. In Table 14.1, we summarize the instruments, the spectral types of the stellar classification, the number of flare events per dataset and the power law slopes of energies with uncertainties.

14.1.1 Observations

Observations of 409 stellar flares have been obtained during 469 hours from 8 UV Ceti-type (cool dwarf) flare stars (CN Leo, UV Cet, Wolf 424 AB, YZ CMi, EQ Peg, EV Lac, AD Leo, and YY Gem) in optical wavelengths (U, B, V band, $\lambda = 3370 - 5550$ Å), observed at the McDonald Observatory (Moffett 1974; Lacy et al. 1976). The flare fluxes (or luminosities) F were calculated for each star separately (Moffett 1974; Lacy et al. 1976). Because Lacy et al. (1976) computed a linear regression fit with inverted x-axis and y-axis for the cumulative distribution function, i.e., $\log(N) = a + b \ \log(E)$, we have to invert the power law slope b and add a value of 1 to obtain the power law slope of the differential distribution function of the cumulative distribution function, which yields $\alpha_F = 1/b + 1$. The mean value was found to be $\alpha_F = 1.56 \pm 0.07$, which is close to the theoretically predicted value by the FD-SOC model, $\alpha_F = 1.80$ (Table 2.1).

A similar dataset of 21 flares occurring on 13 different dMe stars was observed in soft X-rays with EXOSAT (in soft X-ray energies of 0.05 – 2 keV), yielding a power law distribution of the soft X-ray peak flux $\alpha_F = 1.52$ (Collura et al. 1988), and $\alpha_F = 1.70$ (Pallavicini et al. 1990). For the case of Collura et al. (1988), the agreement between the observed value ($\alpha_F = 1.52 \pm 0.42$) and the theoretically predicted value of the flux ($\alpha_F = 1.80$) is commensurable.

An analysis of 16 RS CVn binary star systems (G, F, K-types) was undertaken with Extreme Ultraviolet Explorer (EUVE) photometry, yielding a power law slope with fluxes α_F, where the flare energy is proportional to the time-integrated counts (Osten and Brown 1999). The observed power law slope of $\alpha_F = 1.60 \pm 0.32$, however, is subject to a major uncertainty, as the deviations from a straight power law show in their Figure 12.

Photometric observations of the dM4.5e flare star YZ CMi using the *Hubble Space Telescope* in the ultraviolet wavelength range of (λ = 1,600 – 3,200 Å) detected 54 flare events and yielded an energy distribution with a power law-like slope of $\alpha_F = 2.25 \pm 0.31$ (Robinson et al. 1999). This case is subject to major deviations from ideal power laws also, which may explain the large discrepancy of the observed value and the FD-SOC prediction, $\alpha_F = 1.80$.

Further observations with EUVE targeted two active solar analogs, 47 Cas and EK Dra, gathering 28 flares and a power law slope of $\alpha_F \approx 2.15 \pm 0.15$ (Audard et al. 1999). Twelve datasets of cool dwarf stars, that is, HD 2726, 47 Cas, EK Dra, κ Cet (1994, 1995), AB Dor, ϵ Eri, GJ 411, AD Leo, and CN Leo (1994, 1995), were observed with EUVE, leading to power law slopes in the range of $\alpha_F \approx 1.5 - 2.4$ (Audard et al. 2000).

One study focused on relatively weak flares from cool dwarfs FK Aqr (dM2e type), V1054 Oph (M3Ve type), and AD Leo (M3 V type) and found power law slopes of $\alpha_F \approx 2.03 - 2.74$, using EUVE deep survey data (Kashyap et al. 2002).

Observations with EUVE concentrate on the cool dwarf AD Leo in several studies (Kashyap et al. 2002; Güdel et al. 2003; Arzner and Güdel 2004), applying different statistical models, which converge to similar values for the power law slope in the range of $\alpha_F \approx 2.25 - 2.40$. Two other studies detected soft X-ray emission from stellar flares using the XMM/Newton telescope, performing observations in the Taurus molecular cloud (Arzner et al. 2007; Stelzer et al. 2007).

14.1.2 Size Distributions

Let us have a look at the obtained size distributions of flare durations T, fluxes F, that have been sampled from flares on individual stars, which are compiled in Table 14.1 and plotted in Figure 14.1. Most power law slopes of fluxes are found in the range of $\alpha_F \approx 1.5 - 2.6$, which is significantly higher than measured in solar flares, where we found $\alpha_F \approx 1.5 - 1.7$ in hard X-rays (Table 4.1), $\alpha_F \approx 1.4 - 1.9$ in soft X-rays (Table 5.1) and $\alpha_F \approx 1.4 - 2.4$ in EUV (Table 6.1), while the FD-SOC model predicts a value of $\alpha_F \approx 1.80$, which is indeed matched by solar flare observations. However, several observations found power law slopes of $\alpha_F \approx 1.5 - 1.7$ (Collura et al. 1988; Pallavicini et al. 1990; Osten and Brown 1999) that are close to the predictions of the FD-SOC model ($\alpha_F \approx 1.80$). Almost all size distributions of stellar flares have been characterized as power law functions. The only exception (with an exponential size distribution) has been reported from optical flares of low-mass young stellar objects in the Orion nebula and from the region of ρ Ophiuchi (Akopian 2012a, 2012b, 2013).

We display the results of 75 published datasets in form of scatter plots as a function of the sample size (Figure 14.1). All datasets analyzed here and shown in Figure 14.1 are taken from published literature values. We divide the

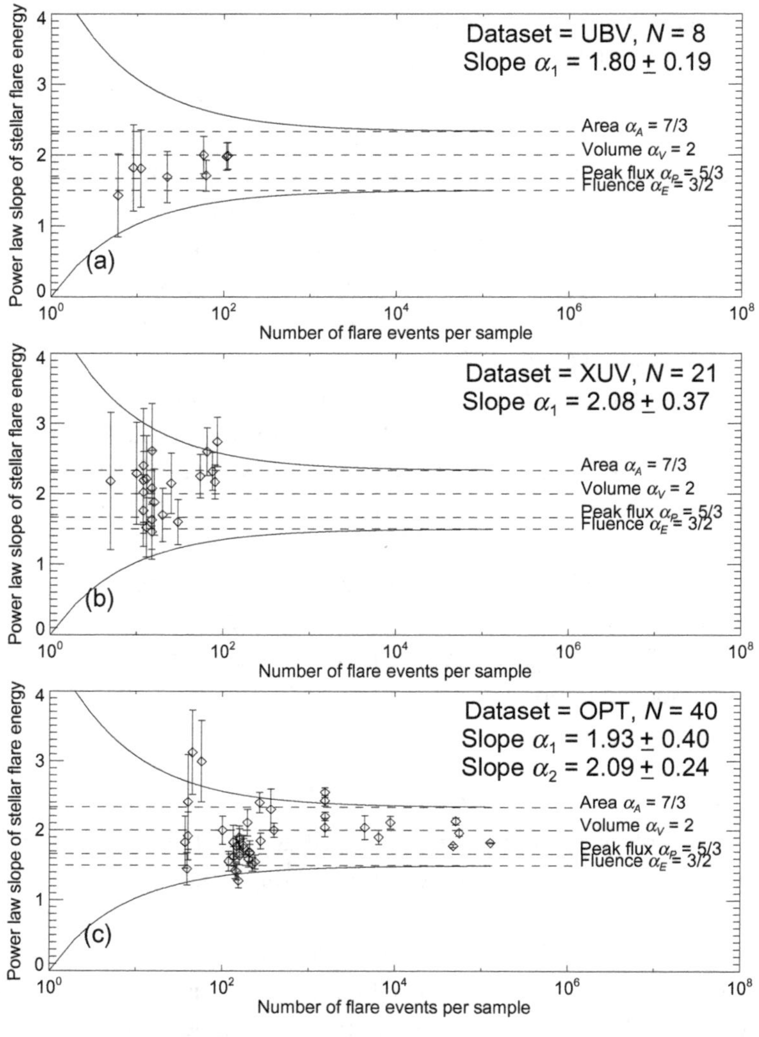

Figure 14.1 The power law slopes (α_F) of flare energy distributions as a function of the number of events per sample are shown separately for three instrument groups or wavelength ranges: (a) UBV, (b) XUV [EXOSAT, EUV, HSP/HST, and XMM-Newton], and (c) OPT [Kepler]. The solid curves bracket the range of statistical uncertainties. The dashed horizontal lines indicate the predicted slopes of four SOC parameters: for the size distribution of the flare volume (α_V), the flare area (α_A), the flux (α_F), and the time-integrated flare energy or fluence (α_F). The slope a_1 includes all datasets, the slope a_2 samples with large statistics only ($n_{ev} \geq 1{,}000$) (Aschwanden and Güdel 2021).

75 datasets into three groups sampled in different wavelength ranges: The dataset with UBV spectroscopy (Figure 14.1a), the dataset with XUV fluences, measured with EXOSAT, EUVE, HSP/HST, XMM-Newton (Figure 14.1b), and the dataset OPT with optical luminosities (integrated over the flare duration) from Kepler

(Figure 14.1c). For these three groups, we find the following means and standard deviations (Figure 14.1):

$$\alpha_{\text{F,UBV}} = 1.80 \pm 0.19, \tag{14.1}$$

$$\alpha_{\text{F,XUV}} = 2.08 \pm 0.37, \tag{14.2}$$

$$\alpha_{\text{F,OPT}} = 1.93 \pm 0.40. \tag{14.3}$$

In order to eliminate small-number samples we select large-size distributions with $n_{ev} > 1{,}000$ events only (i.e., OPT dataset with Kepler data only), and obtain $\alpha_{\text{F,OPT}}$ $(n_{ev} > 1{,}000) = 2.09 \pm 0.24$ (Figure 14.1c), which narrows the uncertainties of α_F-values from 20% to 10%, while their means are consistent within the uncertainties.

The diagram of Figure 14.1c illustrates the convergence of the distribution of α-values as a function of the sample size n_{ev}. The small-sample α_F-values spread over a range of $1.5 \lesssim \alpha_F \lesssim 3$, while the large-sample α_F-values (with $n_{\epsilon v} > 1{,}000$) converge to a range of $1.9 \lesssim \alpha_F \lesssim 2.3$. This convergence clearly demonstrates that the uncertainty of the power law slope scales with the sample size, as predicted by the uncertainty estimate and shown in Figure 14.1 (solid curves).

14.2 Stellar Flares: Kepler Observations

14.2.1 Observations

On 2009 March 7, the Kepler space telescope (Borucki et al. 2010) was launched, which was designed to survey the stars and planets in the Milky Way galaxy, in particular the exoplanets that orbit around other stars than our Sun. It turned out that the studies of light curves reveal numerous stellar flare events, superimposed on the relatively slow rotational modulation of a star (Walkowicz et al. 2011; Nielsen et al. 2013; Notsu et al. 2013, 2015a, 2015b). At the time of writing, the Kepler flare catalog contains 3420 flare stars and 162,262 flare events (Yang and Liu 2019). The number of detected flares strongly depends on the detection algorithm. For instance, the study of Van Doorsselaere et al. (2017) detected 16,850 flares on 6662 stars out of a total of 188,837 in the Kepler field-of-view during the quarter Q15. We provide a compilation of Kepler flare statistics in Table 14.2, based on the studies of Maehara et al. (2012); Shibayama et al. (2013); Balona (2015); Aschwanden (2015, 2019b, 2021); Wu et al. (2015), Lurie et al. (2015); Davenport (2016); Svanda and Karlicky (2016); Aschwanden et al. (2016); and Yang and Liu (2019). These new measurements from Kepler represent an enormous breakthrough in the statistical accuracy of stellar flares, which enables us to determine the power law distributions of stellar flare energies in different stellar types. However, one particular property is the wavelength range of Kepler observations, which is in the optical wavelength regime ($\lambda = 4300 - 8900$ Å), roughly corresponding to white-light flares observed

Table 14.2 *The power law slopes α_F of the frequency distributions of fluxes observed in stellar flares with Kepler, for the differential frequency distribution, and $\alpha_F = \alpha_{F,cum} + 1$ for the cumulative size distributions.*

Star	Instrument	Number of flares and events	Differential slope α_F	Cumulative slope plus 1 $\alpha_{F,cum} + 1$	Reference
G5-stars	Kepler	365	2.30 ± 0.30		Maehara et al. (2012)
G5-stars slow	Kepler	101	2.00 ± 0.20		Maehara et al. (2012)
G5-stars	Kepler	1547	2.20 ± 0.06		Shibayama et al. (2013)
G5-stars slow	Kepler	397	2.00 ± 0.10		Shibayama et al. (2013)
G5-stars	Kepler	1538	2.43 ± 0.08	2.42 ± 0.06	Aschwanden (2015)
G-type stars	Kepler	4494	2.04 ± 0.17		Wu et al. (2015)
K-M,A-F stars	Kepler	209	1.69 ± 0.16	1.71 ± 0.12	Balona (2015), Aschwanden (2015)
KID3557532	Kepler	196	2.11 ± 0.19		Wu et al. (2015)
KID6034120	Kepler	45	3.12 ± 0.60	3.17 ± 0.48	Aschwanden (2015)
KID6697041	Kepler	37	1.83 ± 0.37	1.51 ± 0.25	Aschwanden (2015)
KID6865416	Kepler	147	1.77 ± 0.10		Wu et al. (2015)
KID75264976	Kepler	40	1.92 ± 0.34	1.98 ± 0.32	Aschwanden (2015)
KID7532880	Kepler	159	1.90 ± 0.16		Wu et al. (2015)
KID8074287	Kepler	160	1.87 ± 0.10		Wu et al. (2015)
KID8479655	Kepler	39	1.45 ± 0.23	1.47 ± 0.24	Aschwanden (2015)
KID8547383	Kepler	40	3.41 ± 0.68	2.58 ± 0.41	Aschwanden (2015)
KID9653110	Kepler	158	1.64 ± 0.07		Wu et al. (2015)
KID10422252	Kepler	177	1.75 ± 0.08		Wu et al. (2015)
KID10422252	Kepler	57	2.99 ± 0.58	2.78 ± 0.37	Aschwanden (2015)
KID10745663	Kepler	137	1.63 ± 0.10		Wu et al. (2015)
KID11551430	Kepler	202	1.59 ± 0.06		Wu et al. (2015)
GJ 1243	Kepler	...	1.92 ± 0.01		Lurie et al. (2015)
GJ 1245A	Kepler	683	1.99 ± 0.02		Lurie et al. (2015)
GJ 1245B	Kepler	605	2.03 ± 0.02		Lurie et al. (2015)

G-type	Kepler	...	1.59 ± 0.02		Svanda and Karlicky (2016)
K-type	Kepler	...	1.79 ± 0.06		Svanda and Karlicky (2016)
F-type	Kepler	...	1.77 ± 0.07		Svanda and Karlicky (2016)
A-type	Kepler	...	1.37 ± 0.07		Svanda and Karlicky (2016)
GJ 1243	Kepler	...	2.008 ± 0.002		Silverberg et al. (2016)
Stellar flares	Kepler	208	1.68 ± 0.12		Aschwanden (2019b)
Bolometric flares	Kepler	1537	2.55 ± 0.07		Aschwanden (2019b)
KID10000490	Kepler	241		1.55 ± 0.10	Davenport (2016)
KID10001145	Kepler	271		2.40 ± 0.15	Davenport (2016)
KID10001154	Kepler	119		1.56 ± 0.14	Davenport (2016)
KID10001167	Kepler	147		1.41 ± 0.12	Davenport (2016)
KID10007792	Kepler	225		1.52 ± 0.10	Davenport (2016)
KID10002897	Kepler	155		1.28 ± 0.10	Davenport (2016)
KID10004510	Kepler	142		1.43 ± 0.12	Davenport (2016)
KID10004660	Kepler	135		1.83 ± 0.24	Davenport (2016)
KID10005966	Kepler	175		1.79 ± 0.14	Davenport (2016)
KID10006158	Kepler	279		1.85 ± 0.11	Davenport (2016)
A-type	Kepler	583	1.12 ± 0.08	1.65 ± 0.07	Yang and Liu (2019), Aschwanden (2021)
F-type	Kepler	8869	2.11 ± 0.09	1.85 ± 0.09	Yang and Liu (2019), Aschwanden (2021)
G-type	Kepler	55259	1.96 ± 0.04	1.79 ± 0.06	Yang and Liu (2019), Aschwanden (2021)
K-type	Kepler	47112	1.78 ± 0.02	1.80 ± 0.20	Yang and Liu (2019), Aschwanden (2021)
M-type	Kepler	50439	2.13 ± 0.05	1.84 ± 0.02	Yang and Liu (2019), Aschwanden (2021)
Giants	Kepler	6496	1.90 ± 0.10	2.00 ± 0.34	Yang and Liu (2019), Aschwanden (2021)
All	Kepler	162262	1.823 ± 0.007		Aschwanden and Güdel (2021)
	TESS		1.68 ± 0.03		Pietras et al. (2022)

on the Sun. We have to keep this optical wavelength bias in mind when we compare with observations in soft X-rays and EUV (EXOSAT, EUVE, and XMM-Newton). Also, there may be a possible wavelength dependence in the related power law slopes. The flare events have been detected with Kepler using different methods that are described in Davenport (2016) and Yang and Liu (2019) and are subject to various biases and false-positive signals. Examples of such erroneous signals are periodic flux modulations from RR Lyrae, pulsating giants, eclipsing binaries, or fluctuations from δ Scuti and γ Doradus stars. Size distributions of flare-radiated energies are gathered for flares from the same star (given by the KID identification number in Table 14.2), as well as for different stellar spectral types (A, F, G, K, M-type, Giants) separately.

14.2.2 Size Distributions

In the study of Yang and Liu (2019), the stellar flare catalog of the Kepler mission has been sorted according to the spectral type of the observed stars. This enables us to investigate the power law slopes α_F of the flare size distributions for each stellar spectral type (A, F, G, K, M, and Giants) separately. Each spectral type may have different stellar properties. We show the power law slopes α_F of the various stellar spectral types as a function of the sample size in Figure 14.2. Interestingly, all stellar spectral types have power law slopes near $\alpha_F \approx 2.0$. The largest datasets (with $n_{ev} > 1000$) have the following power law slopes: $\alpha_G = 2.10 \pm 0.21$ for G-type stars (Figure 14.2a), $\alpha_F = 2.36 \pm 0.25$ for F-type stars (Figure 14.2b), $\alpha_M = 1.99 \pm 0.35$ for M-type stars (Figure 14.2c), $\alpha_K = 2.02 \pm 0.33$ for K-type stars (Figure 14.2d), and $\alpha_{Giants} = 1.90 \pm 0.10$ for Giant stars (Figure 14.2f). The only exception is $\alpha_A = 1.12 \pm 0.08$ for A-type stars (Figure 14.2e), as it was noted before (Yang and Liu 2019). This outlier is re-visited in a recent study and was attributed to an ill-defined power law range, while fits with the PM (Pareto) model, the EM (extreme event) model, and the FM (finite system size) model (Fig. 1 in Aschwanden 2021), yield a value of $\alpha_F = 1.65 \pm 0.01$ that is more consistent with the other spectral types, and invalidates the claim of a different physical mechanism made by Yang and Liu (2019). Moreover, the existence of starspots indicates the presence of magnetic fields, but the fact that A stars lack coronae implies that surface convection is a necessary condition for the formation of the corona (Balona et al. 2016).

From the Kepler data shown in Figure 14.1c it appears that the datasets with the largest statistics ($n_{ev} > 1000$) with a power law value of $\alpha_{\mathrm{P,OPT}} = 2.09 \pm 0.24$ seem to cluster around the mean value of $\alpha \approx 2.0$. Given the fact that the same trend holds for all spectral types (Figure 14.2), it is conceivable that all stellar types can be explained in terms of a single general model that predicts an universal

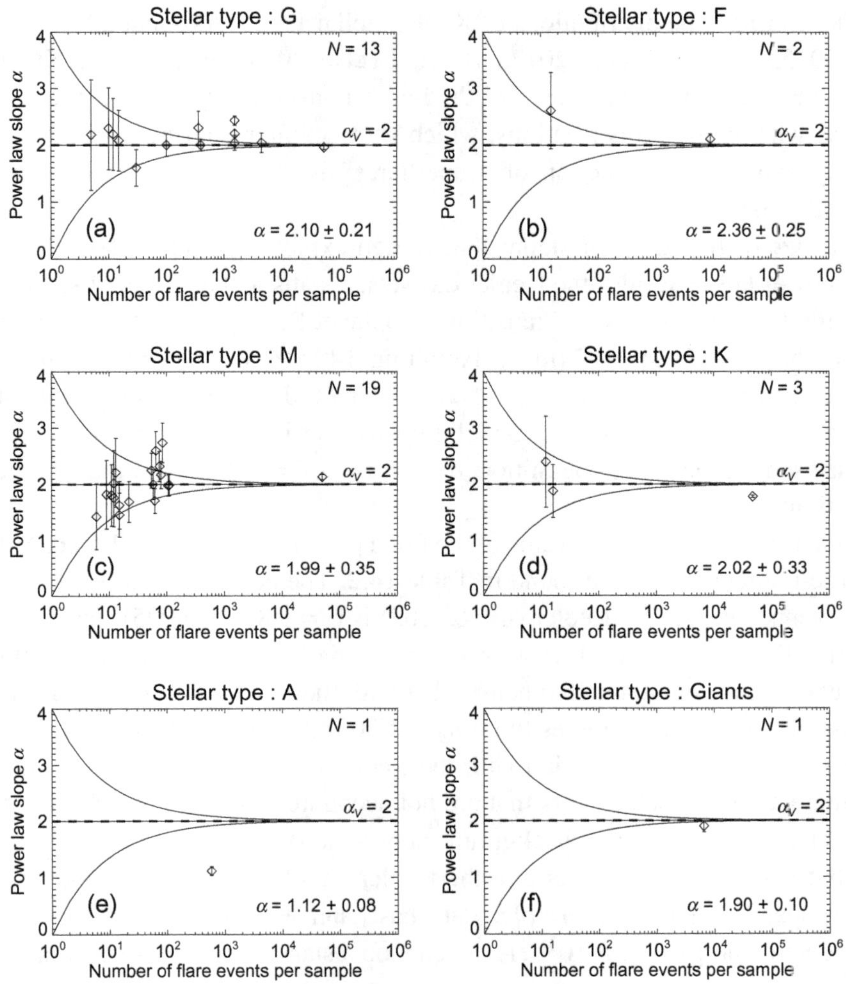

Figure 14.2 Means and standard deviations of power law slopes from datasets grouped by star types (A, F, G, K, M, Giants), based on the classification of the Kepler flare catalog (Yang and Liu 2019) and the star classification given in Tables 14.1 and 14.2. The dashed horizontal lines indicate power law slopes of $\alpha_V \approx 2.0$, while the curves indicate the expected uncertainties. The number of flare datasets per stellar spectral type varies from $N = 1$ (A-type stars) to $N = 13$ (G-type stars) (Aschwanden and Güdel 2021).

value of $\alpha_F \approx 2$. Statistics of the star GJ 1243 was carried out for different data ranges, subdivided by quarters (years) and months, while the mean power law slope remained to be constant, $\alpha_F = 2.008 \pm 0.002$ (Silverberg et al. 2016). The Kepler stellar flare data were found to be multifractal (DeFreitas et al. 2017, 2019a, 2019b), which is consistent with the multiple-fractal approach of our FD-SOC model, which we parameterized with $D_d = (D_{d,min} + D_{d,max})/2 = [(d-1)+d]/2 = d - 1/2$.

The waiting time distributions of Kepler stellar flares (Akopian 2013; Maehara et al. 2015, 2017; Gizis et al. 2017a, 2017b; Li et al. 2018; Tu et al. 2020; Kuznetsov and Kolkotkov 2021) have been studied with nonstationary Poisson processes as well as with Weibull distributions, which show memory during events with high activity, while the production of "superflares" is closer to a stochastic process (Li et al. 2018).

The size distributions of stellar flares often exhibit significant deviations from ideal power laws. Besides the Kepler datasets, a statistical analysis of stellar flares from the first three years of Transiting Exoplanet Survey Satellite (TESS) observations has also been performed, exhibiting 140,000 flare events (Pietras et al. 2022), and yield a power law slope of $\alpha_F = 1.68 \pm 0.03$, which is consistent with the expected value of the FD-SOC model, in particular if the volume-energy relationship is used for the energy definition, $\alpha_V = \alpha_F = 9/5 = 1.80$, rather than the fluence definition $\alpha_E \approx 1.44$.

What is the most accurate value of the fitted power law slopes α_F? A compilation of large-number fits can be found in Table 14.2. The power law slope of all stellar flare events detected by Kepler (n=162,262) is $\alpha = 1.817 \pm 0.005$). Inspecting the uncertainties σ_α in Figure 14.1 we notice that small samples (with $n_{ev} \lesssim 1{,}000$) display uncertainties that are commensurable with their deviation from a mean value of $\alpha_F \approx 2$, while large samples (with $n_{ev} \gtrsim 1{,}000$) display much smaller uncertainties so that they do not overlap with their common mean value, which indicates that there are systematic errors that are not accounted for (such as arbitrary fitting ranges $[x_0, x_2]$ or inaccurate background subtraction).

What is the largest superflare in the Kepler data base? Althukair and Tsiklauri (2023) searched the entire Kepler data base and estimated that a superflare of G-type with energy of $E \approx 10^{35}$ ergs occurs on a star once every 4,360 years.

14.3 Star Formation

The formation of stars is initiated by gravitational collapses of molecular clouds. Such a gravitational collapse can be triggered by collisions of two molecular clouds, by the explosion of a nearby supernova, which ejects shocked matter, or even by galactic collisions, which cause compression of matter and of tidal forces. If there is a critical mass reached, which is quantified by the *Jeans mass criterion*, which mostly depends on the initial size of the unstable galactic fragment, the collapsing molecular cloud will build up a dense core by self-gravity, which forms a star with nuclear burning. Smaller sizes evolve into nonradiating brown dwarfs.

Considering star formation as a SOC process, the situation is similar to the formation of planetesimals and planets, where a critical condition is given by the balance between the forces of self-gravity and diffusion. A collapsing molecular cloud gains kinetic energy from the gravitational potential according to the conservation of angular momentum. However, tidal forces, external gravitational disturbances, and thermal pressure represent forces that contribute to the local diffusion of the molecular cloud. Therefore there is a threshold for the instability of a gravitational collapse, which is self-organizing by the given balance between the opposing forces of contraction and diffusion. This process could possibly also be modeled in terms of a percolation process.

SOC avalanches have a fractal structure, and hence fractals are expected for star-forming regions also. Indeed, fractal and self-similar patterns have been observed in the Milky Way from dense cores to giant molecular clouds in a range of $0.1 < L < 100$ parsec (Elmegreen and Scalo 2004; Bergin and Tafalla 2007), as well as in star-forming regions in the Andromeda nebula M33 (Sanchez et al. 2010). The fractal dimension in the interstellar medium has a value of $D_V \approx 2.3 \pm 0.3$ (Elmegreen and Falgarone 1996), in bright young stars and molecular gas it is $D_A \approx 1.9$, and in fainter stars and HII regions it is $D_V \approx 2.2 - 2.5$. The prediction of the FD-SOC model is $D_V \approx 2.5$, based on the mean value of $D_d = (D_{min} + D_{max})/2 = (2+3)/2 = 2.5$ in Euclidean 3-D space (with dimensionality $d = 3$). The fractal structure has generally been attributed to interstellar turbulence, which however does not exclude a generalized description in terms of a SOC process. It has been argued that the *interstellar mass function* of starbursts is independent of local processes governing star formation and thus can be considered as a universal self-organized criticality process (Melnick and Selman 2000).

14.4 Cataclysmic Variables

Cataclysmic variables (CVs) are close binary systems consisting of a white dwarf and a red-dwarf secondary transferring matter via the Roche lobe overflow. The resultant accretion disk becomes thermally unstable in systems with high mass transfer rates.

A SOC interpretation was suggested for the VY Scl-type cataclysmic variable KR Aurigae (Kato et al. 2002; Dobrotka et al. 2012), UU Aqr (Dobrotka et al. 2012), the broad-line radio galaxy 3C-390.3 (Leighly and O'Brien 1997), galaxy IRAS 13224-3909 (Gaskell 2004), which all exhibit a highly intermittent variability on top of a shot noise background like Cygnus X-1.

14.5 Pulsar Glitches

14.5.1 Theoretical Modeling

A pulsar is a highly magnetized, rapidly rotating neutron star that emits a beam of electromagnetic radiation. Since the beamed emission is aligned with the magnetic axis, we observe rotationally modulated pulses whenever the beam axis points to the Earth (line-of-sight direction) during each period of its rapid rotation. Besides these regular periodic pulses on time scales of milliseconds, which are measured with high accuracy, there occur sporadic glitches in pulse amplitudes and frequency shifts, probably caused by sporadic unpinning of vortices that transfer momentum to the crust (Warszawski and Melatos 2008). Conservation of angular momentum produces then a tiny increase of the angular rotation rate of the neutron star, called *positive spin-ups*.

Theoretical modeling of pulsar glitches has been performed by quantum mechanical simulations of neutron star rotational glitches, triggered by vortex avalanches in the superfluid stellar interior, which reveal that vortices pin non-axis-symmetrically to the crust during the intervals between glitches (Melatos et al. 2015).

For some pulsars, synchronous soft X-ray and radio emission has been discovered, which was interpreted in terms of variations in the accretion rate of material from the interstellar medium through which it is passing (Hermsen et al. 2018).

Kepler K2 observations of the transitional millisecond pulsar PSR J1023+0038 reveal a flaring state and a quiescent state, with flare durations from $T < 2$ min up to several hours. The scale-free flaring behavior is consistent with a SOC mechanism, most likely related to the build-up and release of mass at the inner edge of the accretion disk (Kennedy et al. 2018). A dual waiting time behavior, characterized by flaring states and quiescent states was also noted by Guidorzi et al. (2015).

Few statistically compelling correlations are found in pulsar timing data between the size of a rotational glitch and the time to the preceding glitch (backward waiting time), except for PSR J0537-6810 (Melatos et al. 2018). The lack of a size-duration correlation rules out energy storage-dissipation models (Section 5.3).

One minimal model for pulsar glitches was constructed by introducing a solid-crust potential in the 3-D Gross–Pitaevskii–Poisson equation, which has been used earlier to study gravitationally bound Bose–Einstein condensates, showing signatures of SOC in the observed pulsar glitches (Verma et al. 2020).

The variability of pulsars shows that the primary emission of the pulsars corresponds to log-normal statistics, possibly consistent with the stochastic growth theory, which would imply that the radiation is generated by a linear micro-instability directly via linear mode conversion (Cairns et al. 2004). A secondary component may be the result of scattering by density irregularities, involving non-linear processes (e.g., modulational instability and wave collapse) in a SOC system (Cairns et al. 2004).

Supergiant fast X-ray transients are a subclass of high-mass X-ray binaries, in which the compact object is a neutron star and is accreting materials from its non-generate companion star, which suggests that the high-mass X-ray binaries with similar X-ray flares are likely to belong to a SOC system (Zhang et al. 2022).

Table 14.3 *Frequency distributions measured from giant pulses of pulsars. The size distributions were reported in units of cumulative) pulse energies (Argyle and Gower 1972), in radio flux densities (Lundgren et al. 1995), cumulative pulse amplitudes (Cognard et al. 1996), electric fields (Cairns 2004), and fractional increase of the spin frequency ($\Delta\nu/\nu$) (Melatos et al. 2008). Minimum and maximum values are quoted in brackets.*

Object	Waveband	Number of events	Power law slope of fluxes α_F	References
Crab pulsar	146 MHz	440	3.5	Argyle and Gower (1972)
Crab pulsar	813–1330 MHz	3×10^4	3.06–3.36	Lundgren et al. (1995)
PSR B1937+21	430 MHz	60	2.8 ± 0.1	Cognard et al. (1996)
PSR B1706-44	1.5 GHz		6.4 ± 0.6	Cairns (2004)
Vela pulsar	2.3 GHz		6.7 ± 0.6	Cairns (2004)
PSR B0950+08	0.4 GHz		6.2 ± 0.5	Cairns (2004)
Crab pulsar	0.8 GHz		5.6 ± 0.6	Cairns (2004)
PSR B1937+214	0.4 GHz		4.6 ± 0.2	Cairns (2004)
PSR B1821-24	1.5 GHz		9.0 ± 2.0	Cairns (2004)
PSR 0358+5413		6	2.4 [1.5,5.2]	Melatos et al. (2008)
PSR 0534+2200		26	1.2 [1.1,1.4]	Melatos et al. (2008)
PSR 0537+6910		23	0.42 [0.39,0.43]	Melatos et al. (2008)
PSR 0631+1036		9	1.8 [1.2,2.7]	Melatos et al. (2008)
PSR 0835+4510		17	−0.13 [−0.20,+0.18]	Melatos et al. (2008)
PSR 1341+6220		12	1.4 [1.2,2.1]	Melatos et al. (2008)
PSR 1740+3015		30	1.1 [0.98,1.3]	Melatos et al. (2008)
PSR 1801+2304		9	0.57 [0.092,1.1]	Melatos et al. (2008)
PSR 1825+0935		8	0.36 [−0.30,1.0]	Melatos et al. (2008)
			(1.67)	FD-SOC prediction

14.5.2 Observations

Statistics of pulsar-related sporadic glitches (Table 14.3) exhibit power law distributions of the pulsar fluxes or fluences, such as observed from the Crab pulsar and other pulsars in radio wavelengths (Argyle and Gower 1972; Lundgren et al. 1995; Cognard et al. 1996; Cairns 2004; Melatos et al. 2008), and thus were interpreted in terms of a SOC system (Young and Kenny 1996). While early measurements with extensive statistics exhibit power law distributions with relatively steep slopes of $\alpha_F \approx 3.0$ (Argyle and Gower 1972; Lundgren et al. 1995; Cognard et al. 1996), more recent observations with smaller samples yield a large scatter of power law slopes in the range of $-0.13 \leq \alpha_F \leq 2.4$ (Melatos et al. 2008) and $\alpha_F \approx 4.6 - 9.0$ (Cairns 2004). The latter results with power law slopes of $\alpha_F \geq 3$ appear to be outliers, given the typical ranges of observed SOC values and the maximum Euclidean (and fractal) dimension $d \leq 3$.

Actually, most (16) of the 18 pulsar size distributions listed in Table 14.3 are outside the valid range ($1.44 \leq \alpha_x \leq 3.00$) of the standard FD-SOC model ($d = 3$), which is a strong indication that the FD-SOC model is either not relevant for size distributions of pulsar glitches, or that the data analysis in the measurement of SOC parameters is substantially flawed. The extremely narrow range of the size distribution of pulsar glitches is also visible for the Crab pulsar shown in Figure 15.3, while typical SOC size distributions extend over at least one or two decades. This is a new (perhaps surprising) result that is concluded from data before 2008.

Other recent studies of giant micropulses from pulsars report a log-normal distribution of energies (Johnston and Romani 2002; Cairns et al. 2004), which is only consistent with a power law function at the asymptotic limit of the high-energy tail of a log-normal function. Therefore, the distributions of giant pulses from pulsar glitches do not give rise to a narrow range of power law slopes, and thus are not easy to explain in terms of an universally valid SOC model. Part of the large uncertainties of power law slope measurements is clearly attributable to the small-number statistics in small samples (i.e., 6–30 pulses in the datasets of Melatos et al. 2008). The unusual steepness of reported power law slopes may be associated with finite-size effects in a SOC system, which can cause an exponential-like cutoff at the high-energy end of the size distribution, as it is suspected for giant stellar flares. Furthermore, the size distributions listed in Table 14.3, have been reported in different physical units (i.e., flux, pulse energy, electric field, and frequency decrease ratios), and in form of both differential and cumulative size distributions, which need to be converted into the same energy units in order to make them directly comparable.

14.5.3 Cellular Automaton Models

A cellular automaton model has been developed for pulsar glitches, based on the superfluid vortex unpinning paradigm (Warszawski and Melatos 2008, 2012; Melatos et al. 2008). The lattice grid in this model simulates the collective behavior of up to 10^{16} vortices in the interior of the pulsar. The cellular automaton generates scale-free avalanche distributions with power law slopes of $\alpha_F \approx 2.0 - 4.3$ for avalanche sizes, and $\alpha_T = 2.2 - 5.5$ for avalanche durations. This numerical model produces size distributions that are not consistent with the predictions of the FD-SOC model ($\alpha_F = \alpha_V = 1.80$, $\alpha_T = 2.00$). Moreover, the observed size distributions of pulsar glitches (Table 14.3) suffer from small number statistics and unplausible outliers far off the expected values of the standard FD-SOC model ($1.44 \lesssim \alpha_F \lesssim 1.80$).

14.6 Summary

Among stellar systems, we find a lot of systems with applications of SOC, such as stellar flares or pulsar glitches. Stellar flares occur mostly in the wavelength ranges of ultraviolet (UBV), soft X-rays and UV (XUV), and in visible light. A breakthrough in new stellar data was accomplished with the Kepler spacecraft, which allowed unprecedented detections of exoplanets, while the same light curves could be searched for large stellar flares.

Exploiting these promising new datasets, one finds that most stellar flare datasets exhibit dominant size distributions that converges to a power law slope of $\alpha_F \approx 2.0$, regardless of the star type (e.g., G, F, M, K, A, and giant stars), while the FD-SOC theory predicts a slope of $\alpha_V = \alpha_F = 1.80$ for 3-D isotropic sources, and $\alpha_A \approx 2.33$ for 2-D structures. The size distributions of pulsar glitches are mostly found outside of the valid range ($1.44 \le \alpha_x \le 2.33$) of the Standard FD-SOC model, and thus require a different model.

Power law fits are not always superior to fits with the log-normal function or Weibull function. The reason for this discrepancy between observed and modeled power law slopes in stellar SOC systems is mostly due to small-number statistics of the samples ($N \lesssim 10 - 100$), incomplete sampling near the lower threshold, and due to ill-defined power law fitting ranges, which can cause significant deviations from ideal power laws. Nevertheless, the size distributions of stellar flares is similar to those of solar flares in most cases, and thus largely fit the predictions of the FD-SOC model.

15

Galactic and Black-Hole Systems

15.1 Soft Gamma Ray Repeaters

A host of gamma-ray bursts were detected with the Compton Gamma Ray Observatory (CGRO), the Rossi X-ray Timing Explorer (RXTE), and the International Cometary Explorer (ICE) in hard X-rays $\approx 20 - 40$ keV (a wavelength regime that is also called soft gamma-rays). The events show repeated detections from the same source location and have been dubbed Soft Gamma-Ray Repeaters (SGRs). These gamma-ray bursts are believed to originate from slowly rotating, extremely magnetized neutron stars (magnetars) that are located in supernova remnants (Kouveliotou et al. 1998; 1999), where neutron star crust fractures occur, driven by the stress of an evolving, ultrastrong magnetic field in the order of $B \gtrsim 10^{14}$ G (Thompson and Duncan 1996). We should be aware that repeated bursts from the same source are the exception rather than the rule for gamma-ray bursts (GRBs). GRBs have been proven to be extra-galactic in origin, after a successful identification of an X-ray afterglow (Klebesadel et al. 1973; Beckmann et al. 2003). The recent discovery of a galactic fast radio burst (FRB) associated with a hard X-ray burst from a SGR has been established, at least for some FRBs, that they originate in magnetars (Wei et al. 2021). An open question is whether they are related to supernova explosions.

15.1.1 Size Distributions

The size distributions of the fluences of sources SGR 1900+14 and SGR 1806-20 were found to exhibit power law distributions with slopes of $\alpha_F = 1.66$, Figure 15.1, (Gogus et al. 1999), and $\alpha_F = 1.43$, 1.76, and 1.67, Figure 15.2 (Gogus et al. 2000), extending over a range of about 4 orders of magnitude in fluence. These values are close to the theoretical prediction of the FD-SOC model, which predicts $\alpha_F = 1.80$. The waiting time distributions were found to be consistent with a log-normal distribution (which is approximately a power law function in the high-energy tail). Based on these observational statistics, SGR bursts have been interpreted in terms

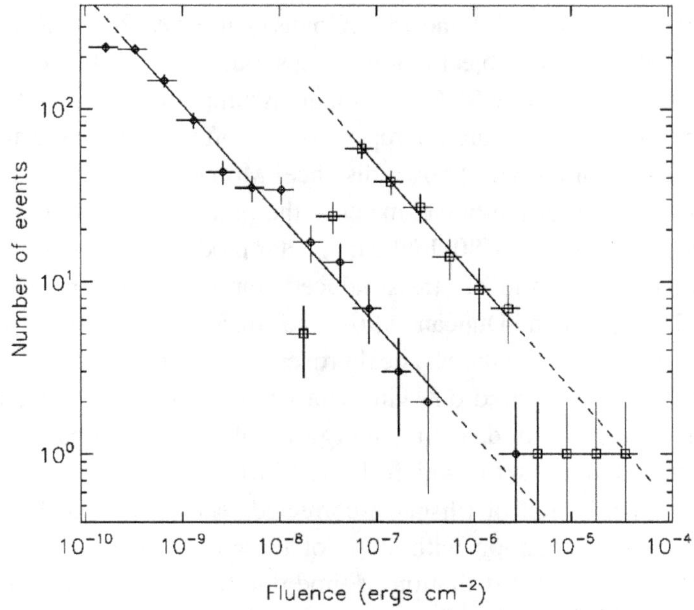

Figure 15.1 Differential size distribution of the fluences of bursts from SGR 1900+14 as measured with RXTE (diamonds) and BATSE.GCRO (squares). The solid lines denote the interval where used in the fit, and the dashed lines are the extrapolations of the model (Gogus et al. 1999).

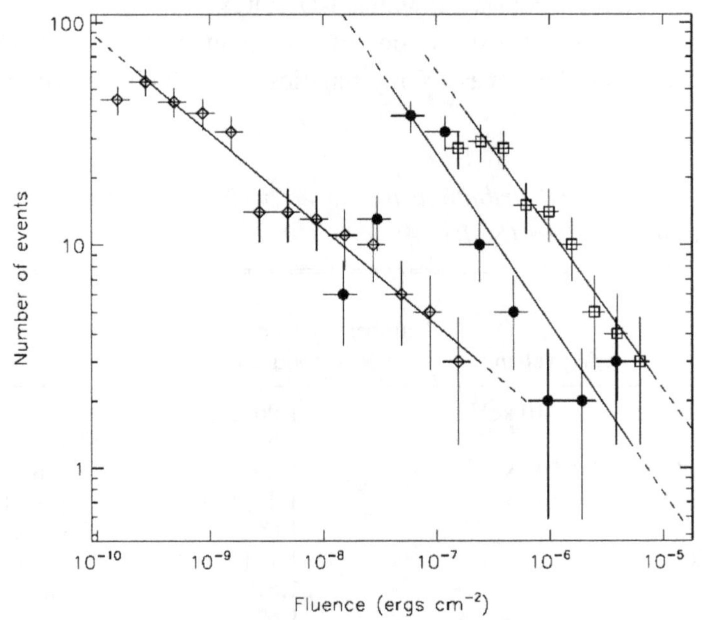

Figure 15.2 Differential fluence distributions of from SGR 1806-20 bursts as seen by RXTE (diamonds), BATSE.GCRO (circles), and ICE (squares). The lines are obtained by fitting a power law model with the maximum likelihood technique. The solid lines show the intervals used in the fit, and the dashed lines are the extrapolation of the model (Gogus et al. 2000).

of a SOC process, Figures 15.1 and 15.2 (Gogus et al. 1999; 2000). Since the source location is identical for an object that produces multiple SGR bursts, we can identify each event with a single SOC system, an assumption that cannot be made for other gamma-ray bursts that are nonrepetitive and often do not have an unambiguous source identification with known distance. Moreover we find that the fluence or energy distribution of the bursts matches the prediction of the fractal-diffusive FD-SOC model, with $\alpha_F = 1.80$. In the magnetar model, the triggering mechanism for SGR bursts is a hybrid of stress-induced starquakes and magnetically powered flares (Thompson and Duncan 1996). The stress-induced trigger in starquakes shares some similarity with the physical process of earthquakes.

A recent study was carried out with data from the *Swift* satellite, which has a rapid response, suitable for detecting afterglows of gamma-ray bursts. In a sample with 83 localized sources for which the redshift was known (and thus the distance), a size distribution of (distance-corrected) energies could be constructed, and a power law distribution with slope of $\alpha_F = 1.06 \pm 0.15$ was found (Wang and Dai 2013). The size distribution of time duration was found to have a slope of $\alpha_T = 1.10 \pm 0.15$. These Swift results were interpreted in terms of a 1-D SOC system (Wang and Dai 2013), for which the generalized FD-SOC model (with dimensionality $d = 1$ instead of $d = 3$) predicts $\alpha_F = 1$ and $\alpha_T = 1$. This 1-D interpretation for GRBs with afterglows appears to be different from normal SGRs (Table 15.1), which are consistent with a 3-D SOC system.

Cumulative luminosity distributions of supergiant fast (hard) X-ray transiens (SFXT) and classical high-mass X-ray binaries (HMXB) have been accumulated

Table 15.1 *Frequency distributions measured from gamma-ray bursts (GRB) and soft gamma-ray repeaters (SGR).*

Object	Waveband	Number of events	Power law slope of fluxes α_F	References
Gamma-ray bursts	0.3–10 keV	83	1.06 ± 0.15	Wang and Dai (2013)
SGR 1806–20			1.6	Chang et al. (1996)
SGR 1900+14	>25 keV		1.66	Gogus et al. (1999)
SGR 1806–20	>21 keV		1.43, 1.76, 1.67	Gogus et al. (2000)
SGRJ1550–5418			1.667 ± 0.078	Cheng et al. (2020)
SGR1806–20			1.741 ± 0.067	Cheng et al. (2020)
SGR1900+14			1.672 ± 0.048	Cheng et al. (2020)
			1.67	Median
			1.80	FD-SOC prediction

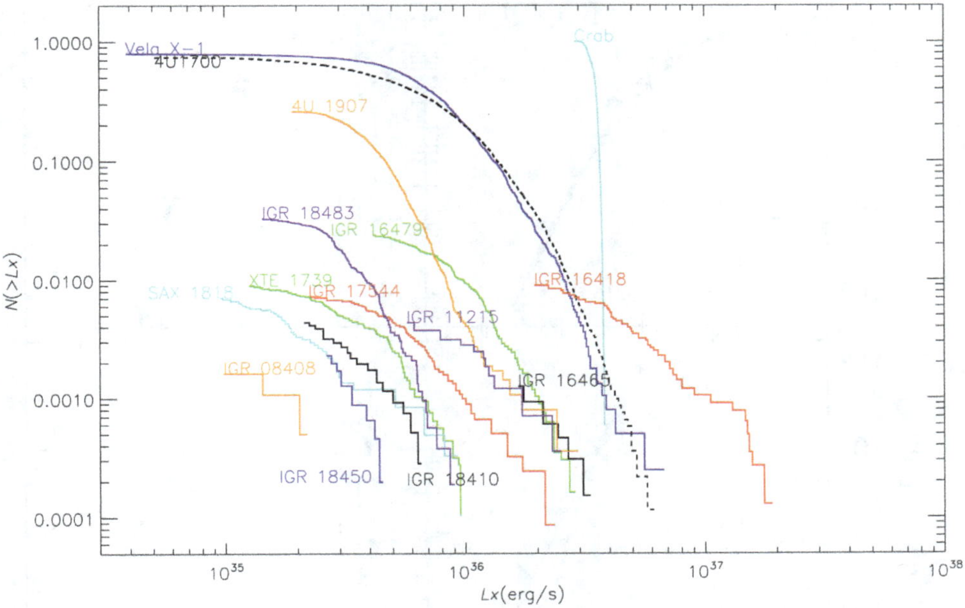

Figure 15.3 Cumulative luminosity distributions $N(>L_x)$ as a function of luminosity $L_x(erg/s)$ of GRBs, for sources in the energy range of 17–30 keV. The events consist of supergiant fast (hard) X-ray transiens (SFXT) and classical high-mass X-ray binaries (HMXB) (Paizis and Sidoli 2014).

from INTEGRAL data in the energy range of 17–100 keV (Paizis and Sidoli 2014), Figure 15.3. The application of power law fits shows plausible power law fits for SFXT, while it can only reproduce the very high-luminosity tail for the classical HMXBs. The slopes of the power law size distributions clusters around $\alpha_F \approx$ 1.3–1.4, 2.2–2.7, and 4.1 (see Table 7 in Paizis and Sidoli 2014), which roughly compares with our standard FD-SOC model for 3-D structures ($\alpha_F = 1.80$), and for 2-D structures ($\alpha_A = 2.33$). A synopsis of all analyzed size distributions is shown in Figure 15.3. The power law fits displayed in Figure 15.4 illustrate the similarity of power law functions and log-normal functions, which often are not significantly different in the high-energy tail.

A statistical study of multi-pulse GRBs, where at least 3 pulses per GRB event are required, produced a sample with 454 pulses in 93 GRBs, and displayed differential frequency distributions with power law slopes of $\alpha_F^{diff} = 1.54 \pm 0.09$, $\alpha_F^{cum} = 1.82 \pm 0.15$, and $\alpha_F^{diff} = 2.09 \pm 0.18$, which agrees to some degree with the theoretical prediction of the FD-SOC model, $\alpha_F = 1.80$, where multiple jet avalanches are possibly responsible for the GRBs, driven by a magnetic instability (Lyu et al. 2020).

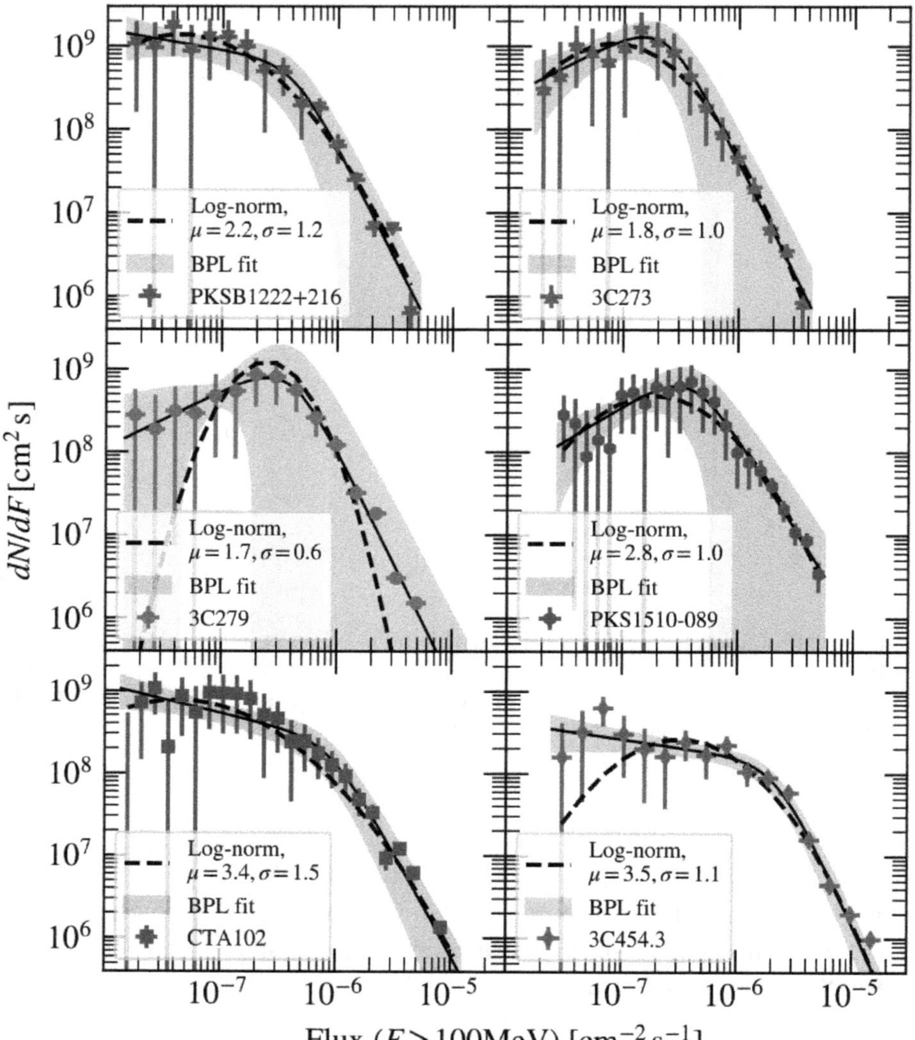

Figure 15.4 Distribution of the fluxes of the weekly 9.5 year gamma-ray light curves. The broken power law fit is shown as a solid line with 1-σ uncertainties (shaded bands). The black dashed line indicates a fit with a log-normal distribution with location and scale parameters μ and σ, as indicated in each panel (Meyer et al. 2019).

15.1.2 Time Variability

Avalanching SOC systems generally produce irregular time series, which can be characterized by stochastic noise, by quasi-periodic oscillations (such as limit cycles in nonlinear systems), or by strictly periodic (Fourier-type) signals. What is the time variability behavior of GRBs and SGRs? Although gamma-ray bursts are normally stochastic, evidence for periodicity (4 pulses with a period of ≈ 8 s) was found in the French SIGNE II MS experiment aboard the Soviet Venera 12 Venus probe in the 130–205 keV energy range (Barat et al. 1979).

Waiting time distribution statistics of type I X-ray bursts in low-mass X-ray binaries, which are caused by unstable nuclear burning of accreted materials, exhibit significant deviations from the predicted power law functions in their fluences, counts, rise times, durations, and waiting times (Wang and Dai 2017).

Gamma-ray observations with *FERMI* revealed extreme gamma-ray outbursts from *flat spectrum radio quasars (FSRQ)*, but there is only weak evidence for variability at time scales less than FERMIs orbit of 95 min, which suggests extremely compact emission regions in the jet (Meyer et al. 2019).

The minimum variation timescale (MVT) of SGR can be used to estimate the emission region in pulsar-like models, as well as to determine the Lorentz factor and radius of the possible relativistic jets in GRB-like models, which in turn can reveal their progenitors, the physical mechanism, and the identification of bursts with a magnetar origin (Xiao et al. 2023a). Energy-dependent time delays of the spectral lag of GRBs can probe temperature changes of the blackbody-emitting plasma with time (Xiao et al. 2023b).

15.1.3 Cellular Automaton Models

Numerical simulations of a SOC system with gamma-ray afterglows have been conducted with a 1-D cellular automaton code, in order to test the existence of 1-D magnetized flows, which was found to be the case (Harko et al. 2015).

A SOC cellular automaton model with recurrent bursts of soft gamma repeaters (SGR) has been developed, where the recurrent bursts are interpreted as avalanches of reconnection in the magnetosphere of neutron stars (Nakazato 2014). This model reproduces the energy distributions observed in SGR 1806-20 and SGR 1900 + 14. This result shows that the SOC of SGRs can be interpreted not only by the crust quake model, but also by the magnetic reconnection model (Nakazato 2014).

A 1-D cellular automaton model in a magnetized grid is applied to gamma-ray burst (GRB) X-ray afterglows by Harko et al. (2015). While SOC models in a real world are described in terms of a 3-D lattice, next-neighbor interactions, a slow external driver, a threshold condition, energy dissipation, and testable size distributions, the model of Harko et al. (2015) is formulated in terms of a 1-D lattice, dual next-neighbor interactions, a stochastically loading driver, boundary dissipation, and size distributions observed in X-ray afterglows of GRBs. The underlying physical mechanism is derived from a discretization of the induction equation in the study of Harko et al. (2015). The observed size distribution is found to have a power law slope of $\alpha_F \gtrsim 1.00$, which is different from the theoretical prediction of the standard SOC model, $\alpha_F = 1.80$ for the dimensionality $d = 3$.

Another study of the peak flux (F) and flaring time (T) duration of the blazar Mrk 421 was carried out from XMM-Newton data by Yan et al. (2018). The size distributions of these two parameters were found to have power law slopes of $\alpha_F = \alpha_T \approx 1.0$, which is the lower limit of the valid parameter range in the generalized FD-SOC model. This extreme value appears to follow a similar trend of a dimensionality of $d = 1$ as in the case of GRBs simulated by Harko et al. (2015). How can we understand this trend? While the spatial structure is assumed to be isotropic in the generalized FD-SOC model, this assumption is obviously not valid in the case of blazars, because they emit a relativistic beam or jet that is aligned or nearly aligned with the line-of-sight direction to Earth, and thus is highly anisotropic. Consequently, this anisotropic geometry of blazars corresponds to a dimensionality of $d = 1$, for which power law slopes of $\alpha_F = \alpha_T = 1.00$ are predicted. This applies to the observed blazar Mrk 421 (Yan et al. 2018) as well as to the simulated blazar (Harko et al. 2015).

15.2 Cygnus X-1

A host of gamma-ray bursts were detected with the Compton Gamma Ray Observatory (CGRO). The first galactic X-ray source that has been identified as a black-hole candidate, Cygnus X-1, emits hard X-ray pulses with a time variability down to 1 ms. These hard X-ray pulses are attributed to inverse Compton scattering of soft photons by hot electrons heading toward the event horizon within the black hole's accretion disk.

Statistics of the fluctuations in the light curve of Cygnus X-1, observed in hard X-rays with *Ginga* and *Chandra*, exhibit complex 1/f noise spectra and size distributions of peak fluxes with very steep power law slopes of $\alpha_F \approx 7.1$ (Negoro et al. 1995; Mineshige and Negoro 1999), which have been interpreted in terms of SOC models applied to accretion disks (Mineshige et al. 1994a, 1994b; Takeuchi et al. 1995; Mineshige and Negoro 1999). However, since the standard FD-SOC model limits the valid range of size distribution slopes to $1.80 \lesssim \alpha_F \lesssim 2.33$ (Table 2.1), which is far off the observed value of $\alpha_F \lesssim 7.1$, and since the power law fitting range extends over less than a decade, an interpretation of pulses from Cygnus X-1 in terms of the FD-SOC model is not plausible.

15.3 Sagittarius (Sgr) A*

In contrast to Cygnus X-1 observations, a total of 39 X-ray flares observed with *Chandra* from Sgr A*, the black hole with 4×10^6 $M_\odot$ at the center of our Galaxy, revealed power law distributions with slopes of $\alpha_F = 1.9 \pm 0.4$ for the luminosity (of the 2-8 keV flux) and $\alpha_E = 1.5 \pm 0.2$ for the fluence (Nielsen et al. 2013), which

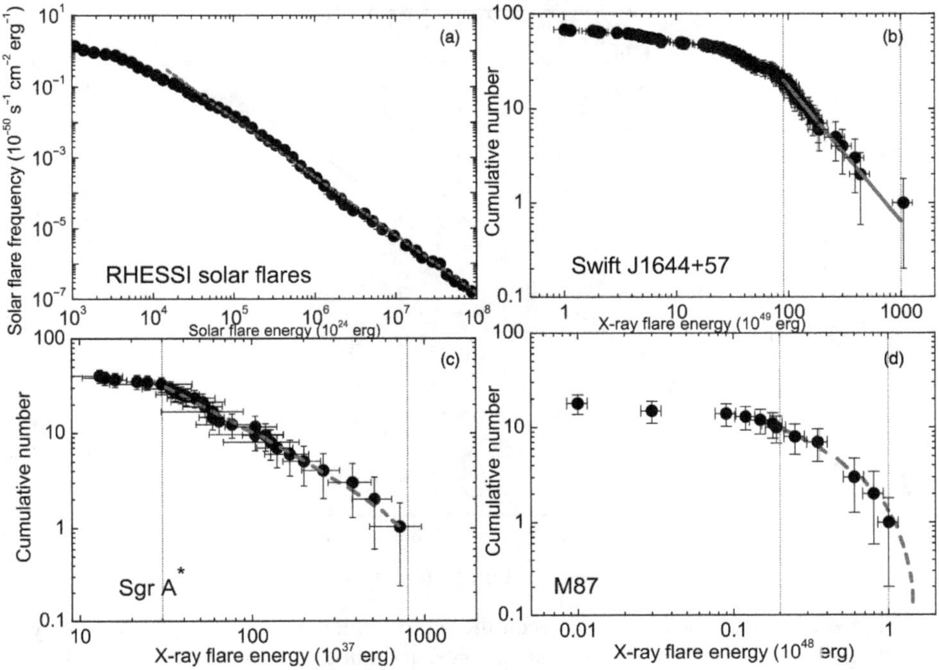

Figure 15.5 Cumulative size distributions of energy for: (a) solar flares observed by RHESSI, (b) the tidal disruption event Swift J1644+57, (c) the supermassive black hole Sgr A*, and (d) galaxy M87 with an active galactic nucleus. The observations are rendered with dots and the cumulative distribution fit with dashed curves. The fitting ranges are indicated with vertical lines (Wang et al. 2015).

is consistent with the predictions of the standard FD-SOC model ($\alpha_F = 1.80$ and $\alpha_E = 1.44$).

A systematic Chandra study (1999–2012) of Sgr A* traces dynamic, high-energy (X-ray) processes in the immediate vicinity of the supermassive black hole (Yuan et al. 2018). Power law size distributions with slopes of $\alpha_F = 1.73 \pm 0.20$ are reported for the flux F, while the theoretical prediction is $\alpha_F = 1.80$. There is no significant correlation between the flux and duration detected, which rules out the energy storage-and-dissipation model.

The galactic center black hole Sgr A* is a variable near-infrared source that exhibits bright flux excursions called “flares.” When flux from Sgr A* is detected, the light curve has been shown to exhibit red noise characteristics and the distributions of flux densities is nonlinear, non-Gaussian, and skewed to higher flux densities. It is concluded that Sgr A* has two states: the bulk of the emission is generated in a log-normal noise process with a well-defined median flux density, and this quiescent emission is supplemented by sporadic flares that create the observed power law extension of the flux distribution (GRAVITY Collaboration 2020).

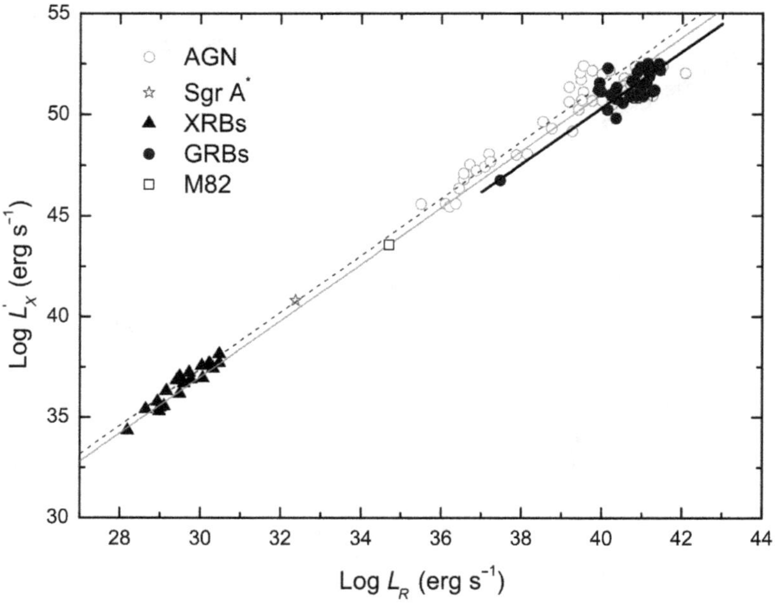

Figure 15.6 The correlation between the radio luminosity (x-axis) and the X-ray luminosity (y-axis) is shown for black holes, assuming a black hole mass of 3 $M_\odot$) (Wang and Dai 2017).

A universal behavior of X-ray flares from black hole systems has been claimed (Figure 15.5), based on the finding of consistent power law slopes in solar flare gamma-ray bursts and galactic gamma-ray bursts (Wang et al. 2015). This universal scaling law of black hole activity is further corroborated by the close correlation between radio luminosity F_R, X-ray luminosity F_X, and the black hole mass (Wang and Dai 2017), Figure 15.6.

15.4 SOC in Magnetars

Kouveliotou et al. (1999) discovered a soft gamma-ray repeater (SGR 1900 + 14) associated with a magnetar. The SGR 1900+14 exhibits a spin-down rate of 1.1×10^{-10} s^{-1} and is attributed to magnetic dipole radiation, possibly accelerated by a quiescent flux, yielding a pulsar magnetic field of $B = (2 - 8) \times 10^{14}$ [G], and confirms that SGRs are magnetars (Kouveliotou et al. 1999). The SGRs have a firm association with supernova remnants (SNRs), based on positional coincidence. For a comprehensive review of the physics behind observations of magnetars see Turolla et al. (2015). What distinguishes these transient soft gamma-ray bursters from the "classical" gamma-ray sources is their recurrence and the much softer spectra of their bursts. SGRs emit up to 40 very short sub-pulses lasting less than 10 ms each. The recent discovery of a galactic fast radio burst (FRB) associated with a hard

X-ray burst from a soft gamma-ray repeater (SGR) J1935+2154 has established the magnetar origin of at least some FRBs (Wei et al. 2021).

Frequency distributions of SGR are provided in Table (15.1) (Gogus et al. 1999; 2000; Cheng et al. 2020). The scale-free behavior of magnetars, SGRs, and FRBs has been evaluated with q-Gaussian distributions and the q-values suggest that they can be well explained within the physical framework of fractal-diffusive FD-SOC systems (Wei et al. 2021). Size distributions have been fitted with a log-normal distribution, which in fact are similar to the fitted power law size distributions.

A Bayesian model was developed for magnetars (Huppenkothen et al. 2015). It was found that the sub-burst variability does not conform to SOC models, and thus a simplified cascade model for the different trigger mechanisms has been suggested.

More recently, further magnetar observations were reported by Enoto et al. (2017), Gogus et al. (2017), Cheng et al. (2020), and Mottez et al. (2020). Broadband X-ray spectra from magnetars were found to be correlated with magnetic fields, using observations from the Suzaku Archive of SGRs and AXP combined with NuSTAR, Swift, and RXTE (Enoto et al. 2017). Burst and outburst characteristics of magnetar 4U 0142+61 are compiled in Gogus et al. (2017). Statistical properties of magnetar bursts and FRB 121102 show approximate agreement with SOC parameters of power law slope in energy, duration, and waiting times (Cheng et al. 2020), but the size distributions display significant deviations from ideal power law functions. The cause of repeating fast radio bursts is explained by small bodies that orbit around a pulsar or a magnetar (Mottez et al. 2020).

15.5 SOC in Blazars

Blazars are very compact objects associated with super-massive black holes in the center of active, giant elliptical galaxies. They represent a subgroup of active galactic nuclei (AGN), which emit a relativistic beam or jet that is aligned or nearly aligned with the line-of-sight direction to Earth. Due to this particular geometry, blazars exhibit highly variable and highly polarized emission in radio and X-ray emission. Optically violent variable (OVV) quasars are a subclass of blazars.

The optical variability of blazar GC 0109+224 was monitored from 1994 onwards and the light curve exhibited a power spectrum $P(\nu) \approx \nu^{-p}$, with $1.57 < p < 2.05$ (Ciprini et al. 2003), which is consistent with the 1/f noise or flicker noise characteristics of SOC avalanches in the BTW model (Bak et al. 1987; Hufnagel and Bregman 1992). The frequency distribution of radio peak fluxes of flaring events from blazar GC 0109+224 was found to be a power law distribution (over about one order of magnitude), $N(F) \propto F^{-1.55}$ (Ciprini et al. 2003), which is consistent with the prediction of the FD-SOC model, that is, $N(F) \propto F^{-1.80}$,

Table 15.2 *Frequency distributions measured from black-hole objects (Cygnus X-1, Sgr A*), and a blazar (GC 0109+224). One case with* $\alpha_F = (7.1)$ *is likely to be an outlier.*

Object	Waveband	Number of events	Power law slope of energies α_F	References
GC 0109+224	Optical		1.55	Ciprini et al. (2003)
Cygnus X-1	1.2–58.4 keV		(7.1)	Mineshige and Negoro (1999)
Sgr A*	2–8 keV		1.5, 1.0	Nielsen et al. (2013)
Swift J1644+57		68	2.4 ± 0.6	Wang et al. (2015)
Sgr A*		39	1.8 ± 0.6	Wang et al. (2015)
M87		18	1.6 ± 0.7	Wang et al. (2015)
CORE		14	1.7 ± 0.5	Yang et al. (2019)
HST-1		19	1.9 ± 0.3	Yang et al. (2019)
			1.67	Median
			1.80	FD-SOC prediction

within the uncertainties of the measurements. Interpreting blazars as a SOC phenomenon, the critical threshold for a pulse is given by the geometric coalignment condition between the emitted beam direction (of accelerated particles producing gyrosynchrotron emission) and the observer's line-of-sight direction from Earth. The intermittency of blazar bursts observed on Earth is believed to be caused by sporadic bursts of energy releases, created by internal shocks that occur within AGN jets.

Another statistical analysis of a blazar is reported from XMM/Newton X-ray flares of Mrk 421 (Yan et al. 2018), with power laws with slopes of $\alpha_F = \alpha_T \approx 1$. This result can be explained with a low dimensionality ($d = 1$) in our generalized SOC model, which implies linear 1-D structures in the emission region of blazars.

Probing shock acceleration in BL Lac jets through X-ray polarimetry has been described in terms of a time-dependent view (Tavecchio et al. 2020a). The distribution of fluxes of gamma-ray blazars may provide hints for a stochastic process (Tavecchio et al. 2020b). SOC avalanches have been interpreted in terms of anisotropic electron populations in BL Lac jets (Tavecchio and Sobacchi 2020c).

15.6 Active Galactic Nuclei

An active galactic nucleus (AGN) is believed to host a super-massive black hole, which may be the central engine that powers an AGN-associated jet. As an example, M87 is a radio galaxy with a central supermassive black hole with a mass of $\approx (3-6) \times 10^9 M_\odot$, and has a collimated jet misalignment of $\approx 30°$ with respect to the line-of-sight (Yang et al. 2019). Statistical analysis of X-ray flares from the nucleus and HST-1 knot in the M87 jet yield support for the SOC interpretation (Yang et al. 2019). Power laws are found to fit the cumulative size distribution of the peak intensity P and the flaring time duration T. Fits of the cumulative size distribution yield a good match to the theoretical predictions of the FD-SOC model, for both the power law and the log-normal functions.

15.6.1 Cellular Automatons

Cellular automaton models were constructed to mimic mass accretion by avalanches that are triggered when the mass density of the disk exceeds some critical value, which could reproduce the 1/f power spectra $N(\nu) \propto \nu^{-1.6}$ and produces size distributions with power law slopes of $\alpha_F = 2.8$ for energies and $\alpha_T = 1.4$ for durations (Mineshige et al. 1994a, 1994b; Yonehara et al. 1997). A BTW-related model produces a flux distribution of $\alpha_F = 1.35$ (Mineshige et al. 1994a; 1994b) that is significantly different from the standard FD-SOC prediction ($\alpha_F = 1.80$). Adding gradual diffusion to the SOC avalanches in the cellular automaton simulations produces a steeper (exponential) energy size distribution that is closer to the observations (Takeuchi et al. 1995).

Further modified cellular automaton models were developed that include reservoirs of different capacities (Negoro et al. 1995), hydrodynamic models of advection-dominated accretion disks (Takeuchi et al. 1997), relativistic effects (Xiong et al. 2000), nonlocal transport of angular momentum in terms of the kinematic viscosity of magnetic loops in the accretion disk corona (Pavlidou et al. 2001), and boson clouds around black holes (Mocanu and Grumiller 2012).

Most of the various cellular automaton models designed to mimic a physical mechanism operating in black-hole objects have difficulty to reproduce the observed steep size distributions, while most of them seem to produce 1/f power spectra without special assumptions. The observed steep size distributions may represent deviations of the accretion disk system from a pure SOC system. The notion of SOC may still be useful to understand the observations in a qualitative way,

As an alternative to power law size distributions, it was explored whether SOC accretion disks can generate a log-normal emission variability in AGN. As a result, a modified SOC model based on cellular automaton mechanism for producing log-normal flux distribution has been formulated (Kunjaya et al. 2011).

15.7 Bosons Clouds

Modern particle physics subdivides subatomic particles into bosons (photons, W^+, Z^0, Gluons, Higgs bosons), hadrons (mesons, pions, kaons, baryons, protons, neutrons), and fermions (electrons, neutrinos). A boson is a subatomic particle whose spin quantum number has an integer value (0, 1, 2, ...). The presence of bosons near a rotating black hole leads to the formation of a boson cloud around the black hole under certain conditions (Mocanu and Grumiller 2012). Once the number of bosons in the cloud becomes too large, the nonlinearities in the boson cloud lead to an unstable behavior and the cloud goes through a Bosenova-type event, which may be compared to an exponentially growing SOC avalanche event. The first condition requires that a boson exists so that its associated Compton wavelength is approximately equal to the characteristic size of the black hole (such as the Schwarzschild radius). The SOC system is slowly driven as a result of threshold dynamics, in form of bosons that are added to the boson cloud with a very slow rate, in analogy to the sand grains in the BTW cellular automaton realization.

Eroencel et al. (2019) discussed a new possible approach to the Higgs hierarchy problem. The model is in part inspired by attempts to model aspects of SOC in condensed matter systems in which it has been hypothesized that some classes of these critical states on the brink of catastrophic failure contain critical exponents that are becoming complex under loading of the system, leading to discrete scale invariance and instabilities.

In high-energy physics, it is proposed that the observed behavior is due to an evolution in global nonequilibrium, leading to SOC. At the confinement point, the initial quark-gluon medium becomes quenched by the vacuum, breaking up into all allowed free hadronic and nuclear mass states without formation of any subsequent thermal hadronic medium (Castorina and Satz 2019).

15.8 SOC in Galactic Structures

What physical mechanism produces galactic structures? A nonlinear theory was proposed in which the structure of spiral galaxies arises from percolation phase transition (Schulman and Seiden 1986a, 1986b; Seiden and Schulman 1990). The differential rotation of the galaxy triggers propagating patterns of star formation. This scenario is very similar to a SOC model, since it has a critical point at the second-order phase transition associated with the percolation threshold, which causes avalanches of star formations. Percolation processes, however, require fine-tuning, in contrast to SOC systems. The process of stochastic self-propagating star formation was simulated with a cellular automaton model that provides a representation of the percolation process operating in spiral galaxies (Seiden and Schulman 1990).

The formation of galaxies has been modeled with two opposite scenarios, the *top-down scenario* that starts with a monolithic collapse of a large cloud (Eggen, Lynden-Bell, and Sandage 1962; Zeldovich 1970), versus the now more widely accepted *bottom-up scenario*, where smaller objects merge and form larger structures that ultimately turn into galaxies (Searle and Zinn 1978; Peebles 1980). The second scenario is more widely accepted now and corresponds also closer to a SOC-driven avalanching scenario. In most models of galaxy formation, thin, rotating galactic disks result as a consequence of clustering of dark matter halos, gravitational forces and disturbances, and conservation of angular momentum. The fractal-like patterns of the universe from galactic down to solar system scales is thought to be a consequence of the gravitational self-organization of matter (Da Rocha and Nottale 2003). Fractal structures are observed throughout the universe (Baryshev and Teerikorpi 2002). It is conceivable that gravitational forces in an expanding universe lead to sporadic density fluctuations or waves that initiate a local instability of self-gravitating matter like an avalanche in a sandpile SOC model, where a critical threshold exists without need of fine-tuning.

The spatial structure of the universe exhibits fractal structures of galaxy clusters out to a redshift of $z = 0.23$ (Figure 15.7) but becomes very homogeneous and isotropic at cosmological scales of the microwave background, with inhomogeneities of $\lesssim 10^{-4}$ according to the latest results of the COBE and WMAP missions. Self-organization and fractal scaling has been applied to some large-scale structures in our universe, such as to the galactic spiral structure (Nozakura and Ikeuchi 1988), the near-Eddington accretion sources in the Seyfert 1 galaxy MCG-6-30-15 (Sivron and Goralski 1998), the extreme narrow-line Seyfert 1 (Sivron 1998), the formation of the interstellar medium (Tainaka et al. 1993), the initial mass function of starbursts (Melnick and Selman 2000), the stellar dynamics in elliptical galaxy formation (Kalapotharakos et al. 2004), or to the gravitational structure formation in general on many scales (Da Rocha and Nottale 2003).

The spatial flatness, homogeneity, and isotropy of the universe at cosmological scales can be considered as a critical point that would require an extreme fine-tuning, unless there is a self-organizing principle that creates such a special state in a natural way. Moffat (1997) proposes that the universe evolves as a SOC system (in the sense of a BTW model), where the Hubble expansion undergoes "punctuated equilibria" like the SOC scenario of intermittent evolution (Bak and Sneppen 1993). The inflationary scenario, which predicts a rapid expansion of the early universe to explain the flatness and the horizon problem, could be the manifestation of a major SOC avalanche, while a SOC scenario would predict many intermittent inflationary phases (Moffat 1997). The critical point of a cosmological system would be the critical density $\Omega = 1$ that discriminates between an open ($\Omega < 1$) and a closed ($\Omega > 1$) universe, independent of the initial conditions and without fine tuning of the parameters. A related SOC concept has also been applied to quantum gravity, with geometries measured in terms of the Compton wavelength and Planck

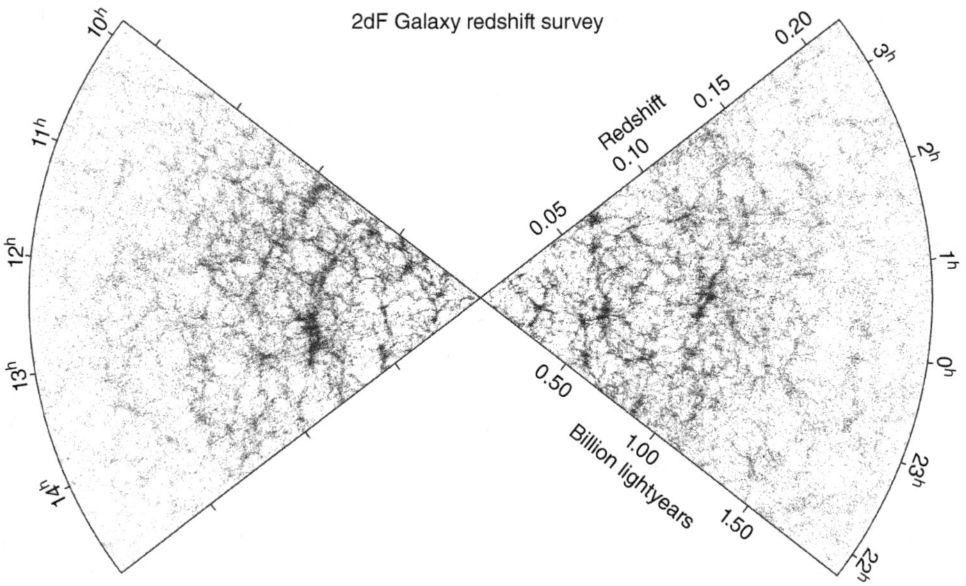

Figure 15.7 The 2dF galaxy redshift survey (2dFGRS), conducted at the Anglo-Australian Observatory, shows a map of the galaxy distribution out to redshifts of $z = 0.23$ or approximately 2 billion lightyears, which includes approximately 250,000 galaxies. Note the fractal large-scale structure of the universe that makes up the galaxy density (Colless et al. 2001).

length (Ansari and Smolin 2008). With the recent advent of string theory and multiverses, we might even consider our universe being only one single avalanche episode in a multiverse SOC scenario.

15.9 SOC and Cosmic Rays

Cosmic rays are high-energetic particles (protons, helium nuclei, or electrons) that originate from within our Milky Way, as well as from extra-galactic space, and are detected when they hit the Earth's atmosphere and produce a shower of high-energy particles (muons). The energy spectrum of cosmic rays extends over a large range of 10^9 eV $\lesssim E \lesssim 10^{21}$ eV, with an approximate power law slope of $\alpha_E \approx 3.0$ (Figure 15.8). A closer inspection reveals a broken power law with a "knee" at $E_{knee} \approx 10^{16}$ eV, which separates the cosmic rays accelerated inside our Milky Way (with a spectral slope of $\alpha_{E1} \approx 2.7$) and in extra-galactic space (with a slope of $\alpha_{E2} \approx 3.3$). The sources of cosmic rays are believed to be supernova remnants, pulsars, pulsar-wind nebulae, active galactic nuclei, and gamma-ray burst sources. The particles with higher energies ($E \gtrsim E_{knee}$) have a uniform and isotropic distribution over the sky and are believed to originate mostly from active galactic

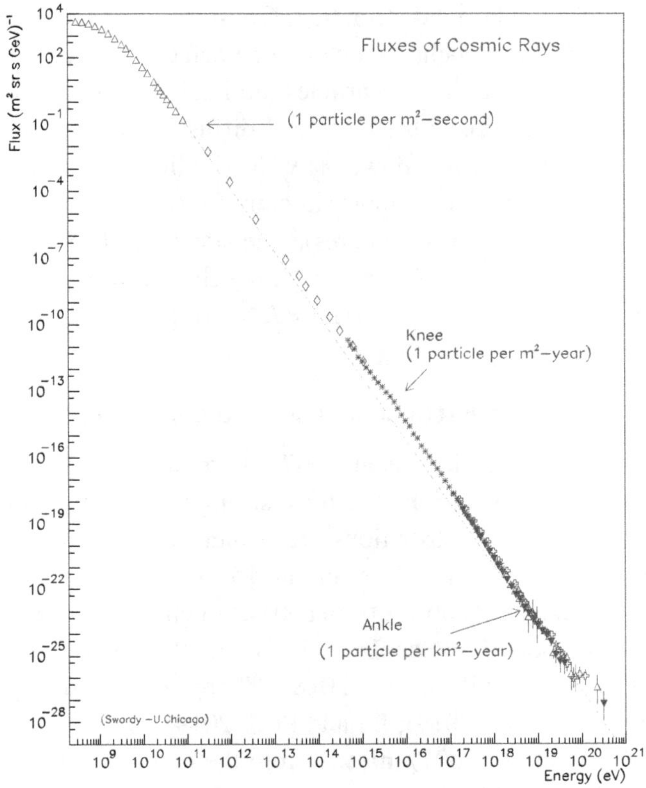

Figure 15.8 Cosmic ray spectrum in the energy range of $E = 10^9 - 10^{21}$ eV, covering 12 orders of magnitude. There is a "knee" in the spectrum around $E \approx 10^{16}$ eV, which separates cosmic rays originating within our galaxy (at lower energies) and those from outside the galaxy (at higher energies) (credit: Simon Swordy, University of Chicago).

nuclei). Localizations of gamma-ray bursts have been obtained from interplanetary networks with six near-Earth and four interplanetary spacecraft (Atteia et al. 1987).

High-energy particles can be accelerated by a number of physical mechanisms, for example, by electric fields, by shock waves, or by stochastic wave–particle interactions, such as by cyclotron resonance, which requires magnetic fields. A first-order Fermi process, as well as a fractal reconnection model, can produce the observed power law spectra of high-energy particles (Nishizuka and Shibata 2013). Cosmic rays, which travel through a large part of the universe, probably undergo many local acceleration processes, and thus their trajectories may look like a diffusive random walk. The acceleration process of cosmic rays has been interpreted in terms of a SOC process (Aschwanden 2014). The critical threshold is the "runaway regime" (e.g., Holman 1985) of a charged particle in a thermal distribution, which is a critical velocity, that is, $v_{crit} \gtrsim 4v_{th}$, that is necessary to enable efficient acceleration out of the thermal distribution. Considering the subsequent acceleration process as a SOC avalanche, which can be achieved by an arbitrary number of localized acceleration steps, the particles are likely to undergo a diffusive random walk, as it is characterized by the fractal-diffusive SOC model. The standard FD-SOC model predicts then a power law distribution for the energy spectrum of accelerated particles, which is approximately fulfilled for cosmic rays (as well as for nonthermal particles in solar flares). The standard FD-SOC model predicts an energy size distribution of $N(E) \approx E^{-1.44}$, which is however different from the observed cosmic ray spectrum, with $N(E) \approx E^{-3.0}$ (Figure 15.8).

15.10 Extragalactic Fast Radio Bursts

Fast radio bursts (FRBs) (Lorimer et al. 2007; Thornton et al. 2013) are millisecond radio pulses with large dispersion measures, strongly suggesting an extra-galactic origin. Some of their size distributions are power law-like, but deviations from straight power laws are clearly visible (Figure 15.9).

Comparisons of size distributions from FRBs, which contain cosmological information, have been reported by Scholz et al. (2016, 2017); Chatterjee et al. (2017); Spitler et al. (2018); Michilli et al. (2018); Zhang et al. (2018b); Zhang et al. (2018c); Wang and Zhang (2019); Lu and Piro (2019); Gourdji et al. (2019); Wei et al. (2021); Zhang et al. (2021), mostly from the repeating source FRB 121102 (Figs. 15.3 and 15.4). FRB burst statistics includes observations from the following radio telescopes: UTMOST, CHIME, ASKAP, Parkes, Arecibo, and GBT (Lu and Piro 2019). Flux and spectral variation characteristics of 3C 454.3 GeV at the GeV band are analyzed in Zhang et al. (2018a). The nature of 3C 454.3 may be consistent with a SOC system, similar to Sagittarius A^*, and thus the outbursts are from plasmoid ejections driven by magnetic reconnection. This may further support the idea that the jet radiation regions are magnetized.

A sample of 41 low-energy bursts from FRB 121102 were observed with the Arecibo and VLA radio telescope (Gourdji et al. 2019). A power law approximation of the cumulative distribution of burst energies yields a slope of $\alpha_E = 1.8 \pm 0.3$, which is consistent with the predictions of the standard FD-SOC model, $\alpha_V = \alpha_F = 1.80$.

A universal energy distribution is invoked for FRB 121102, but substantial deviations from straight power laws are recognizable (Figure 15.9) (Wang and Zhang 2019). Although scale-invariance is observed in the repeating FRB 121102 (Lin and Sang 2020), it is found that the cumulative distributions of fluence, flux density, total energy, and waiting time can be well fitted by broken power laws (Lin and Sang 2020). Energy and waiting time distributions of FRB 121102 were observed by FAST (Zhang et al. 2021). The authors demonstrate that the waiting

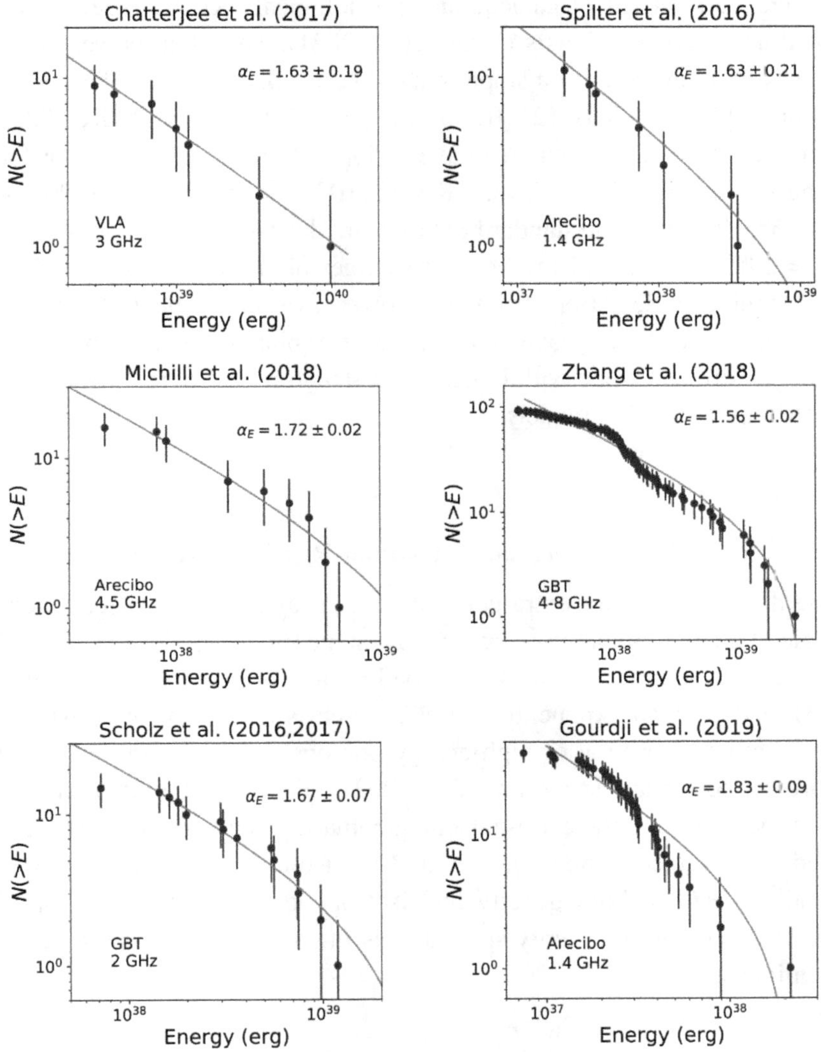

Figure 15.9 Cumulative energy distributions for different samples of FRB 121102. The observed frequencies and instruments are shown. The best-fitting power law index α_E for a power law function with a high-energy cutoff is in a narrow range $1.6 < \alpha_E < 1.8$ (Wang and Zhang 2019).

time distribution is very similar to a Poisson distribution, a log-normal distribution, and a Weibull distribution function, within a maximum uncertainty of $\lesssim 10\%$ (see Fig. 3 in Zhang et al. 2021).

The recent discovery of a galactic FRB associated with a hard X-ray burst from a soft gamma-ray repeater has established the magnetar origin of at least some FRBs (Wei et al. 2021).

The size distributions of galactic and cosmological FRB have been considered as scaled-up solar radio bursts (Wang et al. 2021). The fitted power laws of the energies were found to have a slope of $\alpha_E = 1.51 \pm 0.19$ for the solar radio bursts, $\alpha_E = 1.66 \pm 0.06$ for FRB 121102, and $\alpha_E = 1.82 \pm 0.20$ for SGR 1935+2154. The time durations were found to have a slope of $\alpha_T = 1.69 \pm 0.02$ for the solar radio bursts, $\alpha_T = 1.57 \pm 0.11$ for FRB 121102, and $\alpha_T = 1.78 \pm 0.06$ for SGR 1935+2154. Following the standard FD-SOC model, we expect values of $\alpha_E = 1.44$ and $\alpha_T = 2.00$ (Table 2.1). However, the accuracy of the power law fits is ill-defined because significant deviations from pure power laws are visible in Figures 2 and 3 of Wang et al. (2021). Improvements in the determination of the power law slope could be possible obtained with larger event statistics or different (nonpower law) size distribution function models.

15.11 Extragalactic Gamma-Ray Background

The size distribution of the extra-galactic gamma-ray background generated by the ensemble of all extra-galactic sources has been measured with the Fermi-LAT telescope, and was found to be well described by a power law function (with slope $\alpha \approx 2.3$), and with an exponential cutoff, which is consistent with the existence of the effect of absorption of high-energy photons, assuming an emission that is an unbroken power law spectrum (Lipari 2021). Fermi-LAT measured three estimates of the total flux: the extra-galactic gamma-ray background, the ensemble of resolved extra-galactic point sources, and the isotropic gamma-ray background, which all reflect the homogeneity and isotropy of the universe (Lipari 2021). The size distribution (or energy spectrum) is fitted by a power law function with exponential cutoff,

$$\Phi_\gamma(E) = K_\gamma E^{-\alpha} \exp^{-E/E_{cut}}, \tag{15.1}$$

yielding best-fit parameters of $\alpha = 2.30 \pm 0.02$, $E_{cut} = 330 \pm 70$ GeV.

The standard FD-SOC model predicts a power law slope of $\alpha_A = 7/3 \approx 2.33$ for area-like structures with constant line-of-sight depth, and $\alpha_V = 9/5 \approx 1.80$ for isotropic volume-like structures. Since the extra-galactic gamma-ray background is assumed to be isotropic in all directions, the resulting line-of-sight depth in each direction is invariant and the relevant power law slope is $\alpha_A \approx 2.33$, while the assumption of an isotropic geometry for extra-galactic gamma-ray background fluctuations would predict $\alpha_V = 1.80$. It is possible that relativistic effects could explain this geometry (Xiong et al. 2000).

15.12 Summary

Among black-hole systems, we find a variety of systems with applications of SOC, such as: soft gamma-ray repeaters, magnetars, blazars, black holes in accretion disks, and galactic fast radio bursts. Gamma-ray bursts (GRBs), soft gamma-ray repeaters (SGRs), as well as black hole objects are found to be selfconsistent with the theoretical prediction of the FD-SOC model ($\alpha_F = 9/5 = 1.80$).

Power law fits are not always superior to fits with a log-normal function or a Weibull function. The reason for this discrepancy between observed and modeled power law slopes in stellar (or galactic) SOC systems is mostly due to small-number statistics of the samples ($N \lesssim 10 - 100$), incomplete sampling near the lower threshold, and due to ill-defined power law fitting ranges, which can cause significant deviations from ideal power laws. Nevertheless, the size distributions of stellar flares and galactic systems is similar to those of solar flares in most cases, and thus largely fit the predictions of the FD-SOC model.

Galactic phenomena that possibly have some characteristics in common with SOC models are: (i) fractal galaxy distributions (out to redshifts of $z \lesssim 0.23$); (ii) cosmic ray energy spectrum; (iii) extra-galactic FRB; and (iv) extra-galactic background fluxes.

(i) The fractal structures of galaxy distributions has been modeled in terms of a percolation process, which requires fine-tuning, in contrast to SOC.
(ii) The cosmic ray energy spectrum has been found to fit a double power law distribution, but produces a mean slope ($\alpha_L \approx 3.0$) that agrees with isotropic 3-D (d = 3) structures of the standard FD-SOC model.
(iii) The size distribution of extra-galactic fast radio bursts fits a power law function with a slope of $\alpha_F \approx 1.80$ for the fluxes in the standard FD-SOC model.
(iv) The extra-galactic background flux is isotropic and fits a power law function of $\alpha_A \approx 2.33$, which implies area-like 2-D structures in the FD-SOC model, while the line-of-sight depth is constant.

Part III

Conclusions

16

Conclusions

In this final chapter, we are drawing conclusions of our new physical understanding of avalanche processes in astrophysical systems, governed by the principle of self-organized criticality (SOC). Most fundamentally, a theory or a theoretical model can only justify its existence if it is able to make predictions that are consistent with observed data. According to some scientific philosophers (such as Karl Popper and Kurt Gödel), no theory can be proven, but can only be falsified. In practice, a theory should predict a quantity that can be measured from observations by the numerical values of the corresponding physical parameters, where quantitative comparisons are preferred over qualitative observations. This is exactly the aim of this monograph, where we use quantitative SOC parameters to prove or to falsify self-consistency between theory and observations.

In the present case, we claim the following six testable hypotheses:

(1) Size distributions: The occurrence frequency distribution function (or size distribution function) follows a power law (or power law-like) function for the statistics of SOC events, that is, $N(x) \propto x^{-\alpha_x}$.

(2) Physical scaling laws: The underlying five physical scaling laws are: (i) the mean value of the area-like multifractal dimension $D_A = D_2 = 1.5$ for dimensionality d = 2; (ii) the mean value of the volume-like multifractal dimension $D_V = D_3 = 2.5$ for dimensionality d = 3; (iii) the scaling law of classical diffusion, $T \propto L^{2/\beta}$ for $\beta = 1$; (iv) the volume-flux scaling law, $F \propto V^{\gamma}$, for $\gamma = 1$, and (v) the fluence or energy scaling law, $E = F \times T = L^{(D_V\gamma+2/\beta)}$.

(3) Power law slopes: The power law slopes of various physical SOC parameters have predictable values α_x for each given SOC parameter $x = L, A, V, T, F, E$, such as the slope $\alpha_L = 3$ for the length scale L, the slope $\alpha_A = 7/3 \approx 2.33$ for the event area A, the slope $\alpha_V = 9/5 = 1.8$ for the event volume V, the slope $\alpha_T = 2$ for the time duration T, the slope $\alpha_F = 9/5 = 1.8$ for the flux F, and the slope $\alpha_E = 13/9 \approx 1.44$ for the fluence or energy E);

(4) Energy integral: The energy integral is predicted to converge at the largest energies of the size distribution, which indicates that the largest events are energetically most dominant in the energy budget of the energy size distribution, while the energy contributions of nanoflares is negligible.

(5) Universality: The power law slopes α_x have universal validity, as it was originally suggested by Per Bak.

(6) Paradigm shift: The original Bak, Tang, and Wiesenfeld (BTW) model based on the cellular automaton concept undergoes here a paradigm shift to the *fractal-diffusive FD-SOC* model, without losing its characteristic behavior of SOC avalanche dynamics driven by a system-wide threshold and nonlinear instability.

16.1 Comparing Observations with Theory

In Table 16.1, we present a summary of available quantitative results. The seven columns contain selected astrophysical phenomena (column 1), the observational instruments, mostly from spacecraft (column 2), the median power law slopes $\alpha_{\mathrm{x,median}}$ of the observed size distributions (column 3), the theoretically predicted slopes $\alpha_{\mathrm{FD-SOC}}$ for six SOC parameters (column 4), the Euclidean dimension d of the FD-SOC model (column 5), the normalized difference between the observed median value α_{median} and the theoretically predicted value $\alpha_{\mathrm{FD-SOC}}$ (column 6),

$$\sigma_x = \frac{(\alpha_{\mathrm{median}} - \alpha_{\mathrm{FD-SOC}})}{\alpha_{\mathrm{FD-SOC}}}, \tag{16.1}$$

and literature references for each median case (column 7). Table 16.1 contains 30 datasets, which are subdivided into 5 groups according to their SOC parameters: the power laws of the flux F, the fluence or energy E, the time duration T, the area A, and the length scale L. This dataset of 30 size distributions is drawn from a larger amount of 162 datasets (Aschwanden 2022a; 2022b) that is detailed in the sections for each observation (such as in Tables 4.1, 5.1, ..., 15.1). For example, Table 4.1 contains 20 different observed size distributions of solar flare hard X-ray events, where the median value is $\alpha_{\mathrm{F,median}} = 1.74$ (Crosby et al. 1993), the theoretically predicted value is $\alpha_{\mathrm{F,FD-SOC}} = 1.80$, and the observed power law slopes (Table 4.1) extend over a range of $\alpha_F = 1.58 - 2.00$. The normalized difference for this case is $\sigma_F = (1.74 - 1.80)/1.80 = -0.03$, which means that the observation and the theoretical model agree within 3%. From all 13 cases listed (of the flux) in column 6 of Table 16.1, we find that 11 cases agree within 10% accuracy or uncertainty. Half of the cases agree within 5%. The median value α_{median} is a very robust measure of a representative value (among a range of values), being largely

Table 16.1 *Synopsis of astrophysical phenomena and observed size distributions with power law slopes $\alpha_{\rm median}$, juxtaposed to the theoretical predictions of the FD-SOC model. The 7 columns contain: (1) observed phenomenon; (2) instrument; (3) median values of observed power law slopes; (4) power law slopes of theoretically predictions of FD-SOC model; (5) Euclidean dimension; (6) normalized difference σ_x (Eq. 16.1), which is a measure of the agreement between observations and theory; and (7) references.*

Phenomenon	Instrument	Power law slope $\alpha_{\rm median}$	Power law slope $\alpha_{\rm FD-SOC}$	Eucl. dim. d	Diff.	Ref.
Flux F:						
Solar flare HXR	HXRBS/SMM	1.74	$\alpha_F = 1.80$	3	− 0.03	[1]
Solar flare SXR	GOES	1.83	$\alpha_F = 1.80$	3	+ 0.02	[2]
Solar nanoflares	EIT,TRACE	1.74	$\alpha_F = 1.80$	3	− 0.03	[3]
Solar photosphere	SOT,MDI	1.85	$\alpha_F = 1.80$	3	+ 0.03	[4]
Solar radio bursts	NOAA,NGDC	1.80	$\alpha_F = 1.80$	3	+ 0.00	[5]
Solar CMEs	LASCO/SOHO	1.98	$\alpha_F = 1.80$	3	− 0.08	[6]
SEPs	IMP,OGO	1.32	$\alpha_F = 1.67$	2	− 0.27	[7]
Auroras	UVI/POLAR	1.80	$\alpha_F = 1.80$	3	+ 0.00	[16]
Stellar flares	UBV	1.99	$\alpha_F = 1.80$	3	+ 0.11	[10]
Stellar flares	XUV	2.08	$\alpha_F = 1.80$	3	+ 0.16	[11]
Stellar flares	Kepler	1.93	$\alpha_F = 1.80$	3	+ 0.07	[12]
Soft gamma-repeaters	BATSE/CGRO	1.67	$\alpha_F = 1.80$	3	− 0.07	[13]
Sgr A*	CORE/HST	1.75	$\alpha_F = 1.80$	3	− 0.03	[14]
Fluence E:						
Solar flare HXR	HXRBS/SMM	1.53	$\alpha_E = 1.44$	3	+ 0.06	[1]
Solar flare SXR	GOES	1.75	$\alpha_E = 1.44$	3	+ 0.22	[2]
Solar CMEs	SOHO/LASCO	2.00	$\alpha_E = 1.44$	3	− 0.39	[6]
SEPs	IMP,OGO	1.32	$\alpha_E = 1.28$	2	− 0.08	[7]
Solar wind	ACE,WIND	1.67	$\alpha_E = 1.44$	3	+ 0.16	[8]
Auroras	UVI/POLAR	1.62	$\alpha_E = 1.44$	3	+ 0.13	[16]
Time duration T:						
Solar flare HXR	HXRBS/SMM	2.20	$\alpha_T = 2.00$	3	+ 0.10	[1]
Solar flare SXR	GOES	2.02	$\alpha_T = 2.00$	3	+ 0.01	[2]
Solar nanoflares	EIT,TRACE	2.20	$\alpha_T = 2.00$	3	+ 0.10	[3]
Solar CMEs	SOHO/LASCO	2.87	$\alpha_T = 2.00$	3	− 0.08	[6]
Auroras	UVI/POLAR	2.27	$\alpha_T = 2.00$	3	+ 0.14	[16]
SEPs	IMP,OGO	1.32	$\alpha_F = 1.50$	2	− 0.34	[7]
Solar wind	ACE,WIND	2.20	$\alpha_T = 2.00$	3	+ 0.10	[8]
Event area A:						
Solar nanoflares	EIT,TRACE	2.49	$\alpha_A = 1.67$	3	+ 0.07	[3]
Auroras	UV	1.85	$\alpha_A = 1.67$	3	− 0.21	[16]

Table 16.1 *(cont.)*

Phenomenon	Instrument	Power law slope $\alpha_{\rm median}$	Power law slope $\alpha_{\rm FD-SOC}$	Eucl. dim. d	Diff.	Ref.
Length scale *L*:						
Cosmic rays	Chicago Univ.	3.00	$\alpha_L = 3.00$	1	+ 0.00	[15]
Planetary systems	Ranger 7,8,9	2.96	$\alpha_L = 3.00$	1	− 0.01	[17]

References: [1] = Crosby et al. (1993); [2] = Aschwanden and Freeland (2012); [3] = Aschwanden and Parnell (2002); [4] = Parnell et al. (2009); [5] = Nita et al. (2002); [6] = D'Huys et al. 2016; [7] = Kahler (2013); [8] = Moloney and Davidsen (2011); [9] = Uritsky et al. (2008); [10] = Lacy et al. (1976); [11] = Aschwanden and Güdel (2021); [12] = Aschwanden and Güdel (2021); [13] = Cheng et al. (2020); [14] = Wang et al. (2015); [15] = Simon Swordy, University Chicago; [16] = Uritsky et al. (2002, 2008); Kozelov et al. (2004); [17] = Cross (1966).

insensitive to various outliers that occur for a number of reasons, such as the choice of the power law fitting range, the background subtraction, small number statistics, and different event detection methods. In particular, the median value is less sensitive to outliers than the mean and standard deviation.

16.2 Size Distributions

It is widely agreed by the science community that **power law distribution functions** are the hallmarks of SOC. So we might ask whether power law functions are necessary and sufficient to describe SOC size distributions. Strictly speaking, power law functions are neither necessary nor sufficient for detecting SOC behavior because large deviations from straight power laws have been encountered in many observations presented here, while power laws outside the validity range of SOC behavior, ($1 \le \alpha_x \le 3$), have been detected in the fractal-diffusive SOC (FD-SOC) model also. Mathematically, a large number of nonpower law functions exist and have been applied to SOC data, such as the **log-normal function** (Johnston and Romani 2002; Cairns et al. 2004; Abramenko and Yurchyshyn 2010c; Kunjaya et al. 2011; Abramenko 2014; Alipour and Safari 2015; Verbeek et al. 2019; Zhang et al. 2021), the **Weibull distribution function** (Shapoval et al. 2018; Telloni et al. 2014; Love 2020, 2021; Li et al. 2018; Zhang et al. 2021; **Pareto-type functions** (Aschwanden and McTiernan 2010; Aschwanden and Freeland 2012; Aschwanden and Dudok de Wit 2021); the **Tsallis nonextensive statistics** (Pavlos et al. 2012); **q-Gaussian distributions** (Wei et al. 2021); and even **Gaussian** and the **exponential function** within small ranges, which all have a close similarity on a log–log scale, especially in their "fat tails."

While any choice of this vast arsenal of mathematically available functions is arbitrary, unless a physical model justifies its choice, there exist very few studies that investigate the comparison between individual mathematical distribution functions. An example is presented in Figure 15.4, where log-normal distributions are contrasted with a double power law function. Ultimately, only physical models can justify the choice of a particular size distribution function. In our treatment here, the following two models are based on a physical model: (i) the exponential function (an approximation to the Poissonian function that is used for describing random processes), and (ii) the FD-SOC model, which is the main subject of this monograph. We will refer to the **"standard FD-SOC" model** with an Euclidean dimension of $d = 3$ but call it the **"generalized FD-SOC" model** when the dimension d is a free variable, which allows $d = 1, 2, 3$. Since the FD-SOC model assumes power law distribution functions to begin with, we cannot test power law versus nonpower law alternatives. However, the fact that most of the observed power law slopes fit the FD-SOC model within a few percents (Table 16.1), we can confirm the existence of power law distribution functions within this margin of uncertainty.

More importantly, the validity of a SOC model requires a power law slope that is within the range predicted by the model, that is, ($1 \le \alpha_x \le 3$) for the general SOC model, and the fitted power law range (also called "inertial range") has to extend at least over one or two decades). A poignant example is the power law behavior of pulsar glitches (Section 14.5 and Table 14.3). Although we compiled several published references of pulsar glitch events (Argyle and Gower 1972; Lundgren et al. 1995; Cairns 2004; Melatos et al. 2008), most of these datasets do not pass the criteria of a valid power law range ($1 \le \alpha_x \le 3$) and a sufficiently large inertial range $(x_{max}/x_{min}) \gtrsim 10^2$). This example demonstrates that the observed power law slopes in pulsar glitches cannot be modeled with the FD-SOC model, and thus a different physical model is required. However, most of the astrophysical phenomena listed in Table 16.1, which amount to 11 (out of 13) for the flux parameter α_F, satisfy the valid-range criterion and the inertial-range criterion. This example clearly demonstrates that not all power law size distributions are compatible with a SOC model, such as the phenomenon of pulsar glitches, or the data analysis in terms of power law fits (in the published data) is flawed due to other reasons (background subtraction, choice of fitting range, etc.).

16.3 Physical Scaling Laws

The theoretical FD-SOC model is based on four fundamental physical scaling laws: the definition of multifractals, classical diffusion, the flux–volume correlation, and the fluence definition.

16.3.1 Multifractals

The first scaling law is just the definition of the **fractal (Hausdorff) dimension** D_d, which is $D_A = D_2 = 1.5$ for area-like features with $d = 2$, and $D_V = D_3 = 2.5$ for volume-like features with $d = 3$, respectively. This definition is simply the mean value $D_d = (D_{d,min} + D_{d,max})/2$, which contains the maximum possible fractal dimension and coincides with the Euclidean dimension, $D_{d,max} = d$, and the minimum possible fractal dimension, which is the next lower Euclidean dimension value $D_{d,min} = (d - 1)$. Note that this definition is slightly different from an earlier definition with $D_{d,min} = 1$ (Aschwanden 2012). Nevertheless, since the fractal dimension has a spread of one decade $D_{d,max} - D_{d,min} = 1$, it corresponds to a **multifractal definition**, rather than a **monofractal dimension**.

Multifractal concepts have been postulated in other SOC-related studies (Chang 1999a, 1999b; Macek 2010; Abramenko and Yurchyshyn 2010c; Rypdal and Rypdal 2011; Georgoulis 2011, 2012; Kelty-Stephen et al. 2013; Garcia-Marin et al. 2013; Abramenko 2005, 2014; Giorgi et al. 2015; Golovko and Salakhutdinova 2015; DeFreitas et al. 2017, 2019a, 2019b; Giannattasio and Consolini 2021). The multifractal approach therefore appears to be more consistent with observations than the earlier monofractal approach.

16.3.2 Classical Diffusion

The second scaling law represents the classical diffusion process, which links the spatial length scale L to the time (duration) scale T, expressed with the scaling law, that is, $T \propto L^{2/\beta}$, where β is the diffusion coefficient. This definition of the diffusion coefficient corresponds to subdiffusion for $\beta < 1$, to classical diffusion for $\beta = 1$, to hyper-diffusion or Lévy flight for $\beta > 1$, and to linear propagation for $\beta = 2$. Thus, this diffusion coefficient is chosen to be $\beta = 1$ for the standard FD-SOC model, and $1 \lesssim \beta \lesssim 2$ for the generalized FD-SOC model. Note that this spatiotemporal relationship used here in the FD-SOC model replaces the complex kinematics of next-neighbor interactions in the original BTW model of Bak et al. (1987). In other words, we transform a microscopic cellular automaton process to a macroscopic physical diffusion process. The justification for this transformation is the fact that the resulting size distributions are equivalent in the two approaches, in the sense that we obtain essentially identical power law slopes in the size distributions obtained with the methods of a (microscopic) cellular automaton or a (macroscopic) diffusion process.

There are several studies that analyze the diffusion process in solar data, mostly interpreted in terms of the photospheric magneto-convection process (Lawrence

1991; Lawrence and Schrijver 1993; Cadavid et al. 1999; Grigolini et al. 2002; Bergman 2007; Aschwanden 2012a, 2012b; Giannattasio et al. 2013, 2014a).

16.3.3 The Flux–Volume Correlation

The third physical scaling law that we make use of is the volume–flux scaling law, that is, $F \propto V^{\gamma} \propto L^{D_V \gamma}$, which essentially assumes a correlation between the observed flux of an electromagnetic signal (or possibly another observable signal) F and the size of the emitting volume V. If the emission is constant in every volume element, the flux–volume relationship would be exact, which however is not necessary in the FD-SOC model here. The flux–volume relationship could be nonlinear, in which case the coefficient exceeds $\gamma \geq 1$, as it can be modeled with the "generalized FD-SOC model," while the setting to unity $\gamma = 1$ defines the "standard FD-SOC model."

In the plasma physics of solar radio emission, one generally distinguishes between incoherent emission ($\gamma = 1$) and coherent emission ($\gamma > 1$), where the latter emission type exhibits exponential growth and produces power law size distributions (see derivation from first principles in Section 2.6). A number of electromagnetic emission mechanisms with incoherent behavior includes free–free bremsstrahlung, gyroresonance emission, gyrosynchrotron emission, while electromagnetic emission mechanisms with coherent behavior include type III plasma emission, loss-cone emission, or microwave amplification by stimulated emission of radiation (MASER) emission.

The multifractal approach represents a refined scaling law of the flux–volume relationship $F \propto L^{D_V} \approx L^{2.5}$, while 2-D imaging data yield $F \propto L^{D_A} \approx L^{1.5}$, respectively. The validity of these relationships can be proven by comparing the power law slopes α_x with $x = L, A, V, F$, which is essentially accomplished with the results given in Table 16.1. Other studies are enumerated in Section 16.3.1 on multifractals. Note that the power law slopes for the event volume and the flux is identical ($\alpha_V = \alpha_F = 1.8$) for incoherent emission ($\gamma = 1$).

16.3.4 The Fluence

The fluence is generally defined by the time integral of the flux $f(t)$ over some time interval T, so that the total flux of an event amounts to $E = \int_0^T f(t)\, dt$. This exact fluence definition can be approximated by multiplication of the peak flux F with the event duration T, so that $E = F \times T = L^{(D_V \gamma + 2/\beta)}$. The motivation to calculate this parameter E is to include space and time variables in a common spatiotemporal parameter. The flux F is defined in terms of an energy flux and has the physical units of [energy/time], while the total emitted or radiated (electromagnetic) signal has the physical unit of [energy]. The multiplication of the flux F with time T mimics the time integration of the flux, and thus contains the total emitted energy of a SOC avalanche event. In the standard FD-SOC model the expression for the energy reduces to $E = F \times T = L^{(D_V+2)} = L^{9/2} = L^{4.5}$, for $\gamma = 1$, $\beta = 1$, and $D_V = 2.5$.

The measurement of the fluence is available for six datasets in published studies (second group of datasets in Table 16.1).

16.3.5 Waiting Time Distribution

Besides the widely used size distribution functions, the so-called waiting time distribution function provides useful constraints also, which are defined by the rank-order of monotonically sorted events, that is, $\tau_{i-1} < \tau_i < \tau_{i+1}, \tau_i = \tau_2, ..., \tau_{n-1}$ (see Section 2.5). If the statistics of SOC avalanche events is a random process, the resulting constraint is that the waiting time distribution follows a Poissonian process, which can be approximated by an exponentially decreasing function, $N(\tau) \propto \exp^{-\tau/\tau_0}$. This exponential size distribution function has been used by some authors as an argument for a random process in some SOC datasets (Li et al. 2018; Wheatland 2000a; Lepreti et al. 2001; Grigolini et al. 2002).

In the observed Sun, however, flare events occur in active regions, which have different locations and different magnetic field strengths, leading to nonstationary waiting time distribution functions (Wheatland 2001; 2000b), but depend on the flare rate function. We calculated exact solutions of the waiting time distribution and obtained the following expression (Section 3.3; Eq. 3.22),

$$\alpha_\tau = 2 + \frac{1}{p}, \tag{16.2}$$

where p is the nonlinearity index, which has a value of $p = 1$ for linear flare rates, and a typical value of $p \approx 2$ for (quadratically) growing flare rates $\lambda(t) \propto t^p$. This implies a power law slope of $\alpha_\tau = 3$ for $p = 1$, $\alpha_\tau = 2.5$ for $p = 2$, and $\alpha_\tau = 2$ for $p = \infty$. Thus the validity range for waiting time slopes is $\alpha_\tau = [2, 3]$, which is a useful constraint for the waiting time distribution functions. Consequently, although we have a quantitative understanding of the waiting time distribution function, it does not allow us to make an accurate prediction due to its dependence on the nonlinearity index p. A comparison of the power law function and the Pareto approximation is shown in Figure 3.2.

16.4 Power Law Slopes

The power law slopes of various physical SOC parameters have predictable values α_x for each given SOC parameter $x = L, A, V, T, F, E$, such as the slope $\alpha_L = 3$ for the length scale L, the slope $\alpha_A = 7/3 \approx 2.33$ for the event area A, the slope $\alpha_V = 9/5 = 1.8$ for the event volume V, the slope $\alpha_T = 2$ for the time duration T, the slope $\alpha_F = 9/5 = 1.8$ for the flux F, and the slope $\alpha_E = 13/9 \approx 1.44$ for the fluence or energy E). These theoretically predicted values of power law slopes are tabulated in column 4 of Table 16.1, which can be compared with the median values of all observed size distributions listed in column 3 of Table 16.1. Interestingly, all observed power law slope fits occur in the valid range of $1 \le \alpha_x \le 3$, except for the phenomenon of pulsar glitches (Section 15.4 and Table 15.3). Thus, the FD-SOC model can predict all analyzed size distributions, with exception of a single case that is indicative of a different physical mechanism.

16.5 The Energy Integral

The total integral (or first moment) of a power law distribution function, for instance, the total energy of an occurrence frequency distribution of energies, can be obtained by convolving the variable x with the power law distribution $N(x) = (\alpha - 1)x^{-\alpha}$ over the valid range $x_{min} \le x \le x_{max}$,

$$x_{tot} = \int_{x_{min}}^{x_{max}} x\, N(x) dx = \int_{x_{min}}^{x_{max}} (\alpha - 1) x^{1-\alpha} dx \tag{16.3}$$

$$= \left(\frac{\alpha - 1}{2 - \alpha} \right) \left[x_{max}^{2-\alpha} - x_{min}^{2-\alpha} \right], \tag{16.4}$$

which is only defined for $\alpha \neq 2$. Generally, both boundaries contribute significantly to the total number, unless the power law distribution extends over a very large range, in which case the following approximations can be used,

$$x_{tot} \approx \left(\frac{\alpha - 1}{2 - \alpha} \right) \begin{cases} x_{min}^{2-\alpha} & \text{for } (x_{max} \gg x_{min}) \text{ and } (\alpha > 2) \\ x_{max}^{2-\alpha} & \text{for } (x_{max} \gg x_{min}) \text{ and } (\alpha < 2) \end{cases} . \tag{16.5}$$

The critical value is $\alpha = 2$, which decides whether the integral diverges at the lower bound x_{min} (if $\alpha > 2$) or upper bound x_{max} (if $\alpha < 2$). A far-reaching application of this integral is the total energy contained in the distribution of solar or stellar flares, which is also responsible for heating of the solar or stellar corona, and could be dominated by nanoflares if $\alpha > 2$ applies over a large energy range, as pointed out by Hudson (1991).

Reviewing Table 16.1, we see that the power law size distributions of fluxes F and energies or fluences E are substantially below the critical value, that is, $\alpha_F = 1.8$ and $\alpha_E = 1.44$, and thus the energy budget is dominated by the largest flare events, rather than by nanoflare events.

16.6 Universality

In the textbook *"Self-organized criticality"* by Pruessner (2012), which mostly deals with the cellular automaton approach of SOC systems, the property of universality is described as:

> ... if SOC were generally to be found across a large class of systems, then universality would unify these systems and their internal interactions could be identified and studied. They dominate the long time and large scale behaviour, just like in ordinary critical phenomena with a broken symmetry and could be studied in a simplified experiment on the laboratory scale or a simple computer model, rather than attempting to model the detailed, and perhaps insuperably complex microphysics ... even if SOC is less common than initially expected, it may still help to elucidate the nature of the critical state in transitional critical phenomena. It might even serve as a recipe to make these systems self-tune to a transition.

In this monograph, we demonstrate that the power law slopes of size distributions are invariant in most of the analyzed (nonlinear) phenomena. For instance, the power law slope $\alpha_F = 1.80$ agrees with the observed slopes for solar flares, photospheric transients, coronal mass ejections, solar energetic particles, auroras, stellar flares, gamma-ray bursts, soft gamma-ray repeaters, and bursts from black-hole candidates (see first group of phenomena in Table 16.1. Thus the universality covers not only a large range of astrophysical phenomena analyzed here, but extends also to other disciplines, such as geophysics, biophysics, and sociophysics (see textbook by Aschwanden 2011).

16.7 Paradigm Shift

The classical paradigm of a SOC process has been described in terms of the sandpile toy-model, which more generally is also known as cellular automaton (BTW) model. Sand grains are dropped on the "paradigmal" sandpile, which then triggers large or small avalanches, once a local instability triggers an event. The kinematics or dynamics of sand avalanches can be simulated with next-neighbor interactions that keep the sandpile in balance. The cycle of triggering an event and restoring of the sandpile represents a self-organizing mechanism, which becomes critical at times, and is thus called a "self-organized criticality" model. This process is so intuitive that a single mathematical equation is sufficient to describe the SOC dynamics, as it served as a paradigm over the last 38 years (Pruessner 2012).

In an attempt to obtain a better physical understanding of the SOC phenomenon, the "standard fractal-diffusive self-organized criticality model" (FD-SOC) has been developed over the last decade, beginning with a formulation of a SOC process in terms of physical scaling laws that are universally valid, as demonstrated with comparisons of the observed power law size distributions applied to astrophysical phenomena, with theoretical predictions based on physical scaling laws. This represents a shift from the well-established SOC paradigm. The microscopic cellular automaton concept is not necessary in this new approach, because the complex lattice-grid kinematics can be modeled with a macroscopic fractal-diffusive transport mechanism that reproduces the same observables, in the statistical mean.

References

Abramenko, V. I., Yurchyshyn, V. B., Wang, H., Spirock, T. J., and Goode, P. R. 2003, Signature of avalanche in solar flares as measured by photospheric magnetic fields, *ApJ* 597, 1135–1144.

Abramenko, V. I. 2005, Multifractal analysis of solar magnetograms, *SoPh* 228, 29–42.

Abramenko, V. I., Yurchyshyn, V., Goode, P., and Kilcik, A. 2010a, Statistical distribution of size and lifetime of bright points observed with the new solar telescope, *ApJ* 725, L101–L105.

Abramenko, V. I. and Yurchyshyn, V. 2010b, Magnetic energy spectra in solar active regions, *ApJ* 720, 717–722.

Abramenko, V. I. and Yurchyshyn, V. 2010c, Intermittency and multifractality spectra of the magnetic field in solar active regions, *ApJ* 722, 122–130.

Abramenko, V. I., Yurchyshyn, V. B., Goode, P. R., Kitiashvili, I. N., and Kosovichev, A. G. 2012, Detection of small-scale granular structures in the quiet sun with the new solar telescope, *ApJ* 756, L27.

Abramenko, V. I., Zank, G. P., Dosch, A., Yurchyshyn, V. B., Goode, P. R., Ahn, K., and Cao, W. 2013, Characteristic length of energy-containing structures at the base of a coronal hole, *ApJ* 773, 167.

Abramenko, V. I. 2014, The multifractal nature of solar magnetism and the solar dynamo problem, *Geomagnetism and Aeronomy* 54, 892.

Abramenko, V. I. 2015, Possibilities of predicting flare productivity based on magnetic field power spectra in active regions, *Geomagnetism and Aeronomy* 55, 860.

Abramenko, V. I. and Yurchyshyn, V. B. 2020, Analysis of quiet-sun turbulence on the bases of SDO/HMI and Goode solar telescope data, *MNRAS* 497, 54.

Akabane, K. 1956, Some features of solar radio bursts at around 3000 Mc/s, *PASJ* 8(3–4), 173–181.

Akopian, A. A. 2012a, Frequency distribution of X-ray flares for low-mass young stellar objects in the Orion nebula, *Astrophysics* 55(4), 505–514.

Akopian, A. A. 2012b, Frequency distribution of X-ray flares for young stellar objects in the region of ρ Oph, *Astrophysics* 55(1), 81–91.

Akopian, A. A. 2013, Frequencies of superflares in solar-type stars detected by the Kepler Orbital Observatory, *Astrophysics* 56(4), 488–500.

Aletti, V., Velli, M., Bocchialini, K., Einaudi, G., Georgoulis, M., and Vial, J. C. 2000, Microscale structures on the quiet sun and coronal heating, *ApJ* 544, 550–557.

Alipour, N. and Safari, H. 2015, Statistical properties of solar coronal bright points, *ApJ* 807, 197 (9p).

Alipour, N., et al. 2022, Automatic detection of small-scale EUV brightenings observed by the Solar Orbiter/EUI, *AA* 663, id. A128, 12pp.

Althukair, A. K. and Tsiklauri, D. 2023, Main sequence star super-flare frequency based on Enire KEPLER data, *Research in Astronomy and Astrophysics* 23(8), id. 085017, 21pp.

Anastasiadis, A., Gontikakis, C., Vilmer, N., and Vlahos, L. 2004, Electron acceleration and radiation in evolving complex active regions, *AA* 422, 323–330.

Andrews, M. D. 2003, A search for CMEs associated with big flares, *SoPh* 218, 261–279.

Angelopoulos, V., et al. 1996, Multipoint analysis of a bursty bulk flow event on April 11, 1985, *GRL* 101(A3), 4966–4990.

Angelopoulos, V., Mukai, T., and Kokubun, S. 1999, Evidence for intermittency in Earth's plasma sheet and implications for self-organized criticality, *Physics of Plasmas* 6, 4161–4168.

Ansari, M. H. and Smolin, L. 2008, Self-organized criticality in quantum gravity, *Classical and Quantum Gravity* 25(9), 09016.

Antolin, P., et al. 2021, Reconnection nanojets in the solar corona, *NatAs* 5, 54.

Antonov, N. V., Gulitskiy N. M., Kakin, P. I. and Kochnev, G. E. 2020, Effects of turbulent environment of self-organized critical behavior: Isotropy vs. Anisotropy, *Universe* 6(9), 145.

Antonov, N. V., Gulitskij, N. M., Kakin, P. I., and Serov, V. D. 2021, Effects of turbulent environment and random noise on self-organized critical behavior: Universality versus nonuniversality, *Physical Review E* 103(4), article ID 042106.

Antonova, E. E. 2004, Magnetostatic equilibrium and current systems in the Earth's magnetosphere, *Advances in Space Research* 33, 752.

Argyle, E. and Gower, J. R. R. 1972, The pulse-height distribution for NP 0532, *ApJ* 175, L89–L91.

Arzner, K. and Güdel, M. 2004, Are coronae of magnetically active stars heated by flares? III. Analytical distribution of superposed flares, *ApJ* 602, 363–376.

Arzner, K., Güdel, M., Briggs, K., Telleschi, A., and Audard, M. 2007, Statistics of superimposed flares in the Taurus molecular cloud, *AA* 468, 477–484.

Aschwanden, M. J., Dennis, B. R., and Benz, A. O. 1998, Logistic avalanche processes, elementary time structures, and frequency distributions in solar flares, *ApJ* 497, 972–993.

Aschwanden, M. J. and Nitta, N. 2000, The effect of hydrostatic weighting on the vertical temperature structure of the solar corona, *ApJ* 535, L59–L62.

Aschwanden, M. J., Nightingale, R., Tarbell, T., and Wolfson, J. 2000a, Time variability of the quiet Sun observed with TRACE. I. Instrumental effects, event detection, and discrimination of EUV nanoflares, *ApJ* 535, 1027–1046.

Aschwanden, M. J., Tarbell, T., Nightingale, R., Schrijver, C. J., Title, A., Kankelborg, C. C., Martens, P. C. H., and Warren, H. P. 2000b, Time variability of the quiet Sun observed with TRACE, II. Physical parameters, temperature evolution, and energetics of EUV nanoflares, *ApJ* 535, 1047–1065.

Aschwanden, M. J. and Charbonneau, P. 2002a, Effects of temperature bias on nanoflare statistics, *ApJ* 566, L59–L62.

Aschwanden, M. J. and Parnell, C. E. 2002, Nanoflare statistics from first principles: Fractal geometry and temperature synthesis, *ApJ* 572, 1048–1071.

Aschwanden, M. J. and Güdel, M. 2008, Scaling laws of solar and stellar flares, *ApJ* 672, 659–673.

Aschwanden, M. J. and Aschwanden P. D. 2008a, Solar flare geometries: I. The area fractal dimension, *ApJ* 574, 530–543.

Aschwanden, M. J. and Aschwanden P. D. 2008b, Solar flare geometries: II. The volume fractal dimension, *ApJ* 574, 544–553.

Aschwanden, M. J. and McTiernan, J. M. 2010, Reconciliation of waiting time statistics of solar flares observed in hard X-rays, *ApJ* 717, 683–692.
Aschwanden, M. J. 2011a, *Self-Organized Criticality in Astrophysics: The Statistics of Nonlinear Processes in the Universe*, Springer-Praxis, New York, 416p.
Aschwanden, M. J. 2011b, The state of self-organized criticality of the Sun during the last 3 solar cycles. I. Observations, *SoPh* 274, 99–117.
Aschwanden, M. J. 2011c, The state of self-organized criticality of the Sun during the last 3 solar cycles. II. Theoretical model, *SoPh* 274, 119–129.
Aschwanden, M. J. 2012a, A statistical fractal-diffusive avalanche model of a slowly-driven self-organized criticality system, *AA* 539, A2 (15p).
Aschwanden, M. J. 2012b, The spatio-temporal evolution of solar flares observed with AIA/SDO: Fractal diffusion, sub-diffusion, or logistic growth? *ApJ* 757, 94.
Aschwanden, M. J. and Freeland, S. L. 2012, Automated solar flare statistics in soft X-rays over 37 years of GOES observations: The invariance of self-organized criticality during three solar cycles, *ApJ* 754, 112.
Aschwanden, M. J. 2013a, A nonlinear force-free magnetic field approximation suitable for fast forward-fitting to coronal loops. I. Theory, *SoPh* 287, 323–344.
Aschwanden, M. J. 2013b, in Theoretical models of SOC systems (Chapter 2), in *Self-Organized Criticality Systems* (ed. Aschwanden, M. J.), Open Academic Press: Berlin, Warsaw, pp. 21–67.
Aschwanden, M. J. 2013c, Self-organized criticality systems in astrophysics (Chapter 13), in *Self-Organized Criticality Systems* (ed. Aschwanden, M. J.), Open Academic Press, Berlin, Warsaw, pp. 439–483.
Aschwanden, M. J. 2019a, *New Millennium Solar Physics*, Springer Nature, Switzerland AG, Science Library 458.
Aschwanden, M. J. 2019b, Self-organized criticality in solar and stellar flares: Are extreme events scale-free?, *ApJ* 880, 105 (16pp).
Aschwanden, M. J. 2021, Finite system-size effects in self-organizing criticality systems, *ApJ* 909, 69.
Aschwanden, M. J. 2022a, Reconciling power law slopes in solar flare and nanoflare size distributions, *ApJL* 934, L3.
Aschwanden, M. J. 2022b, The fractality and size distributions of astrophysical self-organized criticality systems, *ApJ* 934, 33 (27pp).
Aschwanden, M. J. and Shimizu, T. 2013, Multi-wavelength observations of the spatio-temporal evolution of solar flares with AIA/SDO: II. Hydrodynamic scaling laws and thermal energies, *ApJ* 776, 132.
Aschwanden, M. J. 2014, A macroscopic description of a generalized self-organized criticality systems: Astrophysical applications, *ApJ* 782, 54.
Aschwanden, M. J., Wuelser, J. P., Nitta, N. V., Lemen, J. R., Freeland, S., and Thompson, W. T. 2014, STEREO/Extreme Ultraviolet Imager (EUVI) event catalog 2006–2012, *SoPh* 289(3), 919–938.
Aschwanden, M. J. 2015, Thresholded power law size distributions of instabilities in astrophysics, *ApJ* 814, 19 (25pp).
Aschwanden, M. J. 2016, Global energetics of solar flares. IV. Coronal mass ejection energetics, *ApJ* 2016, 831, 105 (34pp).
Aschwanden, M. J., Crosby, N., Dimitropoulou, M., Georgoulis, M. K., Hergarten, S., McAteer, J., Milovanov, A., Mineshige, S., Morales, L., Nishizuka, N., Pruessner, G., Sanchez, R., Sharma, S., Strugarek, A., and Uritsky, V. 2016, 25 years of self-organized criticality: Solar and astrophysics, *Space Science Reviews* 198, 47–166.
Aschwanden, M. J., Caspi, A., Cohen, C. M. S., Holman, G. D., Jing, J., Kretzschmar, M., Kontar, E. P., McTiernan, J. M., O'Flannagain, A., Richardson, I. G., Ryan, D., Warren,

H. P., and Xu, Y. 2017, Global energetics of solar flares: V. Energy closure, *ApJ* 836, 17 (17pp).

Aschwanden, M. J. and Peter, H. 2017, The width distribution of solar coronal loops and strands – Are we hitting rock bottom? *ApJ* 840, 4 (24pp).

Aschwanden, M. J. and Scholkmann, F. 2017, Exoplanet predictions based on harmonic orbit resonances, *Galaxies* 5(4), 56.

Aschwanden, M. J., et al. 2018, Order out of randomness: Self-organization processes in astrophysics, *SSRv* 214, 55.

Aschwanden, M. J. and Güdel, M. 2021, Self-organized critically in stellar flares, *ApJ* 910, 41 (16pp).

Aschwanden, M. J. and Johnson, J. R. 2021, The solar memory from hours to decades, *ApJ* 921, 82.

Aschwanden, M. J., Johnson, J. R., and Nurhan, Y. 2021, The Poissonian origin of power laws in solar flare waiting time distributions, *ApJ* 921, 166 (14pp).

Aschwanden, M. J. and Dudok de Wit, T. 2021, Correlation of the sunspot number and the waiting-time distribution of solar flares, coronal mass ejections, and solar wind switchback events observed with the Parker Solar Probe, *ApJ* 912, 94 (11pp).

Aschwanden, M. J. and Nhalil, N. V. 2022, Interface region imaging spectrograph (IRIS) observations of the fractal dimension in the solar atmosphere, *Frontiers in Astronomy and Space Sciences* 9, id. 999319.

Aschwanden, M. J. and Nhalil, N. V. 2023, The universality of power law slopes in the photosphere and transition region observed with HMI and IRIS, *Frontiers in Astronomy and Space Sciences* 10, id. 1099346, doi 10.3389/fspas.2023.1099346.

Atteia, J. L., et al. 1987, A second catalog of gamma-ray-bursts: 1978–1980 localizations from the interplanetary Network, *ApJSS* 64, 305.

Audard, M., Güdel, M., and Guinan, E. F. 1999, Implications from extreme-ultraviolet observations for coronal heating of active stars, *ApJ* 513, L53–L56.

Audard, M., Güdel, M., Drake, J. J., and Kashyap, V. L. 2000, Extreme-ultraviolet flare activity in late-type stars, *Astrophysical Journal* 541, 396–409.

Bai, T. 1993, Variability of the occurrence frequency of solar flares as a function of peak hard X-ray rate, *ApJ* 404, 805–809.

Baiesi, M., Paczuski, M., and Stella, A. L. 2006, Intensity thresholds and the statistics of the temporal occurrence of solar flares, *Physical Review Letters* 96, 051103.

Bak, P., Tang, C., and Wiesenfeld, K. 1987, Self-organized criticality – An explanation of 1/f noise, *Physical Review Letters* 59(27), 381–384.

Bak, P., Tang, C., and Wiesenfeld, K. 1988, Self-organized criticality, *Physical Review A, Atomic, Molecular, and Optical Physics* 38(1), 364–374.

Bak, P. and Chen, K. 1989, The physics of fractals, *Physica D: Nonlinear Phenomena* 38, 5–12.

Bak, P. and Chen, K. 1991, Self-organized criticality, *Scientific American* 264(1), 46–53 pp.

Bak, P. and Sneppen, K. 1993, Punctuated equilibrium and criticality in a simple model of evolution, *Physical Review Letters* 71(24), 4083–4086.

Bak, P. and Paczuski, M. 1995, Complexity, contingency, and criticality, *Proceedings of the National Academy of Sciences of the United States of America* 92, 6689–6696.

Bak, P. 1996, How Nature Works, Copernicus, Springer Verlag, New York.

Bak, P. and Weissman, M. 1997, How nature works: The science of self-organized criticality, *American Journal of Physics* 65, 579–580.

Baker, D. N., Pulkkinen, T. I., Angelopoulos, V., Baumjohann, W., McPherron, R. L., 1996, Neutral line model of substorms: Past results and present view, *JGR* 101(A6), 12975, S13010.

Balasis, G., Daglis, I. A., Anasasiadis, A., Papadimitriou, C., Mandea, M., and Eftaxias, K. 2011, Universality in solar flare, magnetic storm and earthquake dynamics using Tsallis statistical mechanics, *Physica A: Statistical Mechanics and Its Applications* 390(2), 341–346.

Balke, A. C., Schrijver, C. J., Zwaan, C., and Tarbell, T. D. 1993, Percolation theory and the geometry of photospheric magnetic flux concentrations, *SoPh* 143, 215–227.

Balona, L. A., Baran, A. S., Daszynska-Daszkiewicz, J., and De Cat P. 2015, Analysis of Kepler B stars: Rotational modulation and Maia variables, *MNRAS* 451(2), 1445–1459.

Balona, L. A., Svanda, M., and Karlicky, M. 2016, Differential rotation, flares and coronae in A to M stars, *MNRAS* 463, 1740–1750.

Banerjee, A., Bej, A., Chatterjee, T. N., and Majumdar, A. 2019, An SOC approach to study the solar wind-magnetosphere, *Energy Coupling Earth and Space Science* 6, 565–576.

Barat, C., et al. 1979, An intense gamma-ray burst with possible associated optical transient, *ApJ* 286, L5–L9.

Barat, C., Trottet, G., Vilmer, N., Dezalay, J. P., Talon, R., Sunyaev, R., Terekhov, O., and Kuznetsov, A. 1994, Evidence for intense coronal prompt gamma-ray line emission from a solar flare, *ApJ* 425, L109–L112.

Bargatze, L. F., Baker, D. N., McPherron, R. L., and Hones, E. W. 1985, Magnetospheric impulse response for many levels of geomagnetic activity, *JGR* 90(A7), 6387–6394.

Baring, M. G. 1994, Gamma-ray spectral breaks and source beaming, *ApJSS* 90, 899–903.

Baryshev, Y. and Teerikorpi, P. 2002, *Discovery of Cosmic Fractals*, World Scientific Publishing, Co. Pte. Ltd, Singapore.

Bhattacharya, S. 1997, How nature works: The science of self-organized criticality, by Per Bak, *Physics Today* 50, 71–72.

Beckmann, V., Borkowski, J., Courvoisier, T. J. L., Goetz, D., Hudec, R., Hroch, F., Lund, N., Mereghetti, S., Shaw, S. E., von Kienlin, A., and Wigger, C. 2003, Time resolved spectroscopy of GRB 030501 using INTEGRAL, *AA* 411, L327.

Beim Graben, P., Zhou, C., Thiel, M., and Kurths, J. (eds.). 2008, *Lectures in Supercomputational Neuroscience: Dynamics in Complex Brain Networks*, Springer, Berlin.

Belanger, E., Vincent, A., and Charbonneau, P. 2007, Predicting solar flares by data assimilation in avalanche models. I. Model design and validation, *SoPh* 245, 141–165.

Belov, A., Kurt, V., Mavromichalaki, H., and Gerontidou, M. 2007, Peak-size distributions of proton fluxes and associated soft X-ray flares, *SoPh* 246, 457–470.

Bendjoya, Ph., Petit, J. M., and Spahn, E. 1993, Wavelet analysis of the Voyager data on planetary rings. I. Description of the method, *Icarus* 105, 385.

Berger, M. A., and Asgari-Targhi, M. 2009, Self-organized braiding and the structure of coronal loops, *ApJ* 705, 347–355.

Berger, M. A., Asgari-Targhi, M., and DeLuca, E. E. 2015, Self-organized braiding in solar coronal loops, *Journal of Plasma Physics* 81, 395810404.

Bergin, E. A. and Tafalla, M. 2007, Cold dark clouds: The initial conditions for star formation, *ARAA* 45, 339.

Berrilli, F., Del Moro, D., Russo, S., Consolini, G., and Straus, Th. 2005, Spatial clustering of photospheric structures, *ApJ* 632, 677.

Bershadskii, A. and Sreenivasan, K. R. 2003, Multiscale self-organized criticality and powerful X-ray flares, *European Physical Journal B* 35(4), 513–515.

Benz, A. O. and Krucker, S. 2002, Energy distribution of microevents in the quiet solar corona, *ApJ* 568, 413–421.

Berger, M. A. and Asgari-Targhi, M. 2009, Self-organized braiding and the structure of coronal loops, *ApJ* 705:347–355.

Berger, M. A., Asgari-Targhi, M., and Deluca, E. E. 2015, Self-organized braiding in solar coronal loops, *Journal of Plasma Physics*, 81, 395810404.

Bergman, M. 2007, Strong bursts and diffusion in avalanching systems, *Planetary and Space Science* 55(15), 2228–2232.

Berghmans, D., Clette, F., and Moses, D. 1998, Quiet Sun EUV transient brightenings and turbulence. Spatial clustering of photospheric structures, *ApJ* 632, 677.

Berghmans, D. and Clette, F. 1999, Active region EUV transient brightenings – First results by EIT of SOHO JOP80, *SoPh* 186, 207–229.

Biesecker, D. A., Ryan, J. M., and Fishman, G. J. 1993, A search for small solar flares with BATSE, *Lecture Notes in Physics* 432, 225–230.

Biesecker, D. A. 1994, *On the occurrence of solar flares observed with the Burst and Transient Source Experiment (BATSE)*, PhD Thesis, University of New Hampshire.

Biesecker, D. A., Ryan, J. M., and Fishman, G. J. 1994, *Observations of small solar flares with BATSE*, in *High-energy solar phenomena – A new era of spacecraft measurements* (eds. Ryan, J. M. and Vestrand, W. T.), American Institute Physics, New York, pp. 183–186.

Bingert, S. and Peter, H. 2013, Nanoflare statistics in an active region 3D MHD coronal model, *AA* 550, A30.

Boffetta, G., Carbone, V., Giuliani, P., Veltri, P., and Vulpiani, A. 1999, Power laws in solar flares: Self-organized criticality or turbulence? *PRL* 83, 4662–4665.

Bottke, W. F., Durda, D. D., Nesvorny, D., Jedicke, R., Morbidelli, A., Vokrouhlicky, D., and Levison, H. 2005, The fossilized size distribution of the main asteroid belt, *Icarus* 175, 111–140.

Bolzan, M. J. A. 2018, A modeling substorm dynamics of the magnetosphere using self-organized criticality approach, *Physica A Statistical Mechanics and Its Applications* 503, 1182–1188.

Borucki, W. J., et al. 2010, Kepler planet-detection mission: Introduction and first results, *Science* 327(5968), 977.

Bovelet, B. and Wiehr, E. 2001, A new algorithm for pattern recognition and its application to granulation and limb faculae, *SoPh* 201, 13.

Boynton, R. j., Balikhin, M. A., Billings, S. A., and Sharma, A. S. 2011, Amariutei, Data derived NAREMAX Dst model, *Annales Geophysicae* 29(6), 965–971.

Bregman, M. 2007, Strong bursts and diffusion in avalanching systems, *Planetary and Space Science* 55(15), 2228–2232.

Bristow, W. 2008, Statistics of velocity fluctuations observed by SuperDARN under steady interplanetary magnetic field conditions, *JGR (Space Physics)*, 113(A11), CiteID A11202.

Bromund, K. R., McTiernan, J. M., and Kane, S. R. 1995, Statistical studies of ISEE3/ ICE observations of impulsive hard X-ray solar flares, *ApJ* 455, 733–745.

Bronstein, I. and Semendjajew, K. 1960, *Taschenbuch der Mathematik*, Verlag Harry Deutsch, Zurich, Switzerland.

Brown, J. C. 1971, The deduction of energy spectra of non-thermal electrons in flares from the observed dynamic spectra of Hard X-Ray bursts, *SoPh* 18, 489–502.

Bruno, R., D'Amicis, R., Bavassano, B., Carbone, V., and Sorriso-Valvo, L. 2007, Scaling laws and coherent structures in the solar wind, *Planetary and Space Science*, 55, 2233.

Buchanan, M. 2015, SOC revisited, *Nature Physics* 11, 442.

Buchlin, E., Vial, J. C., and Lemaire, P. 2006, A statistical study of SUMER spectral images: Events, turbulence, and intermittency, *AA* 451, 1091–1099.

Cadavid, A. C., Lawrence, J. K., and Ruzmaikin, A. A. 1999, Anomalous diffusion of solar magnetic elements, *ApJ* 521, 844.

Cadavid, A. C., Miralles, M. P., and Romich, K. 2019, Comparison of the scaling properties of EUV intensity fluctuations in coronal hole and quiet-sun regions, *ApJ* 886, 143.

Cairns, I. H. and Robinson, P. A. 1999, Strong evidence for stochastic growth of Langmuir-like waves in Earth's foreschock, *PRL* 82(15), 3066–3069.

Cairns, I. H. 2004, Properties and interpretations of giant micropulses and giant pulses from pulsars, *ApJ* 610, 948–955.

Cairns, I. H., Johnston, S., and Das, P. 2004, Intrinsic variability and field statistics for pulsars B1641–45 and B0950+08, *MNRAS* 353, 270.

Candelaresi, S., Pontin, D. I., Yeates, A. R., Bushby, P. J., and Hornig, G. 2018, Estimating the rate of field line braiding in the solar corona by photospheric flows, *ApJ* 864, 157.

Carbone, V., et al. 2002, To what extent can dynamical models describe statistical features of turbulent flows? *EPL (Europhysics Letters)*, 58(3), 349.

Cargill, P. J., Bradshaw, S. J., and Klimchuk, J. A. 2012, Enthalpy-based thermal evolution of loops. III. Comparison of Zero-dimensional models, *ApJ* 758, 5.

Caseres, G. A., et al. 2019a, Autoregressive planet search: Methodology, *ApJ* 158, 57.

Caseres, G. A., et al. 2019b, Autoregressive planet search: Application to the Kepler mission, *ApJ* 158, 58.

Cassak, P. A., Mullan, D. J., and Shay, M. A. 2008, From solar and stellar flares to coronal heating: theory and observations of how magnetic reconnection regulates coronal conditions, *ApJ* 676, L69–L72.

Castorinoa P. and Satz, H. 2019, High energy hadron production as self-organized criticality, *International Journal of Modern Physics E*, 28(4), id. 1950025.

Catanzarite, J. and Shao, M. 2011, The occurrence rate of Earth analog planets orbiting sun-like stars, *ApJ* 738, 151 (10pp).

Chang, H. K., Chen, K., Fenimore, E. E., and Ho, C. 1996, Spectral studies of magnetic photon splitting in the March 5 event and SGR 1806-20, *AIP Conference Proceedings* 384, 921–925.

Chang, T. 1999a, Self-organized criticality, multi-fractal spectra, and intermittent merging of coherent structures in the magnetotail, *Astrophysics and Space Science*, 264, 303–316.

Chang, T. 1999b, Self-organized criticality, multi-fractal spectra, sporadic localized reconnections and intermittent turbulence in the magnetotail, *Physics of Plasmas* 6, 4137–4145.

Chang, T. 2000, Forced and/or self-organized criticality in space plasma processes, *Physica Scripta Volume T* 84, 12–13.

Chang, T., Chapman, S., and Klimas, A. 2001, Forced and/or Self-Organized Criticality (FSOC) in space plasmas, *JASTP* 63, 1359–1359.

Chang, T. and Wu, C. C. 2002, Complexity and anomalous transport in space plasmas, *Physics of Plasmas* 9(9), 3679–3684.

Chang, T., Tam, S. W. Y., Wu, C. C., and Consolini, G. 2003, Complexity, forced and/or self-organized criticality, and topological phase transitions in space plasmas, *SSRv* 107, 425–445.

Chang, T., Tam, S. W. Y, and Wu, C. C. 2004, Complexity induced anisotropic bimodal intermittent turbulence in space plasmas, *Physics of Plasmas* 11(4), 1287–1299.

Chang, T. S. 1992, Low-dimensional behavior and symmetry breaking of stochastic systems near criticality – Can these effects be observed in space and in the laboratory, *IEEE Transaction on Plasma Science* 20(6), 691–694.

Chang, T. S. 1998a, Sporadic, localized reconnections and multiscale intermittent turbulence in the magnetotail, in *Geospace Mass and Energy Flow* (Eds. Horwitz, J. L., Gallagher, D. L., and Peterson, W. K.), AGU Geophysical Monograph 104, American Geophysical Union, Washington DC, p. 193.

Chang, T. S. 1998b, Multiscale intermittent turbulence in the magnetotail, in *Proceedings of 4th International Conference on Substorms* (eds. Kamide, Y. et al.), Kluwer Academic Publishers, Dordrecht, and Terra Scientific Company, Tokyo, p. 431.

Chapman, S. C., Watkins, N. W., Dendy, R. O., Helander, P., and Rowlands, G. 1998, A simple avalanche model as an analogue for magnetospheric activity, *GRL* 25(13), 2397–2400.

Chapman, S. C., Dendy, R. O., and Rowlands, G. 1999, A sandpile model with dual scaling regimes for laboratory, space and astrophysical plasmas, *PhPl* 6(11), 4169–4177.

Chapman, S. C. and Watkins, N. 2001, Avalanching and self-organised criticality, a paradigm for geomagnetic activity? *SSRv* 95, 293–307.

Chapman, S. C., Rowlands, G., and Watkins, N. W. 2009, Macroscopic control parameter for avalanche models for bursty transport, *Physics of Plasmas* 16, 012303.

Chapman, S. C. and Nicol, R. M. 2009, Generalized similarity in finite range solar wind magnetohydrodynamic turbulence, *Physical Review Letters*, 103, 241101.

Chapman, S. C. and Watkins, N. W. 2009, Avalanching systems under intermediate driving rate, *Plasma Physics and Controlled Fusion* 51(12), id. 124006, 9pp.

Charbonneau, P., McIntosh, S. W., Liu, H. L., and Bogdan, T. J. 2001, Avalanche models for solar flares, *SoPh* 203, 321–353.

Charbonneau, P. 2013, Self-organized criticality and solar flares (Chapter 12), in "*Self-Organized Criticality Systems*" (ed. Aschwanden, M. J.), Open Academic Press GmbH Co, Berlin; Warsaw, 167–210.

Charbonneau, P. 2017, *Natural Complexity: A Modeling Handbook*, Princeton University Press, Princeton.

Chatterjee, S., Law, C. J., Wharton, R. S., Burke-Spolaor, S., Hessels, J. W. T., et al. 2017, A direct localization of a fast radio bursts and its host, *Nature* 541(7636), 58–61.

Chen, J. and Sharma, A. S. 2006, Modeling and prediction of the magnetospheric dynamics during intense geospace storms, *JGR* 111(A4), A94209.

Chen, J., Sharma, A. S., Edwards, J., Shao, X., and Kamide, Y. 2008, Spatio-temporal dynamics of the magnetosphere during geospace storms: Mutual information analysis, *JGR* 113, A05217.

Chen, Y. and Hu, Q. 2022, Small-scale magnetic flux ropes and their properties based on in situ measurements from the Parker Solar Probe, *ApJ* 924, 43 (12pp).

Cheng, J. X., Qiu, J., Ding, M. D., and Wang, H. 2012, Solar flare hard X-ray spikes observed by RHESSI: A statistical study, *AA* 547, A73 (8pp).

Cheng, Y., Zhang, G. Q., and Wang, F. Y. 2020, Statistical properties of magnetar bursts and FRB 121102, *MNRAS* 491, 1498–1505.

Chesny, D. L., Oluseyi, H. M., and Orange, N. B. 2016, Dynamic flaring non-potential fields on quiet sun network scales, *ApJ* 822, 72.

Choe, W., Kim, H. J., Rostoker, G., and Kamide, Y. 2002, Nonlinear time series analysis of interpeak intervals of AL data, *JGR* 107(A11), CiteID 1392.

Chou, Y. P. 1999, What affects the power-law distribution of the X-ray solar flares? A theoretical study based on a model of uniform normal field, *ApJ* 527, 958–966.

Chou, Y. P. 2001, The effect of helicity dissipation on the critical state of an avalanche model for solar flares, *SoPh* 199, 345.

Ciprini, S., Fiorucci, M., Tosti, G., and Marchili, N. 2003, The optical variability of the blazar GV 0109+224. Hints of self-organized criticality, in *High energy blazar astronomy*, ASP Conference Proceedings 229 (eds. L. O. Takalo and E. Valtaoja), ASP, San Francisco, p. 265.

Christe, S., Hannah, I. G., Krucker, S., McTiernan, J., and Lin, R. P. 2008, RHESSI microflare statistics. I. Flare-finding and frequency distributions, *ApJ* 677, 1385–1394.

Christensen, K. and Olami, Z. 1993, Sandpile models with and without an underlying spatial structure, *Physics Review E* 48(5), 3361–3372.

Cirtain, J. W., et al. 2013, Energy release in the solar corona from spatially resolved magnetic braids. *Nature* 493, 7433, 501.

Clar, S., Drossel, B., and Schwabl, F. 1996, Review article: Forest fires and other examples of self-organized criticality, *Journal of Physics: Condensed Matter* 8(37), 6803–6824.

Clauset, A., Young, M., and Gleditsch, S. 2007, Power-law distributions in empirical data, *Journal of Conflict Resolution* 51, 58.

Clauset, A., Shalizi, C. R., and Newman, M. E. J. 2009, Power-law distributions in empirical data, *SIAM Review* 51(4), 661–703.

Cliver, E., Reames, D., Kahler, S., and Cane, H. 1991, Size distribution of solar energetic particle events, International Cosmic Ray Conference 22nd, Dublin, LEAC A92-36806 15–93, NASA, Greenbelt, p. 2:1–4.

Cliver, E. W., Ling, A. G., Belov, A., and Yashiro, S. 2012, Size distributions of solar flares and solar energetic particle events, *ApJL* 756, L29 (4pp).

Cliver, E. and D'Huys, E. 2018, Size distributions of solar proton events and their associated soft X-ray flares, *ApJ* 864, 48(1), 11.

Cognard, I., Shrauner, J. A., Taylor, J. H., and Thorsett, S. E. 1996, Giant radio pulses from a millisecond pulsar, *ApJ* 457, L81–L84.

Colless, M., et al. 2001, The 2dF galaxy redshift survey: Spectra and redshifts, *MNRAS* 328(4), 1039–1063.

Collura, A., Pasquini, L., and Schmitt, J. H. M. M. 1988, Time variability in the X-ray emission of DM stars observed by EXOSAT, *AA* 205, 197–206.

Consolini, G. 1997, *Sandpile cellular automata and magnetospheric dynamics*, in (Proc, Cosmic Physics in the year 2000 (Eds. Aiello, S., Iucci, N., Sironi, G., Treves, A., and Villante, U.), SIF, Bologna, Italy, Vol. 58, pp. 123–126.

Consolini, G. and Lui, A. T. Y. 1999, Sign-singularity analysis of current disruption, *GRL* 26(12), 1673–1676.

Consolini, G. 2001, Complexity and criticality of the magnetospheric dynamics *Memorie della Societa Astronomica Italiana*, 72, 605.

Consolini, G. and Chang, T. S. 2001, Magnetic field topology and criticality in geotail dynamics: Relevance to substorm phenomena, *SSRv* 95, 309.

Consolini, G. and DeMichelis, P. 2001, A revised forest-fire cellular automaton for the nonlinear dynamics of the Earth's magnetotail, *JASTR* 63(13), 1371–1377.

Consolini, G. 2002, Self-organized criticality: A new paradigm for the magnetotail dynamics, *Fractals* 10, 275–283.

Consolini, G. and Kretzschmar M. 2007, Thermodynamics of rare events and impulsive relaxation events in the magnetospheric substorm dynamics, *Planetary and Space Science* 55, 2244.

Corral, A. and Font-Clos, F. 2013, Criticality and self-organization in branching processes (Chapter 5), in *Self-Organized Criticality Systems* (ed. Aschwanden, M. J.), Open Academic Press GmbH Co, Berlin; Warsaw, 167–210.

Greenhough, J., Chapman, S. C., Dendy, R. O., Nakariakov, V. M. and Rowlands, G. 2003, Statistical characterisation of full-disk EUV/XUV solar irradiance and correlation with solar activity, *AA* 409, L17–L20.

Crosby, N. B., Aschwanden, M. J., and Dennis, B. R. 1993, Frequency distributions and correlations of solar X-ray flare parameters, *SoPh* 143, 275–299.

Crosby, N. B. 1996, *Contribution à l'Etude des Phénomènes Eruptifs du Soleil en Rayons à partir des Observations de l'Expérience WATCH sur le Satellite GRANAT*, PhD Thesis, Université Paris VII, Meudon, Paris.

Crosby, N., Vilmer, N., Lund, N., and Sunyaev, R. 1998, Deka-keV X-ray observations of solar bursts with WATCH/GRANAT: frequency distributions of burst parameters, *AA* 334, 299–313.

Crosby, N. B., Meredith, N. P., Coates, A. J., and Iles, R. H. A. 2005, Modelling the outer radiation belt as a complex system in a self-organised critical state, *Nonlinear Processes in Geophysics* 12, 993–1001.

Crosby, N. B. 2011, Frequency distributions: From the Sun to the Earth, *Nonlinear Processes in Geophysics* 18(6), 791–805.

Crosby, N. B. 2009, Solar extreme events 2005–2006: Effects on near-Earth space systems and interplanetary systems, *Adv. Spac. Res.* 43/4, 559–564.

Crosby, N. B. 2013, *Introduction (Chapter 1)*, in *Self-Organized Criticality Systems* (ed. Aschwanden, M. J.), Open Academic Press GmbH Co , Berlin; Warsaw, 1–20.

Cross, C. A. 1966, The size distribution of lunar craters, *MNRAS* 134, 245–252.

Daei, F., Safari, H., and Dadashi, N. 2017, Complex network for solar active region, *ApJ* 845, 36 (8pp).

Da Rocha, D. and Nottale, L. 2003, Gravitational structure in scale relativity, *Chaos, Solitons, and Fractals* 16(4), 565–595.

Das, T. K., Tarafdar, G., and Sen, A. K. 1997, Validity of power law for the distribution of intensity of radio bursts, *SoPh* 176, 181–184.

Datlowe, D. W., Elcan, M. J., and Hudson, H. S. 1974, OSO-7 observations of solar X-rays in the energy range 10–100 keV, *SoPh* 39, 155–174.

Davenport, J. R. A. 2016, The Kepler catalog of stellar flares, *ApJ* 829(1), article ID 23, 12pp.

DeArcangelis, L., Godano, C., Lippiello, E., and Nicodemi, M. 2006, Universality in solar flare and earthquake occurrence, *Physical Review Letters*, 96, 051102.

DeFreitas, D. B., et al. 2017, New Suns in the cosmos. IV. The multifractal nature of stellar magnetic activity in Kepler cool stars, *ApJ* 843, 103.

DeFreitas, D. B., et al. 2019a, Multifractal detrended moving average analysis of Kepler stars with surface differential rotation traces, *MNRAS* 488, 32.

DeFreitas, D. B., et al. 2019b, New suns in the cosmos. V. Stellar rotation and multifractality in active Kepler stars, *ApJ* 880, 151.

DeMenech, M. and Stella, A. L. 2002, Turbulent self-organized criticality, *Physica A Statistical Mechanics and Its Applications* 309, 289–296.

Dendy, R. O., Chapman, S. C., and Paczuski, M. 2007, Fusion, space and solar plasmas as complex systems, *Plasma Physics and Control Fusion* 49, A95–A108.

Dennis, B. R. 1985, Solar hard X-ray bursts, *SoPh* 100, 465–490.

Dennis, B. R. and Zarro, D. M. 1993, The Neupert effect: What can it tell us about the impulsive and gradual phases of solar flares, *SoPh* 146, 177–190.

Dennis, B. R., Veronig, Z., Schwartz, R. A., Sui, L., Tolbert, A. K. Zarro, D. M. and the RHESSI Team 2003, The Neupert effect and new RHESSI measures of the total energy accelerated in solar flares, *AdSpR* 32, 2459.

Dhar, D. and Ramaswamy, R. 1989, Exactly solved model of self-organized critical phenomena, *Physical Review Letters* 63(16), 1659–1662.

Dhar, D. and Majumdar, S. N. 1990, Abelian sandpile model of the Bethe lattice, *Journal of Physics A* 23(19), 4333–4350.

Dhar, D. 1999, The Abelian sandpile and related models, *Physica A* 263, 4–25.

D'Huys, E., Berghmans, D., Seaton, D. B., and Poedts, S. 2016, The effect of limited sample sizes on the accuracy of the estimated scaling parameter for power-law-distributed solar data, *SoPh* 291, 1561.

Dimitropoulou, M., Isliker, H., Vlahos, L., and Georgoulis, M. K. 2011, Simulating flaring events in complex active regions driven by observed magnetograms, *AA* 529, A101.

Dimitropoulou, M., Isliker, H., Vlahos, L., and Georgoulis, M. K. 2013, Dynamic data-driven integrated flare model based on self-organized criticality, *AA* 553, A65.

Dinkelaker, A. N. and MacKinnon, A. L. 2013a, Wavelets, intermittency and solar flare hard X-rays. I. Local intermittency measure in cascade and avalanche scenarios, *SoPh* 282, 471–481.

Dinkelaker, A. N. and MacKinnon, A. L. 2013b, Wavelets, intermittency and solar flare hard X-rays. II. LIM Analysis of high time resolution BATSE data, *SoPh* 282, 483–501.

Dobrotka, A., Mineshige, S., and Casares, J. 2012, A flickering study of nova-like systems KR Aur and UU Aqr, *MNRAS* 420(3), 2467–2474.

Drake, J. F. 1971, Characteristics of soft solar X-ray bursts, *SoPh* 16, 152–185.

Du, Z. and Du, S. 2006, The relationship between the amplitude and descending time of a solar cycle, *SoPh* 238(2), 431–437.

Dudok de Wit, T., Kopp, G., Shapiro, A., Witzke, V., and Kretzschmar, M. 2018, Response of solar irradiance to sunspot-area variations, *ApJ* 853, 197.

Dudok de Wit, T., et al. 2020, Switchbacks in the near-Sun magnetic field: Long memory and impact on the turbulence cascade, *ApJS* 246, 39.

Dwyer, J. R. and Smith, D. M. 2005, A comparison between Monte Carlo simulations of runaway breakdown and terrestrial gamma-ray flare observations, *GRL* 32(22), L22804.

Eastwood, J. P., Wheatland, M. S., Hudson, H. S., Krucker, S., Bale, S. D., Maksimovic, M., Goertz, K., and Bougeret, J. L. 2010, On the brightness and waiting time distributions of a type III radio storm observed by STEREO/Waves, *ApJL* 706(2), L95–L99.

Eggen, O. J., Lynden-Bell, D., and Sandage, A. R. 1962, Evidence from the motions of old stars that the galaxy collapsed, *ApJ* 136, 748.

Elmegreen, B. G. and Scalo, J. 2004, Interstellar turbulence I. Observations and processes, *ARAA* 42, 211.

Elmegreen, B. G. and Falgarone, E. 1996, A fractal origin for the mass spectrum of interstellar clouds, *ApJ* 471, 816.

Enoto, T., et al. 2017, Magnetar broadband X-ray spectra correlated with magnetic fields: Suzaku Archive of SGRs and AXP combined with NuSTAR, Swift, and RXTE, *ApJSS* 231(1), id. 8, 21pp.

Eroencel, C., Hubisz, J., and Rigo, G. 2019, Self-organized Higgs criticality, *Journal of High Energy Physics*, 2019(3), article ID. 46, 33pp.

Farhang, N., Wheatland, M. S., and Safari, H. 2018, Principle of minimum energy in magnetic reconnection in a self-organized critical model for solar flares, *ApJ* 859, 41.

Farhang, N., Wheatland, M. S., and Safari, H. 2019, Energy balance in avalanche models for solar flares, *ApJ* 883, L20.

Feder, J. 1988, *Fractals*, Plenum Press, New York, 283p.

Feldman, U., Doschek, G. A., and Klimchuk, J. A. 1997, The occurrence rate of soft X-ray flares as a function of solar activity, *ApJ* 474, 511–517.

Fermi, E. 1954, Galactic magnetic fields and the origin of cosmic radiation, *ApJ* 119, 1.

Finnigan, J. 2001, How nature works; The science of self-organized criticality, *Agricultural and Forest Meteorology* 108 241–243.

Fisher, G. H., Canfield, R. C., and McClymont, A. N. 1985, Flare loop radiative dydrodynamics V. Response to thick-target heating. IV. Chromospheric evaporation due to heating by nonthermal electrons. VII. Dynamics of the thick-target heated model, *ApJ* 289(1), 1985, 414–441.

Fishman, G. D. et al. 1994, Discovery of intense gamma-ray flashes of atmospheric origin, *Science* 264, 1313–1316.

Fitzenreiter, R. J., Fainberg, J., and Bundy, R. B. 1976, Directivity of low frequency solar type III radio bursts, *SoPh* 46, 465–473.

Fraser, W. C., Kavelaars, J. J., Holman, M. J., Pritchet, C. J., Gladman, B. J., Grav, T., Jones, R. L., Macwilliams, J., and Petit, J. M. 2008, The Kuiper belt luminosity function from $m_R = 21$ to 26, *Icarus* 195, 827–843.

Fraser, W. C. and Kavelaars, J. J. 2008, A derivation of the luminosity function of the Kuiper belt from a broken power-law size distribution, *Icarus* 198(2), 452–458.

Freeman, M. P., Watkins, N. W., and Riley, D. J. 2000, Power law distributions of burst duration and interburst interval in the solar wind: Turbulence or dissipative self-organized criticality? *Physical Review E*, 62, 8794.

French, R. G. and Nicholson, P. D. 2000, Saturn's rings. II. Particle sizes inferred from stellar occultation data, *Icarus* 145, 502–523.

Fuentes, C. I. and Holman, M. J. 2008, A Subaru archival search for faint trans-Neptunian objects, *AJ* 136, 83–97.

Gabriel, S. B. and Feynman, J. 1996, Power-law distribution for solar energetic proton events, *SoPh* 165, 337–346.

Gabriel, S. B. and Patrick, G. J. 2003, Solar energetic particle events: Phenomenology and prediction, *SSRv* 107(1), 55–62.

Galam, S. 2012, *Sociophysics. A physicist's modeling of psycho-political phenomena*, Springer, New York, 439p.

Garcia-Marin, A. P., Estevez, J., Jimenez-Hornero, F. J., and Ayuso-Munoz, J. L. 2013, Multifractal analysis of validated wind speed time speed series, *Chaos: An Interdisciplinary Journal of Nonlinear Science* 23(1), id. 013133.

Gaskell, C. M. 2004, Lognormal X-ray flux variations in an extreme narrow-line Seyfert 1 galaxy, *ApJ* 612, L21–L24.

Georgoulis, M. K. and Vlahos, L. 1996, Coronal heating by nanoflares and the variability of the occurrence frequency distribution in solar flares, *ApJ* 469, L135–L138.

Georgoulis, M. K. and Vlahos, L. 1998, Variability of the occurrence frequency of solar flares and the statistical flare, *AA* 336, 721–734.

Georgoulis, M. K., Vilmer, N., and Croby, N. B. 2001, A comparison between statistical properties of solar X-ray flares and avalanche predictions in cellular automata statistical flare models, *AA* 367, 326–338.

Georgoulis, M. K., Rust, D. M., Bernasconi, P. N., and Schmieder, B. 2002, Statistics, morphology, and energetics of Ellerman bombs, *ApJ* 575, 506–528.

Georgoulis, M. K. 2005, Turbulence in the solar atmosphere: Manifestations and diagnostics via solar image processing, *SoPh* 228, 5–28.

Georgoulis, M. K. 2011, Are solar active regions with major flares more fractal, multifractal, or turbulent than others?, *SoPh* 276, 161–182.

Georgoulis, M. K. 2012, Are solar active regions with major flares more fractal, multifractal, or turbulent than others? *SoPh* 276, 161–182.

Georgoulis, M. K., Nindos, A., and Zhang, H. 2019, The source and engine of coronal mass ejections, *Philosophical Transactions of the Royal Society A* 377, 20180094.

Gerontidou, M., Vassilaki, A., Mavromichalaki, H., and Kurt, V. 2002, Frequency distributions of solar proton events, *Journal of Atmospheric and Solar-Terrestrial Physics* 64(5/6), 489–496.

Gheibi, A., Hossein, S., and Javaherian, M. 2017, The solar flare complex network, *ApJ* 847, 115 (12pp).

Giannattasio, F., et al. 2013, Diffusion of solar magnetic elements up to supergranular spatial and temporal scales, *ApJ* 770, 36.

Giannattasio, F., Stangalini, M., Berrilli, F., Del Moro, D., and Bellot Rubio, L. 2014a, Diffusion of magnetic elements in a supergranular cell, *ApJ* 788, 137.

Giannattasio, F., et al. 2014b, Pair separation of magnetic elements in the quiet Sun, *AA* 569, A121.

Giannattasio, F., Consolini, G., Berrilli, F., and Del Moro, D. 2019, The complex nature of magnetic element transport in the quiet Sun: The Levy-walk character, *ApJ* 878, 33.

Giannattasio, F., Consolini, G., and Berrilli, F. 2020, Magnetic energy balance in the quiet Sun on supergranular spatial and temporal scales, *ApJ* 904, 7.

Giannattasio, F., and Consolini, G. 2021, The complex nature of magnetic element transport in the quiet Sun: The multiscaling character, *ApJ* 908, 142.

Giles, D. E., Feng, H., and Godwin, R. T. 2011, On the bias of the maximum likelihood estimator for the two-parameter Lomax distribution, Econometrics Workshop Paper EWP1104, ISSN 1485–6441.

Giorgi, F., Ermolli, I., Romano, P., Stangalini, M., Zuccarello, F., and Criscuoli, S. 2015, The signature of flare activity in multifractal measurements of active regions observed by SDO/HMI, *SoPh* 290, 507.

Giuliani, P., Carbone, V., Veltri, P., Boffetta, G. and Vilpiani, A. 2000, Waiting time statistics in solar flares, *Physica A: Statistical Mechanics and Its Applications*, 280, 75–80.

Gizis, J. E., et al. 2017a, K2 Ultracool dwarfs survey. II. The white light flare rate of young brown dwarfs, *ApJ* 845, 33.

Gizis, J. E., Paudel, R. R., Schmidt, S. J., Willians, K. G. and Burgasser, A. J. 2017b, K2 Ultracool dwarfs survey. I. Photometry of an L Dwarf superflare, *ApJ* 838, 22 (6pp).

Gogus, E., Woods, P. M., Kouveliotou, C., van Paradijs, J., Briggs, M. S., Duncan, R. C., and Thompson, C. 1999, Statistical properties of SGR 1900+14 bursts, *ApJ* 526, L93–L96.

Gogus, E., Woods, P. M., Kouveliotou, C., and van Paradijs, J. 2000, Statistical properties of SGR 1806-20 bursts, *ApJ* 532, L121–L124.

Gogus, E., et al. 2017, Burst and outburst characteristics of magnetar 4U 0142+61, *ApJ* 835, 68 (8pp).

Goldstein, M. L., Morris, S. A., and Yen, G. G. 2004, Problems with fitting to the power-law distribution, *European Physical Journal B* 41, 255.

Golovko, A. A. and Salakhutdinova, I. I. 2015, Evolution of solar active regions: Detecting the emergence of new magnetic field through multifractal segmentation, *Astronomy Reports* 59, 776.

Golovko, A. A. and Salakhutdinova, I. I. 2018, Detecting the solar new magnetic flux regions on the base of vector magnetograms, *Journal of Atmospheric and Solar-Terrestrial Physics* 179, 120–127.

Gourdji, K., Michilli, D., Spitler, L. G., Hessels, J. W. T., Seymour, A., Cordes, J. M., and Chatterjee, S. 2019, A sample of low-energy bursts from FRB 12102, *ApJL* 877(2), L19, 12pp.

Gravity Collaboration, et al. 2020, The flux distribution of Sgr A*, *AA* 638, id. A2, 12pp.

Greco, A., Matthaeus, W. H., Servidio, S., and Dmitruk, P. 2009 Waiting-time distributions of magnetic discontinuities: Clustering or Poisson process? *Physical Review E*, 80, 046401.

Greco, A., Perri, S., Servidio, S., Yordanova, E., and Veltri, P. 2016, The complex structure of magnetic field discontinuities in the turbulent solar wind, *ApJ* 823, L39.

Greco, A., et al. 2018, Partial variance of increments method in solar wind observations and plasma simulations, *SSRv* 214, 1.

Grigolini, P., Leddon, D., and Scaffetta, N. 2002, Diffusion entropy and waiting time statistics of hard X-ray solar flares, *PhRvL E*, 65(4), id. 046203.

Güdel, M., Benz, A. O., Schmitt, J. H. M. M., and Skinner, S. L. 1996, The Neupert effect in active stellar coronae: Chromospheric evaporation and coronal heating in the dMe flare star binary UV Ceti, *ApJ* 471, 1002–1014.

Güdel, M., Audard, M., Kashyap, V. L., and Guinan, E. F. 2003, Are coronae of magnetically active stars heated by flares? II. Extreme Ultraviolet and X-ray flare statistics and the differential emission measure distribution, *ApJ* 582, 423–442.

Guerreiro, N., Haberreiter, M., Hansteen, V., and Schmutz, W. 2015, Small-scale heating events in the solar atmosphere. I. Identification, selection, and implications for coronal heating, *ApJ* 813, 61 (11pp).

Guerreiro, N., Haberreiter, M., Hansteen, V., and Schmutz, W. 2017, Small-scale heating events in the solar atmosphere. II. Lifetime, total energy, and magnetic properties, *AA* 603, A103.

Guidorzi, C., Dichiara, S., Frontera, F., Marguitti, R., Baldeschi, A., Amati, L. 2015, A common stochastic process rules gamma-ray burst prompt emission and X-ray flares, *ApJ* 801, id. 57, 11pp.

Haisch, B. M., Johnson, H. M., and Davidson, G. T. 1983, Creation of photometric star catalogs using UBV data and model stellar atmospheres, *Journal of the Astronautical Sciences* 31, 473–506.

Harko, T., Mocanu, G., and Stroia, N. 2015, Self-organized criticality in an one dimensional magnetized grid. Application to GRB X-ray afterglows, *Astrophysics and Space Science* 357(1), id. 84, 9pp.

Hathaway, D. H., Beck, J. G., Bogart, R. S., Bachmann, K. T., Khatri, G., Pelitto, J. M., Han, S., and Raymond, J. 2000, The photospheric convection spectrum, *SoPh* 193, 299–312.

Hergarten, S. 2002, *Self-organized criticality in Earth systems*, Springer, New York, 272p.

Hergarten, S. 2013a, Self-organized criticality in landslides (Chapter 11), in *Self-Organized Criticality Systems* (ed. Aschwanden, M. J.), Open Academic Press GmbH Co, Berlin, Warsaw, 329–348.

Hergarten, S. 2013b, Wildfires and the forest-fire model (Chapter 10), in *Self-Organized Criticality Systems* (ed. Aschwanden, M. J.), Open Academic Press GmbH Co.

Hermsen, W., et al. 2018, Discovery of synchronous X-ray and radio modeling of PSR B0823+26, *MNRAS* 480, 3655–3670.

Hirzberger, J., Vazquez, M., Bonet, J. A., Hanslmeier, A., and Sobotka, M. 1997, Time series of solar granulation images: I. Differences between small and large granules in quiet regions, *ApJ* 480, 406–419.

Hnat, B., Chapman, S. C., Kiyani, K., Rowlands, G., and Watkins, N. W. 2007, On the fractal nature of the magnetic field energy density in the solar wind, *GRL* 34, 15108.

Holman, G. D. 1985, Acceleration of runaway electrons and Joule heating in solar flares, *ApJ* 293, 584–594.

Hood, A. W., Cargill, P. J., Browning, P, K., and Tam, K. V. 2016, An MHD avalanche in a multi-threaded coronal loop, *ApJ* 817, 5 (7pp).

Horbury, T. S., Balogh, A., Forsyth, R. J., and Smith, E. J. 1997, Ulysses observations of intermittent heliospheric turbulence, *Advances in Space Research* 19(6), 847–850.

Horton, W. and Doxas, I. 1996, A low-dimensional energy-conserving state space model for substorm dynamics, *JGR* 101(A2), 27, 223–27, 238.

Hoshino, M., Nishida, A., Yamamoto, T., and Kokubun, S. 1994, Turbulent magnetic field in the distant magnetotail: Bottom-up process of plasmoid formation? *GRL* 21(25), 2935–2938.

Hosking, J. R. M. and Wallis, J. R. 1987, Parameter and quantile estimation for the generalized Pareto distribution, *Technometrics* 29(3), 339–349.

Howes, G. G., Cowley, S. C., Dorland, W., Hammett, G. W., Quataert, E., and Schekochihin, A. A. 2008, A model of turbulence in magnetized plasmas: Implications for the dissipation range in the solar wind, *JGR (Space Physics)* 113(A4), CiteID A05103.

Huberman, B. A. and Adamic, L. A. 1999, Internet: Growth dynamics of the World-Wide Web, *Nature*, 401(6749), 131.

Hudson, H. S., Peterson, L. E., and Schwartz, D. A. 1969, The hard X-Ray spectrum observed from the third orbiting solar observatory, *ApJ* 157, 389–415.

Hudson, H. S. 1978 Threshold effect in second-stage acceleration, *SoPh* 57, 237–240.

Hudson H. S. 1991, Solar flares, microflares, nanoflares, and coronal heating, *SoPh* 133, 357–369.

Hudson, H. S. 2020, Solar flare build-up and release, *SoPh* 295, 132.

Hufnagel, B. R. and Bregman, J. N. 1992, Optical and radio variability in blazars, *ApJ* 386, 473–484.

Hughes, D., Paczuski, M., Dendy, R. O., Helander, P., and McClements, K. G. 2003, Solar flares as cascades of reconnecting magnetic loops, *Physical Review Letters* 90, 131101.

Huppenkothen, D., et al. 2015, Dissecting magnetar variability with Bayesian hierarchical models, *ApJ* 10(1), id. 66, 21pp.

Ilachinski, A. 2001, *Cellular Automata*, World Scientific, New Jersey, 840p.

Ioshpa, B. A., Obridko, V. N., and Rudenchik, E. A. 2008, Fractal properties of solar magnetic fields, *Astronomy Letters* 34, 210.

Isliker, H., Vlahos, L., Benz, A. O., and Raoult, A. 1998, A stochastic model for solar type III bursts, *AA* 336, 371–380.

Isliker, H. and Benz, A. O. 2001, On the reliability of peak-flux distributions, with an application to solar flares, *AA* 375, 1040–1048.

Ivezic, Z., et al. (SDSS Collaboration) 2001, Solar system objects observed in the Sloan Digital Sky Survey commissioning data, *The Astronomical Journal* 122, 2749–2784.

Iwai, K., Masuda, S., Miyoshi, Y., Tsuchiya, F., Morioka, A., and Misawa, H. 2013, Peak flux distributions of solar radio type I bursts from highly resolved spectral observations, *ApJ* 768, L2 (4pp).

Janssen, K., Voegler, A., and Kneer, F. 2003, On the fractal dimension of small-scale magnetic structures in the Sun, *AA* 409, 1127–1134.

Javaherian, M., Safari, H., Dadashi, N., and Aschwanden, M. J. 2017, Statistical properties of photospheric magnetic elements observed by the helioseismic and magnetic imager onboard the solar dynamics observatory, *SoPh* 292, 164.

Jedicke, T. S. and Metcalfe, T. S. 1998, The orbital and absolute magnitude distributions of main belt asteroids, *Ikarus* 131(2), 245–260.

Jedicke, R., Larsen, J., and Spahr, T. 2002, Observational selection effects in asteroid surveys, in *Asteroids III* (eds. Bottke, W., et al.), University Arizona Press Tuscon, AZ, 71–87.

Jensen, H. J. 1998, *Self-organized criticality: Emergent Complex Behavior in Physical and Biological Systems*, Cambridge University Press, Cambridge UK, 153p.

Jensen, H. J. 2023, *Complexity science: A Study of Emergence*, Cambridge University Press, Cambridge.

Jewitt, D. C., Trujillo, C., and Lu, J. X. 2000, Population and size distribution of small Jovian Trojan asteroids, *ApJ* 120(2), 1140–1147.

Johnston, S. and Romani, R. W. 2002, A search for giant pulses in Vela-like pulsars, *MNRAS* 332, 109–115.

Joulin, V., Buchlin, E., Solomon, J., and Guennou, C. 2016, Energetic characterisation and statistics of solar coronal brightenings, *AA* 591, A148.

Kadanoff, L. P. 1991, Complex structures from simple systems, *Physics Today* 44, 9–11.

Kadanoff, L. P., Nagel, S. R., Wu, L., and Zhou, S. M. 1991, Scaling and universality in avalanches, *Physical Review A* 39(12), 6524–6537.

Kahler, S. W. 2013, Does a scaling law exist between solar energetic particle events and solar flares? *ApJ* 769, 35.

Kakinuma, T., Yamashita, T., and Enome, S. 1969, A statistical study of solar radio bursts a microwave frequencies, Proceedings of the Research Institute of Atmospherics, Nagoya University Japan, Vol. 16, 127–141.

Kalapotharakos, C., Voglis, N., and Contopoulos, G. 2004, Chaos and secular evolution of triaxial N-body galactic models due to an imposed central mass, *AA* 428, 905–923.

Kanazir, M. and Wheatland, M. S. 2010, Time-dependent stochastic modeling of solar active region energy, *SoPh* 266, 301–321.

Kanella, C. and Gudiksen, B. V. 2017, Identification of coronal heating events in 3D simulations, *AA* 603, A83.

Kashyap, V. L., Drake, J. J., Güdel, M., and Audard, M. 2002, Flare heating in stellar coronae, *ApJ* 580, 1118–1132.

Kato, T., Ishioka, R., and Uemura, M. 2002, Photometric study of KR Aurigae during the high state in 2001, *PASJ* 54, 1033–1036.

Katz, J. 1986, A model of propagating brittle failure in heterogeneous media, *JGR* 91, 10412.

Kaw, P. K., Milikh, G. M., Sharma, A. S., Guzdar, P. N., and Papadopoulos., K. 2001, Gamma ray flashes produced by plasma effects in the middle atmosphere, *Physics of Plasmas* 8, 4954.

Kawai, T. and Imada, S. 2022, Factors that determine the power-law index of an energy distribution of solar flares, *ApJ* 931, 113.

Kelty-Stephen, D. G., Palatinus, K., Saltzman, E., and Dixon, J. A. 2013, A tutorial on multifractality, cascades, and interactivity for empirical time series in ecological science, *Ecological Psychology* 25, 1–62.

Kennedy, M. R., Clark, C. J., Voisin, G., and Breton, R. P. 2018, Kepler K2 observations of the transitional millisecond pulsar PSR J1023+0038, *MNRAS* 477, 1120–1132.

Klebesadel, R. W., Strong, I. B., and Olson, R. A. 1973, Observations of gamma-ray bursts of cosmic origin, *ApJ* 182, p. L85.

Klimas, A. J., Vassiliadis, D., Baker, D. N., and Roberts, D. A. 1996, The organized nonlinear dynamics of the magnetosphere, *JGR* 101(A6), 13089–13114.

Klimas, A. J., Valdivia, J. A., Vassiliadis, D., Baker, D. N., Hesse, M., and Takalo, J. 2000, Self-organized criticality in the substorm phenomenon and its relation to localized reconnection in the magnetosphere plasma sheet, *JGR* 105(A8), 18,765–18,780.

Klimas, A. J., Uritsky, V. M., Vassiliadis, D., and Baker, D. N. 2004, Reconnection and scale-free avalanching in a driven current-sheet model, *Journal of Geophysical Research (Space Physics)* 104(A2), 2218.

Klimas, A. J., Uritsky, V., and Donovan, E. 2010, Multiscale auroral emission statistics as evidence of turbulent reconnection in Earth's midtail plasma sheet, *JGR* 115(A6), CiteID A06202.

Klimas, A. J. and Uritsky, V. M. 2017, Criticality and turbulence in a resistive magnetohydrodynamic current sheet, *PhRvE* 95, 3209.

Knizhnik, K. J., Uritsky, V. M., Klimchuk, J. A., and DeVore, C. R. 2018, Power-law statistics of driven reconnection in the magnetically closed corona, *ApJ* 853, 82, 14pp.

Kokubo, E. and Ida, S. 1998, Oligarchic growth of protoplanets, *Icarus* 131, 171–178.

Kolmogorov, A. N. 1941, The local structure of turbulence in incompressible viscous fluid for very large Reynolds numbers, *Doklady Akademia NAUK SSR*, 30, 301–305.

Kou, Y. K., Jing, Z. C., Cheng, X., Fan, W. Q., Liu, Y., Li, C., and Ding, M. D. 2020, What determines solar flares producing interplanetary type III radio bursts? *ApJ* 898, L24 (6pp).

Kouveliotou, C., Dieters, S., Strohmayer, T., van Paradijs, J., Fishman, G. J., Meegan, C. A., Hurley, K., Kommers, J., Smith, I., Frail, D., and Muakami, T. 1998, An X-ray pulsar with a superstrong magnetic field in the soft γ-ray repeater SGR 1806-20, *Nature* 393, 235–237.

Kouveliotou, C., Strohmayer, T., Hurley, K., van Paradijs, J., Finger, M. H., Dieters, S., Woods, P., Thomson, C., and Duncan, R. C. 1999, Discovery of a magnetar associated with the soft gamma ray repeater SGR 1900+14, *ApJ* 510, L115–L118.

Kovacs, P, Carbone, V., and Voeroes, Z. 2001, Wavelet-based filtering of intermittent events from geomagnetic time-series, *Planetary and Space Science*, 49, 1219.

Kozelov, B. V., Uritsky, V. M., and Klimas, A. J. 2004, Power law probability distributions of multiscale auroral dynamics from ground-based TV observations, *GRL* 31, 20804.

Kraichnan, R. H. 1974, On Kolmogorov's inertial-range theories, *Journal of Fluid Mechanics* 62, 305–330.

Krucker, S. and Benz, A. O. 1998, Energy distribution of heating processes in the quiet solar corona, *ApJ* 501, L213.

Kundu, M. R. 1965, *Solar Radio Astronomy*, Interscience Publication, New York, 660p.

Kunjaya, C., Mahasena, P., Vierdayanti, K., and Herlie, S. 2011, Can self-organized critical accretion disks generate a log-normal emission variability in AGN? *Astrophysics and Space Science* 336(2), 455–460.

Kurt, V. G. 1990, Electrons and X-ray emission of solar flares, in *Basic Plasma Processes in the Sun, Proceedings of 142 Symposium IAU* (ed. Priest, E. R. and Krishan, V.), Kluwer, Dordrecht.

Kuznetsov, S. N. and Kurt, V. G. 1991, Flare-generated protons of different energies, in *Dynamics of Solar Flares* Proceedings (eds. Schmieder, B. and Priest, E.), Observatoire de Paris, DASOP, Paris, France, pp. 63–64.

Kuznetsov, A. A. and Kolkotkov, D. Y. 2021, Stellar superflares observed simultaneously with Kepler and XMM-Newton, *ApJ* 912/1, id.81, 16pp.

Lacy, C. H., Moffett, T. J., and Evans, D. S. 1976, UV Ceti stars: Statistical analysis of observational data, *ApJSS* 30, 85–96.

Lamy, P. L., FLoyd, O., Boclet, B., Wojak, J., Gilardy, H., and Barlyaeva, T. 2019, Coronal mass ejections over solar cycles 23 and 24, *SSRv* 215, 39.

Lawrence, J. K. 1991, Diffusion of magnetic flux elements on a fractal geometry, *SoPh* 135, 249–259.

Lawrence, J. K. and Schrijver, C. J. 1993, Anomalous diffusion of magnetic elements across the solar surface, *ApJ* 411, 402–405.

Lee, T. T., Petrosian, V., and McTiernan, J. M. 1993, The distribution of flare parameters and implications for coronal heating, *ApJ* 412, 401–409.

Lee, T. T., Petrosian, V., and McTiernan, J. M. 1995, The Neupert effect and the chromospheric evaporation model for solar flares, *ApJ* 418, 915–924.

Lei, W. H., et al. 2020, Do the solar flares originating from individual active region follow a random process or a memory-dependent correlation? *MNRAS* 494, 975.

Leighly, K. M. and O'Brien, P. T. 1997, Evidence for nonlinear X-Ray variability from the broad-line radio galaxy 3C 390.3, *The Astrophysical Journal* 481, L15.

Leighton, R. B. 1964, Transport of magnetic fields on the Sun, *ApJ* 140, 1574.

Lepreti, F., Carbone, V. and Veltri, P. 2001, Solar flare waiting time distribution: Varying-rate Poisson or Levy function? *ApJ* 555, L133–L136.

Lepreti, F., Carbone, V., Giuliani, P., Sorriso-Valvo, L., and Veltri, P. 2004, Statistical properties of dissipation bursts within turbulence: Solar flares and geomagnetic activity, *Planetary and Space Science* 52, 957–962.

Li, C., Zhong, S. J., Xu, Z. G., He, H., Yan, Y., Chen, P. F., and Fang, C. 2018, Waiting time distributions of solar and stellar flares: Poisson process or with memory? *MNRAS* 479, L139–L142.

Li, C. Z., Yang, J. H., Li, C., Tian, Y., Yang, Y., and Wei, H. Q. 2016a, Self-organized criticality of the solar eruptions during solar cycle 23, *Sciencia Sinica Physics*, 46(2), 029501.

Li, Y. P., Gan, W. Q., and Su, Y. 2009, Double hard X-ray peaks in RHESSI flares as evidence of chromospheric evaporation and implications for modifying the Neupert effect, *RAA* 9(19), 1155–1164.

Li, Y. P., Gan, W. Q., and Feng, L. 2012, Statistical analyses on thermal aspects of solar flares, *ApJ* 747, 133.

Li, Y. P., Gan, W. Q., Feng, L., Liu, S. M., and Struminsky, A. 2013, The breakdown of the power-law frequency distributions for the hard X-ray peak count rates of solar flares, *Research in Astronomy and Astrophysics* 13, 1482.

Li, Y. P., Feng, L., Zhang, P., Liu, S. M., and Gan, W. Q. 2016b, On the power-law distributions of X-ray fluxes from solar flares observed with GOES, *Research in Astronomy and Astrophysics* 16, 161.

Lin, H. N. and Sang, Y. 2020, Scale-invariance in the repeating fast radio burst 121102, *MNRAS* 491, 2156–2161.

Lin, R. P., Schwartz, R. A., Kane, S. R., Pelling, R. M., and Hurley, K. C. 1984, Solar hard X-ray microflares, *ApJ* 283, 421–425.

Lin, R. P., Feffer, P. T., and Schwartz, R. A. 2001, Solar hard X-Ray bursts and electron acceleration down to 8 keV, *ApJ* 557, L125–L128.

Lion, S., Alexandrova, O., and Zaslavsky, A. 2015, Coherent events and spectral shape at ion kinetic scales in the fast solar wind turbulence, *ApJ* 824, 47 (13pp).

Lipari, P. 2021, The origin of the power-law form of the extragalactic gamma-ray flux, *Astroparticle Physics* 125, id. 102507.

Lippiello, E., de Arcangelis, L., and Godano, C. 2008, Different triggering mechanisms for solar flares and coronal mass ejections, *AA* 488, L29–L32.

Litvinenko Y. 1996, Particle acceleration in reconnecting current sheets with a non-zero magnetic field, *ApJ* 462, 997.

Longbottom, A. W., Rickard, G. J., Craig, I. J. D. and Sneyd, A. D. 1998. Magnetic flux braiding: Force-free equilibria and current sheets, *ApJ* 500, 471–482.

Longcope, D. W. and Noonan, E. J. 2000, Self-organized criticality from separator reconnection in solar flares, *ApJ* 542, 1088–1099.

Longcope, D. W., Ravindra, B., and Barnes, G. 2007, Determining the source of coronal helicity through measurements of braiding and spin helicity fluxes in active regions, *ApJ* 668, 571–585.

Lopez Fuentes, M. C., and Klimchuk, J. A. 2006, Coronal loops as self-organized critical systems, *Boletin de la Asociacion Argentina de Astronomia La Plata Argentina* 49, 108.

Lopez Fuentes M. C., and Klimchuk, J. A. 2010, A simple model for the evolution of multi-stranded coronal loops, *ApJ* 719, 591–601.

Lopez Fuentes, M. C., and Klimchuk, J. A. 2015, Two-dimensional cellular automaton model for the evolution of active region coronal plasmas, *ApJ* 799, 128.

Lopez Fuentes, M. C., and Klimchuk, J. A. 2016, A nanoflare-based cellular automaton model and the observed properties of the coronal plasma, *ApJ* 828, 86.

Loreto, V., Pietronero, L., Vespignani, A., and Zapperi, S. 1995, Renormalization group approach to the critical behavior of the forest-fire model, *Physical Review Letters* 75(3), 465–468.

Lorimer, D. R., Bailes, M., McLaughlin, M. A. 2007, A bright millisecond radio burst of extragalactic origin, *Science* 318, 777.

Love, J. J. 2020, Some experiments in extreme-value statistical modeling of magnetic superstorm intensities, *Space Weather* 18(1), id. e02255.

Love, J. J. 2021, Extreme event magnetic storm probabilities derived from rank statistics of historical Dst intensities for solar cycles 14–24, *Space Weather* 1901579.

Lu, E. T. and Hamilton, R. J. 1991, Avalanches and the distribution of solar flares, *Astrophysical Journal* 380, L89–L92.

Lu, E. T., Hamilton, R. J., McTiernan, J. M., and Bromund, K. R. 1993, Solar flares and avalanches in driven dissipative systems, *ApJ* 412, 841–852.

Lu, E. T. 1995, Avalanches in continuum driven dissipative systems, *PhRvL* 74, 2511–2514.

Lu, E. T. 1995a, Constraints on energy storage and release models for astrophysical transients and solar flares, *ApJ* 447, 416.

Lu, E. T. 1995b, The statistical physics of solar active regions and the fundamental nature of solar flares, *ApJ* 446, L109–L112.

Lu, W. and Piro, A. L. 2019, Implications from ASKAP fast radio burst statistics, *ApJ* 883(1), id. 40, 8pp.

Lui, A. T. Y., Lopez, R. W., Krimigis, S. M., McEntire, R. W., Zanetti, L. J., and Potemra, T. A. 1988, A case study of magnetotail current sheet disruption and diversion, *GRL* 15(7), 721–724.

Lui, A. T. Y., Chapman, S. C., Liou, K., Newell, P. T., Meng, C. I., Brittnacher, M., and Parks, G. K. 2000, Is the dynamic magnetosphere an avalanching system? *GRL* 27(7), 911–914.

Lundgren, S. C., Cordes, J. M., Ulmer, M., Matz, S. M., Lomatch, S., Foster, R. S., and Hankins, T. 1995, Giant pulses from the Crab pulsar: A joint radio and gamma-ray study, *ApJ* 453, 433–445.

Lurie, J. C., Davenport, J. R. A., Hawley, S. H., Wilkinson, T. D., Wisniewski, J. P., Kowalski, A. F., and Hebb, L. 2015, Kepler flares III: Stellar activity on GJ 1245A and B, *ApJ* 800, 95, 14pp.

Lyon, G. J. 2000, The solar wind-magnetosphere-ionosphere system, *Science* 288, 1987–1991.

Lyu, F., Li, Y. P., Hou, S. J., Wei, J. J., Geng, J. J., and Wu, X. F. 2020, Self-organized criticality in multi-pulse gamma-ray bursts, *Frontiers of Physics* 16(1), id. 14501.

Macek, W. M. and Szczepaniak, A. 2008, Generalized two-scale weighted Cantor set model for solar wind turbulence, *GRL* 35, 2108.

Macek, W. M. and Wawrzaszek, A. 2009, Evolution of asymmetric multifractal scaling of solar wind turbulence in the outer heliosphere, *JGR (Space Physics)*, 114, 3108.

Macek, W. M. 2010, Chaos and multifractals in the solar wind, *Advances in Space Research*, 46, 526.

Maehara, H., Shibayama, T., Notsu, S., Notsu, Y., Nagao, T., Kusaga, S., Honda, S., Nogami, D., and Shibata, K. 2012, Superflares on solar-type stars, *Nature* 485, 478–481.

Maehara, H., Shibayama, T., Notsu, Y., Notsu, S., Honda, S., Nobami, D., and Shibata, K. 2015, Statistical properties of superflares on solar-type stars based on 1-min cadence data, *Earth, Planets and Space* 67, 59.

Maehara, H., et al. 2017, Starspot activity and superflares on solar-type stars, *PASJ* 69, 41.

Main, I. G. and Burton, P. W. 1984, Physical links between crustal deformation, seismic moment and seismic hazard for regions of varying seismicity, *Geophysics Journal* 79(2), 469–488.

Mandelbrot, B. B. 1977, Fractals: Form, Chance, and Dimension, Translation of *Les objects fractals*, W. H. Freeman, San Francisco.

Mandelbrot, B. B. 1983, The fractal geometry of nature, W. H. Freeman, San Francisco.

Mandelbrot, B. B. 1985, Self-affine fractals and fractal dimension, *Physica Scripta* 32, 257–260.

Marek, M. and Schreiber, R. 2017, Is the AKR cyclotron maser instability a self-organized criticality system? in *Planetary Radio Emissions VIII* (ed. Fischer, G.), Academy of Sciences Press, Vienna 2017, pp. 269–277.

Marek, M. and Schreiber, R. 2022, Can the auroral kilometric radiation be a self-organized criticality system? *Earth and Space Science* 8(5), id. e02148.

Markovic, D. and Gros, C. 2014, Power laws and self-organized criticality in theory and nature, *Physics Reports* 536, 41–74.

Martin, E., Shreim, A., and Paczuski, M. 2010, Activity-dependent branching ratios in shocks, solar x-ray flux, and the Bak-Tang-Wiesenfeld sandpile model, *Physical Review E* 016109-1:6

Matthaeus, W. H. and Goldstein, M. L. 1986, Low-frequency 1/f noise in the interplanetary magnetic field, *Physical Review Letters* 57(4), 495–408.

McAteer, R. T. J., Gallagher, P. T., Ireland, J., and Young, C. A. 2005, Automated boundary-extraction And region-growing techniques applied to solar magnetograms, *SoPh* 228, 55–66.

McAteer, R. T. J., Young, C. A., Ireland, J., and Gallagher, P. T. 2007, The bursty nature of solar flare X-ray emission, *ApJ* 662, 691–700.

McAteer, R. T. J. 2013, Self-organized criticality and fractal geometry (Chapter 3), in *Self-Organized Criticality Systems* (ed. Aschwanden, M. J.), Open Academic Press GmbH Co.

McAteer, R. T. J., et al. 2016, 25 Years of self-organized criticality: numerical detection methods, *SSRv* 198, 217–266.

McFadden, L. A. and Binzel, R. P. 2007, Near-earth objects, in *Encyclopedia of the Solar System* (eds. McFadden, L.A., Weissman, P.R., and Johnson, T.V.), Elsevier, p. 293.

McIntosh, S. W. and Charbonneau, P. 2001, Geometric effects in avalanche models of solar flares: Implications for coronal heating, *ApJ* 563, L165–168.

McIntosh, S. W., Charbonneau, P., Bogdan, T. J., Liu, H. L., and Norman, J. P. 2002, Geometrical properties of avalanches in self-organized critical models of solar flares, *PhRvE* 65, 6125.

McIntosh, S. W. and Gurman, J. B. 2005, Nine years of EUV bright points, *SoPh* 228, 285–299.

McTiernan, J. M., Fisher, G. H., and Li, P. 1999, The solar soft X-ray differential emission measure and the Neupert effect at different temperatures, *ApJ* 514, 472–483.

Melatos, A., Peralta, C., and Wyithe, J. S. B. 2008, Avalanche dynamics of radio pulsar glitches, *ApJ* 672, 1103–1118.

Melatos, A., Douglass, J. A., and Simula, T. P. 2015, Persistent gravitational radiation from glitching pulsars, *ApJ* 807(123), 12.

Melatos, A., Howitt, G., and Fulgenzi, W. 2018, Size-waiting-time correlations in pulsar glitches, *ApJ* 863, 196.

Melnick, J. and Selman, F. J. 2000, Self-organized criticality and the IMF of starbursts, in *Cosmic Evolution and Galaxy Formation: Structure, Interactions, and Feedback*, ASP Conference Series 215 (eds. Franco, J., Terlevich, E. , Lopez-Cruz, O., and Aretxaga, I.), Astronomical Society of the Pacific, San Francisco.

Mendoza, B., Melendez, R., Miroshnichenko, L. I., and Perez Enriquez. R. 1997, Frequency distributions of solar proton events, Proceeding of 25th International Cosmic Ray Conference, Potchefstroom University, Vol. 1, p. 81.

Mercier, C. and Trottet, G. 1997, Coronal radio bursts: A signature of nanoflares? *ApJ* 484, 920–926.

Meunier, N. 1999, Large-scale dynamics of active regions and small photospheric magnetic features, *ApJ* 527, 967–976.

Meunier, N. 2004, Complexity of magnetic structures: Flares and cycle phase dependence, *AA* 420, 333–342.

Meyer, M., Scargle, J. D., and Blandford, R. D. 2019, Characterizing the gamma-ray variability of the brightest flat spectrum radio quasars observed with the Fermi LAT, *ApJ* 877, 39.

Meyrand, R., and Galtier, S. 2010, A universal law for solar-wind turbulence at electron scales, *ApJ* 721, 1421–1424.

Michilli, D., et al. 2018, An extreme magneto-ionic environment associated with the fast radio burst source FRB 121102, *Nature* 553(7687), 182–185.

Milikh, G. M., Guzdar, P. N., and Sharma, A. S. 2005, Gamma ray flashes due to plasma processes in the atmosphere: Role of whistler waves, *JGR* 110(A2), A02308.

Milovanov, A. V., Zelenyi, L. M., Zimbardo, G., and Veltri, P. 2001, Self-organized branching of magnetotail current system near the percolation threshold, *JGR* 106(A4), 6291–6308.

Milovanov, A. V. and Zelenyi, L. M. 2002, Nonequilibrium stationary states in the earth's magnetotail: Stochastic acceleration processes and nonthermal distribution functions, *Advances in Space Research*, 30, 2667.

Milovanov, A. V. 2013, Percolation models of self-organized critical phenomena (Chapter 4), in *Self-Organized Criticality Systems* (ed. Aschwanden, M. J.), Open Academic Press GmbH Co.

Mineshige, S. and Negoro, H. 1999, Accretion disks in the context of self-organized criticality: How to produce 1/f fluctuations? in *High Energy Processes in Accreting Black Holes*, ASP Conference Series 161, 113–128.

Mineshige, S., Takeuchi, M., and Nishimori, H. 1994a, Is a black hole accretion disk in a self-organized critical state? *ApJ* 435, L125–L128.

Mineshige, S., Ouchi, B., and Nishimori, H. 1994b, On the generation of 1/f fluctuations in X-rays from black-hole objects, *PASJ* 46, 97–105.

Miroshnichenko, L. I., Mendoza, B., and Perez-Enriquez R. 2001, Size distributions of the >10 MeV solar proton events, *SoPh* 202, 151–171.

Miroshnichenko, L. I. and Nymmik, R. A. 2014, Extreme fluxes in solar energetic particle events: Methodological and physical limitations, *Radiation Measurements* 61, 6–15.

Mocanu, G. and Grumiller, D. 2012, Self-organized criticality in boson clouds around black holes, *Physical Review D* 85(10), is.105022.

Moffett, J. J. 1974, UV Ceti flare stars observational data, *ApJS* 273, 29, 1–42.

Moffat, J. W. 1997, Stochastic gravity and self-organized critical cosmology, in *Very High Energy Phenomena in the Universe*, Morion Workshop (eds. Giraud-Heraud, Y. and Than Van, J. T.), p. 353.

Moloney, N. R. and Davidsen, J. 2011, Extreme bursts in the solar wind, *GRL* 38(4), CiteID L14111.

Moloney, N. R. and Davidsen, J. 2014, Stationarity of extreme bursts in the solar wind, *Physical Review E*, 89(5), id 052812.

Monte-Moreno, E. and Hernandez-Pajares, M. 2014, Occurrence of solar flares viewed with GPS: Statistics and fractal nature, *JGR Space Physics* 119(11), 9216–9227.

Moon, Y. J., Choe, G. S., Wang, H., and Park, Y. D. 2003, Sympathetic coronal mass ejections, *ApJ* 588, 1176–1182.

Moradhaseli, M. A., Javaherian, M., Fathalian, N., and Safari, H. 2021, Empirical scaling relations for the photospheric magnetic elements of the flaring and non-flaring active regions, *Acta Astronomia* 71, 163.

Morales, L. and Charbonneau, P. 2008, Self-organized critical model of energy release in an idealized coronal loop, *ApJ* 682, 654–666.

Morales, L. F. and Charbonneau, P. 2009, Geometrical properties of avalanches in a pseudo-3D coronal loop, *ApJ* 698, 1893–1902.

Morales, L. F., Dmitruk, P., and Gomez, D. O. 2020, Energy dissipation in coronal loops: Statistical analysis of intermittent structures in magnetohydrodynamic turbulence, *ApJ* 894, 90–95.

Morina, D., Serra, I., Puig, P., and Corral, A. 2019, Probability estimation of a Carrington-like geomagnetic storm, *Scientific Reports* 9, id. 2393.

Mottez, F., Zarka, P., and Voison, G. 2020, Repeating fast radio bursts caused by small bodies orbiting a pulsar or a magnetar, *AA* 644, id. A156.

Mullan, D. J., Mathioudakis, M., Bloomfield, D. S., and Christian, D. J. 2006, A comparative study of flaring loops in active stars, *ApJ*SS 164, 173–201.

Nagler, J., Hauert, C., and Schuster, H. G. 1999, Self-organized criticality in a nutshell, *Physical Review E* 60 2706–2709.

Nakazato, K. 2014, Self-organized criticality in a spherically closed cellular automaton: Modeling soft gamma repeater bursts driven by magnetic reconnection, *Physical Review D* 90(4), id. 043010.

Negoro, H., Kitamoto, S., Takeuchi, M., and Mineshige, S. 1995, Statistics of X-ray fluctuations from Cygnus X-1: Reservoirs in the disk? *ApJ* 452, L49–L52.

Nelson, K. P. 2020, Independent approximates enable closed-form parameter estimation of heavy-tailed distributions, arXiv:2012.11026.

Neupert, W. M. 1968, Comparison of solar X-ray line emission with microwave emission during flares, *ApJ* 153, L59–L64.

Newman, M. E. J. 2011, Complex systems: A survey, *American J. Physics* 79/8, 800–810.

Nhalil, N. V., Nelson, C. J., Mathioudakis, M., and Doyle, G. J. 2020, Power-law energy distributions of small-scale impulsive events on the active Sun: Results from IRIS, *MNRAS* 499, 1385–1394.

Nicol, R. M., Chapman, S. C., and Dendy, R. O. 2009, Quantifying the anisotropy and solar cycle dependence of 1/f solar wind fluctuations observed by advanced composition explorer, *ApJ* 703, 2138.

Nielsen, M. B., Gizon, L., Schunker, H., Karoff, C. 2013, Rotation periods of 12,000 main-sequence Kepler stars: Dependence on stellar spectral type and comparison with v sin i observations, *AA* 557 id. L10.

Ning, Z., Wu, H., and Meng, X. 2007, Frequency distributions of microwave pulses for the 18 March 2003 solar flare, *SoPh* 242, 101–109.

Ning, Z. and Cao, W. 2010a, Investigation of the Neupert effect in the various intervals of solar flares, *SoPh* 264, 329–344.

Ning, Z. and Cao, W. 2010b, Investigation of chromospheric evaporation in a Neupert-type solar flare, *ApJ* 717, 1232–1242.

Nishizuka, N., Asai, A., Takasaki, H., Kurokawa, H., and Shibata, K. 2009, The power-law distribution of flare kernels and fractal current sheets in a solar flare, *ApJ* 694, L74–L78.

Nishizuka, N. and Shibata, K. 2013, Fermi acceleration in plasmoids interacting with fast shocks of reconnection via fractal reconnection, *Physics Reviews and Letters* 110, id. 051101.

Nita, G. M., Gary, D. E., Lanzerotti, L. J., and Thomson, D. J. 2002, The peak flux distribution of solar radio bursts, *ApJ* 570, 423–438.

Notsu, Y., Shibayama, T., Maehara, H., Notsu, S., Nagao, T., Honda, S., Ishi, T. T., Nogami, D., and Shibata, K. 2013, Superflare on solar-type stars observed with KEPLER II. Photometric variability of superflare-generating stars: A signature of stellar rotation and starspots, *ApJ* 771, 127.

Notsu, Y., Honda, S., Maehara, H., Notsu, S., Shibayama, T. Nogami, D., and Shibata, K. 2015a, High dispersion spectroscopy of solar-type superflare stars, I. Temperature, surface gravity, metallicity, and v sin i, *PASJ* 67(3), id. 32, 24pp.

Notsu, Y., Honda, S., Maehara, H., Notsu, S., Shibatama, T., Nogami, D., and Shibta, K. 2015b, High dispersion spectroscopy of solar-type superflare stars. II. Stellar rotation, Starspots, and chromospheric activities, *PASJ* 67(3), id.33, 14pp.

Nozakura, T. and Ikeuchi, S. 1988, Spiral patterns on a differentially rotating galactic disk – Self-organized structures in galaxies, *ApJ* 333, 68–77.

Nurhan, Y. I., Johnson, J. R., Homan, J. R., Wing, S., and Aschwanden, M. J. 2021, Role of the solar minimum in the waiting time distribution throughout the heliosphere, *GRL* 48, 16.

Ofman, L. 2010, Wave modeling of the solar wind, living reviews in solar physics, 7, 4.

Osten, R. A. and Brown, A. 1999, Extreme Ultraviolet Explorer photometry of RS Canum Venaticorum systems" Four flaring megaseconds, *ApJ* 515, 746–761.

Ostgaard, N., Gjesteland, T., Hansen, R. S., Collier, A. B., and Carlson B. 2012, The true fluence distribution of terrestrial gamma flashes at satellite altitude, *JGR* 117(A3), CiteID A03327.

Paczuski, M. and Hughes D. 2004, A heavenly example of scale free networks and self-organized criticality, *Physica A* 342(1), 158–163.

Paizis, A. and Sidoli, L. 2014, Cumulative luminosity distributions of supergiant fast X-ray transients in hard X-rays, *MNRAS* 439(4), 3439–3452.

Paniveni, U., Krishan, V., Singh, J., and Srikanth, R. 2005, On the fractal structure of solar supergranulation, *SoPh* 231, 1.

Paniveni, U., Krishan, V., Singh, J., and Srikanth, R. 2010, Activity dependence of solar supergranular fractal dimension, *MNRAS* 402, 424.

Pankine, A. A. and Ingersoll, A. P. 2004a, Interannual variability of Mars global dust storms: An example of self-organized criticality? *Ikarus* 170, 514–518.

Pankine, A. A. and Ingersoll, A. P. 2004b, Interannual variability of Mars global dust storms: An example of self-organized criticality? American Geophysical Union, Fall Meeting, abstract #P31B-0993.

Pallavicini, R., Tagliaferri, G., and Stella L. 1990, X-ray emission from solar neighbourhood flare stars a comprehensive survey of EXOSAT results, *AA* 228, 403–425.

Papadopoulos, K., Sharma, A. S., and Valdivia, J. A. 1993, Is the magnetosphere a lens for MHD waves? *GRL* 20(24), 2809–2812.

Parker, E. N. 1958, Dynamics of the interplanetary gas and magnetic fields, *ApJ* 128, 664.

Parker, E. N. 1988, Nanoflares and the solar X-ray corona, *ApJ* 330, 474–479.

Parnell, C. E. and Jupp, P. E. 2000, Statistical analysis of the energy distribution of nanoflares in the quiet Sun, *ApJ* 529, 554–569.

Parnell, C. E., DeForest, C. E., Hagenaar, H. J., Johnston, B. A., Lamb, D. A., and Welsch, B. T. 2009, A power-law distribution of solar magnetic fields over more than five decades in flux, *ApJ* 698, 75–82.

Pavan, J. and Vinas, A. F. 2019, Temperature fluctuation at the Sun and large-scale electric field in the solar wind: A challenge for the Parker Solar Probe Mission, *ApJ* 882, 28 (6pp).

Pavlidou, V., Kuijpers, J., Vlahos, L., and Isliker, H. 2001, A cellular automaton model for the magnetic activity in accretion disks, *AA* 372, 326–337.

Pavlos, G. P., Karakatsanis, L. P., and Xenakis, M. N. 2012, Tsallis non-extensive statistics, intermittent turbulence, SOC and chaos in the solar plasma. I. Sunspot dynamics, *Physica A: Statistical Mechanics and Its Applications* 391, 6287.

Pearce, G., Rowe, A. K., and Yeung, J. 1993, A statistical analysis of hard X-ray solar flares, *ApJ*SS 208, 99.

Peebles, P. J. E. 1980, *The Large-Scale Structure of the Universe*, Princeton University Press, Princeton, NJ, 435p.

Peng, F. K., et al. 2023, Self-organized criticality in solar GeV flares, *MNRAS* 518(3), 3959–3965.

Perez-Enriquez, R. and Miroshnichenko, L. I. 1999, Frequency distributions of solar gamma ray events related and not related with SPEs 1989–1995, *SoPh* 188, 169–185.

Perez-Enriquez, R., Carrillo, A., and Rodriguez, C. 2006, Fractal character of G index of IPS data for the period 1991–1994, obtained from multiscale wavelet analysis, *Advanced Journal of Social Science* 38(8), 1819–1823.

Pietras, M., Falewicz, R., Siarkovski, M., Bicz, K., and Pres, P. 2022, Statistical analysis of stellar flares from the first three years of TESS observations, *ApJ* 935, 143.

Pietronero, L., Vispignani, A., and Zapperi, S. 1994, Renormalization scheme for self-organized criticality in sandpile models, *Physical Review Letters* 72(11), 1690–1693.

Podesta, J. J. 2006a, Radial decay law for large-scale velocity and magnetic field fluctuations in the solar wind, *JGR (Space Physics)*, 111, 8103.

Podesta, J. J., Roberts, D. A., and Goldstein, M. L. 2006b, Self-similar scaling of magnetic energy in the inertial range of solar wind turbulence, *JGR (Space Physics)*, 111, 9105.

Podesta, J. J., Roberts, D. A., and Goldstein, M. L. 2007a, Spectral exponents of kinetic and magnetic energy spectra in solar wind turbulence, *ApJ* 664, 543–548.

Podesta, J. J. 2007b, Self-similar scaling of kinetic energy density in the inertial range of solar wind turbulence, *JGR (Space Physics)*, 112, 11104.

Pontin, D. I. and Hornig, G. 2015, The structure of current layers and degree of field-line braiding in coronal loops, *ApJ* 805(1), article ID 47, 12.

Price, C. P. and Newman, D. E. 2001, Using the /R/S statistic to analyze/AE data, *JASTR* 63(13), 1387–1397.

Pruessner, G. 2012, *Self-organised Criticality: Theory, Models and Characterisation*, Cambridge University Press: Cambridge.

Pruessner, G. 2013, Self-organized criticality computer simulations (Chapter 7), in *Self-Organized Criticality Systems* (ed. Aschwanden, M. J.), Open Academic Press GmbH Co.

Purkhart, S. and Veronig, A. M. 2022, Nanoflare distributions over solar cycle 24 based on SDO/AIA differential emission measure observations, *AA* 661, id. A149, 13pp.

Raup, D. M. 1997, How nature works: The science of self-organized criticality, *Complexity*, 2(6), 30–33.

Reames, D. V. 2013, The two sources of solar energetic particles, *SSRv* 175(1–4), 53–92.

Rieutord, M., et al. 2008, Solar supergranulation revealed by granule tracking, *AA* 479, L17.

Rieutord, M., Roudier, T., and Rincon, F., 2010, On the power spectrum of solar surface flows, *AA* 512, A4.

Rincon, F., Roudier, T., Schekochihin, A. A., and Rieutord, M. 2017, Supergranulation and multiscale flows in the solar photosphere. Global observations versus a theory of anisotropic turbulent convection, *AA* 599, A69.

Robbrecht, E., Patsourakos, S., and Vourlidas, A. 2009a, No trace left behind: STEREO observations of a coronal mass ejection without low coronal signatures, *ApJ* 701, 283–291.

Robbrecht, E., Berghmans, D., and Van der Linden, R. A. M. 2009b, Automated LASCO CME catalog for solar cycle 23: Are CMEs scale invariant? *ApJ* 691, 1222–1234.

Robinson, R. D., Carpenter, K. G., and Percival, J. W. 1999, A search for microflaring activity on dMe flare stars. II. Observations of YZ Canis Minoris, *ApJ* 516, 916–923.

Rosa, R. R., Sharma, A. S., and Valdicia, J. A. 1998, Characterization of localized turbulence in plasma extended systems, *Physica A*, 257,1–4, 509–514.

Rosa, R. R., Sharma, A. S., and Valdicia, J. A. 1999, Characterization of asymmetric fragmentation patterns in spatially extended systems. *International Journal of Modern Physics C*, 10(01), 147–163.

Rosenshein, E. B. 2003, Applicability of complexity theory to Martian fluvial systems: A preliminary Analysis in 34*th* Annual Lunar and Planetary Science Conference, League City, Texas, abstract no. 1660.

Rosner, R. and Vaiana, G. S. 1978, Cosmic flare transients: Constraints upon models for energy storage and release derived from the event frequency distribution, *ApJ* 222, 1104.

Rosner, R., Tucker, W. H., and Vaiana, G. S. 1978, Dynamics of quiescent solar corona, *ApJ* 220, 643.

Rostoker, G. 1984, Implications of the hydrodynamic analogue for the solar terrestrial interaction and the mapping of high latitude convection pattern into the magnetotail, *Geophysical Research Letters* 11, 251.

Roudier, T. and Muller, R. 1986, Structure of the solar granulation, *SoPh* 107, 11.

Rybczynski, M., Wodarczyk, Z., and Wilk, G. 2001, Self-organized criticality in atmospheric cascades, *Nuclear Physics B Proceedings Supplements* 97, 81–84.

Rypdal, M. and Rypdal, K. 2010, Stochastic modeling of the AE index and its relation to fluctuations in $_z$ of the IMF on time scales shorter than substorm duration, *JGR* 115(A11), CiteID A11216.

Rypdal, M. and Rypdal, K. 2011, Discerning a linkage between solar wind turbulence and ionospheric dissipation by a method of confined multifractal motions, *JGR* 116, A02202, doi:10.1029/2010JA015907.

Ryan, D. F., Dominique, M., Seaton, D., Stegen, K., and White, A. 2016, Effects of flare definitions on the statistics of derived flare distributions, *AA* 592, A133.

Sachs, M. K., et al. 2013, Self-organizing complex earthquakes: Scaling in data, models, and forecasting (Chapter 9), in *Self-Organized Criticality Systems* (ed. Aschwanden, M. J.), Open Academic Press GmbH Co.

Sahraoui, F., Goldstein, M. L., Robert, P., and Khotyaintsev, Y. V. 2009, Evidence of a cascade and dissipation of solar-wind turbulence at the electron gyroscale, *Physical Review Letters* 102(23), id. 231102.

Saichev, A., Malevergne, Y., and Sornette, D. 2009, *Theory of Zipf's Law and Beyond, Lecture Notes in Economics and Mathematical Systems*, Vol. 632, Springer, Berlin.

Sanchez, R., Newman, D. E., and Carreras, B. A. 2002, Waiting-time statistics of self-organized criticality systems, *Physical Review Letters* 88(6), 068302.

Sanchez, N., Anez, N., Alfaro, E., and Odekon, M. C. 2010, The fractal dimension of star-forming regions at different spatial scales in M33, *ApJ* 720, 541–547.

Sanchez, R. and Newman, D. 2018, A primer on complex systems, *Lecture Notes in Physics* 943, S.

Savin, S., et al. 2002, Multi-spacecraft tracing of turbulent boundary layer, *Advances in Space Research* 30(12), 2821–2830.

Scholkmann, F. 2017, Harmonic orbital resonances and orbital long-range order of the TRAPPIST-1 exoplanetary system, *Progress in Physics* 2017, 9, 85–89.

Schulman, L. S. and Seiden, P. E. 1986a, Percolation and galaxies, *Science* 425–431.

Schulman, L. S. and Seiden, P. E. 1986b, Hierarchical structure in the distribution of galaxies, *ApJ* 311, 1–5.

Scholz, P., Spitler, L. G., Hessels, J. W. T., Chatterjee, S. 2016, The repeating fast radio burst FRB 121102: Multi-wavelength observations and additional bursts, *ApJ* 833(2), id. 177, 17pp.

Scholz, P., et al. 2017, Simultaneous X-ray, gamma-ray, and radio observations of the repeating fast radio burst FRB 121102, *ApJ* 846(1), id. 80, 10pp.

Schrijver, C. J., et al. 2012, Estimating the frequency of extremely energetic solar events, based on solar, stellar, lunar, and terrestrial records, *JGR (Space Physics)* 117, 8103.

Schroeder, M. 1991, *Fractals, Chaos, Power Laws: Minutes from an Infinite Paradise*, Freeman: New York.

Schwartz, R. A., Dennis, B. R., Fishman, G. J., Meegan, C. A., Wilson, R. B., Paciesas, W. S. 1992, BATSE flare observations in solar cycle 22, in *The Compton Observatory Science Workshop* (eds. Shrader, C. R., Gehrels, N., and Dennis, B. R.), NASA CP 3137, NASA, Washington DC, 457.

Scott, A. C. 2007, *The Nonlinear Universe: Chaos, Emergence, Life*, Springer, Berlin, 364p.

Searle, L. and Zinn, R. 1978, Compositions of halo clusters and the formation of the galactic halo, *ApJ* 225, 357–379.

Seiden, P. E. and Schulman, L. S. 1990. Percolation model of galactic structure, *Advanced Physics* 39(1), 1–54.

Shao, X., Sitnov, M. I., Sharma, A. S., Papadopoulos, K., Goodrich, C. C., Guzdar, P. N., Milikh, G. M., Wiltberger, M. J., and Lyon, J. G. 2003, Phase transition-like behavior of substorms: Global MHD simulations results, *Journal of Geophysical Research* 108(A1), 1037.

Shapoval, A., Le Mouel, J. L., Shnirman, M., and Courtillot, V. 2018, Observational evidence in favor of scale-free evolution of sunspot groups, *AA* 618, A183.

Sharma, A. S., Vassiliadis, D., and Papadopoulos, K. 1993, Reconstruction on low-dimensional magnetospheric dynamics by singular spectrum analysis, *Research Letters* 20(5), 335–338.

Sharma, A. S. 1995, Assessing the magnetospheres nonlinear behavior – its dimension is low, its predictability high *Reviews of Geophysics* 33 (Suppl.), 645–650.

Sharma, A. S. and Kaw, P. K. 2005, *Nonequilibrium Phenomena in Plasmas* (eds. Sharma, A.S. and Kaw, K.), e-book, Dordrecht.

Sharma, A. S., Sitnov, M. I., and Papadopoulos, K. 2001, Substorms as nonequilibrium transitions of the magnetosphere, *JASTR* 63(13), 1399–1406.

Sharma, A. S., Baker, D. N., Bhattacharyya, A., Bunde, A., Dimri, V. P., Gupta, H. K., Gupta, V. K. , Lovejoy, S., Main, I. G., Schertzer, D., von Storch, H., and Watkins, N. W. 2012, Complexity and extreme events in geosciences: An overview, in *Complexity and Extreme Events in Geosciences*, Geophysical Monograph Series, Vol. 196 (eds., Sharma, A. S., Dimri, V. P., Bunde, A., and Baker, D. N.), American Geophysical Union, Washington, pp. 1–16.

Sharma, A. S., Aschwanden, M. J., Crosby, N. B., Klimas, A. J., Milovanov, A. V., Morales, L., Sanchez, R., and Uritsky, V. 2016, 25 Years of self-organized criticality: Space and laboratory plasmas, *SSRv* 198, 167–216.

Sheikh, M. A., Weaver, R. L., and Dahmen, I. A. 2016, Avalanche statistics identity intrinsic stellar processes near criticality in KIC 8462852, *PhRvL* 117, 1101.

Sheppard, S. S. and Trujillo, C. A. 2010, The size distribution of the Neptune Trojans and the missing intermediate-sized planetesimals, *ApJ* 723, L233–L237.

Shibayama, T., Maehara, H., Notsu, S., Notsu, Y., Nagao, T., Honda, S., Ishii, T. T., Nogami, D., and Shibata, K. 2013, Superflares and solar-type stars observed with Kepler. I. Statistical properties of superflares, *ApJSS* 209, 5 13pp.

Shimizu, T. 1995, Energetics and occurrence rate of active-region transient brightenings and implications for the heating of the active-region corona, *PASJ* 47, 251–263.

Shimojo, M. and Shibata, K. 1999, Occurrence rate of microflares in an X-ray bright point within an active region, *ApJ* 516, 934.

Shuo, X., et al. 2023, Discovery of the linear energy dependence of the spectral lag of X-ray bursts from SGR J1935+2154, *MNRAS* 521(4), 5308–5333.

Silverberg, S. M., Kowalski, A. F., Davenport, J. R. A., Wisniewski, J. P., Hawley, S. L., and Hilton, E. J. 2016, Kepler flares. IV. A Comprehensive analysis of the activity of the dM4e star GJ 1243, *ApJ* 829.2, id. 129, 11pp.

Siscoe G., 1991, The magnetosphere: A uniform of interdependent parts, *EOS* 72, S494–S495.

Sitnov, M. I., Sharma, A. S., Papadopoulos, K., Vassiliadis, D., Valdivia, J. A., Klimas, A. J., and Baker, D. N. 2000, Phase transition-like behavior of the magnetosphere during substorms, *JGR* 12955.

Sitnov, M. I., Sharma, A. S., Papadopoulos, K., and Vassiliadis, D. 2002, Modeling substorm dynamics of the magnetosphere: From self-organization and self-organized criticality to nonequilibrium phase transitions, *Physical Review* 65, 016116.

Sivron, R. and Goralski, E. A. 1998, Organized criticality and variability of structure with data from MCG-6-30-15, 193 AAS Meeting, abstract no. 57.04, Vol. 30, p. 1333.

Sivron, R. 1998, Self-organized criticality in compact plasmas, *ApJ* 503, L57.

Smith, D. M., Lopez, L. I., Lin, R. P., and Barrington-Leigh, C. P. 2005, Terrestrial gamma-ray flashes observed up to 20 MeV, *Science* 307, 5712, 1085–1088.

Smirnova, N. A. and Hayakawa, M. 2007, Fractal characteristics of the ground-observed ULF emission in relation to geomagnetic and seismic activities, *JASTP* 69(15), 1833–1841.

Spitler, L. G., et al. 2018, Detection of bursts from FRB 121102 with the Effelsberg 100 m radio telescope at 5 GHz and the Role of scintillation, *ApJ* 863(2), id. 150, 8pp.

Sornette, D. 2004, *Critical Phenomena in Natural Sciences: Chaos, Fractals, Self-Organization and Disorder: Concepts and Tools*, Springer, Heidelberg, 528p.

Sornette, D. 2009, *Dragon-Kings, black swans and the prediction of crises*, preprint arXiv:0907. 4290.

Sornette, D. and Ouillon, G. 2012, *Dragon-Kings: Mechanisms, statistical methods and empirical evidence*, preprint arXiv:1205.1002.

Souza, A. M., Echer, E., Bolzan, M. J. A., and Hajra, R. 2016, A study on the main periodicities in interplanetary magnetic field B_z component and geomagnetic AE index during HILDCAA events using wavelet analysis, *JASTP* 149, 81–86.

Stelzer, B., Flaccomio, E., Briggs, K., Micela, G., Scelsi, L, Audard, M., Pillitteri, I., and Güdel, M. 2007, A statistical analysis of X-ray variability in pre-main sequence objects of the Taurus molecular cloud, *AA* 468, 463–475.

Strugarek, A. and Charbonneau, P. 2014, Predictive capabilities of avalanche models for solar flares, *SoPh* 289, 4137.

Stumpf, M. P. H. and Porter, M. A. 2012, Critical truths about power laws, *Science* 335, February 10, 2012.

Su, Y., Gan, W. Q., and Li, Y. P. 2006, A statistical study of RHESSI flares, *SoPh* 238, 61–72.

Svanda, M. and Karlicky M. 2016, Flares on A-type stars: Evidence for heating of solar corona by nanoflares, *ApJ* 831(9), 7.

Tainaka, K. I., Fukazawa, S., and Mineshige, S. 1993, Spatial pattern formation of an interstellar medium, *PASJ* 45, 57–64.

Takalo, J., Timonen, J., and Koskinen, H. 1993, Correlation dimension and affinity of AE data and bicolored noise, *GRL* 20(15), 1527–1530.

Takalo, J., Timonen, J., Klimas, A. J., Valdivia, J. A., and Vassiliadis, D. 1999a, A coupled-map model for the magnetotail current sheet, *GRL* 26, 2913.

Takalo, J., Timonen, J., Klimas, A., Valdivia, J., and Vassiliadis, D. 1999b, Nonlinear energy dissipation in a cellular automaton magnetotail field model, *GRL* 26, 1813.

Takalo, J., Timonen, J., Klimas, A. J., Valdivia, J. A., and Vassiliadis, D. 2001, A coupled map as a model of the dynamics of the magnetotail current sheet, *JASTP* 63(13), 1407–1414.

Takeuchi, M., Mineshige, S., and Negoro, H. 1995, X-ray fluctuations from black-hole objects and self organization of accretion disks, *Publications of the Astronomical Society of Japan* 47, 617–627.

Takeuchi, M., Mineshige, S., and Negoro, H. 1997, X-ray fluctuations from advection-dominated accretion disks with a critical behavior, *ApJ* 486, 160–168.

Tam, S. W. Y., Chang, T., Chapman, S. C., and Watkins, N. W. 2000, Analytical determination of power-law index for the Chapman et al. sandpile (FSOC) analog for magnetospheric activity – renormalization-group analysis, *GRL* 27(9), 1367–1370.

Tavecchio, F., Landoni, M., and Sironi, L. 2020a, Probing shock acceleration in BL Lac jets through X-ray polarimetry: the time-dependent view, *MNRAS* 498(1), 599–608.

Tavecchio, F., Bonnoli, G., and Galanti, G. 2020b, On the distribution of fluxes of gamma-ray blazars: hints for a stochastic process? *MNRAS* 497(1), 1294–1300.

Tavecchio, F. and Sobacchi, E. 2020c, Anisotropic electron populations in BL Lac jets: Consequences for the observed mission, *MNRAS* 491(2), pp. 2198–2204.

Telloni, D., Carbone, V., Lepreti, F., and Antonucci, E. 2014, Stochasticity and persistence of solar coronal mass ejections, *ApJ* 781, L1 (5pp).

Thalmann, J. K., Tiwari, S. K., and Wiegelmann, T. 2014, Force-free field modeling of twist and braiding-induced magnetic energy in an active-region Corona, *ApJ* 780, 102.

Thibeault, C., Strugarek, A., Charbonneau, P., and Tremblay, B. 2022, Forecasting solar flares by data assimilation in sandpile models, *SoPh* 297, 9, 125.

Tindale, E., Chapman, S. C., Moloney, N. R., and Watkins, N. W. 2018, The dependence of solar wind burst size on burst duration and its invariance across solar cycles 23 and 24, *JGR* 123(9), 7196–7210.

Thompson, C. and Duncan, R. C. 1996, The soft gamma repeaters as very strongly magnetized neutron stars. II. Quiescent neutrino, X-ray, and Alfvén wave emission, *ApJ* 473, 322–342.

Tlatov, A. G. and Pevtsov, A. A. 2014, Bimodal distribution of magnetic fields and areas of sunspots, *SoPh* 289, 1143.

Toledo-Roy, J. C., Rivera, A. L., and Frank, A. 2019, Symmetry, criticality and complex systems, *AIPC* 2150, 0014.

Tomczak, M. 1999, Yohkoh observations of the Neupert effect, *AA* 342, 583–591.

Thronton, D., et al. 2013, A population of fast radio bursts at cosmological distances, *Science* 341(6141), 53–56.

Tranquille, C., Hurley, K., and Hudson, H. S. 2009, The Ulysses catalog of solar hard X-ray flares, *SoPh* 258, 141–166.

Tsurutani, B. T., Gonzalez, W. D., Guarnieri, F., Kamide, Y., Zhou, X., and Arballo, J. K. 2004, Are high-intensity long-duration continuous AE activity (HILDCAA) events substorm expansion events? *JASTP* 66(2), 167–176.

Tsurutani, B., Suguira, M., Iyemori, T., Goldstein, B. E., Gonzalez, W. D., Akasofu, S.-I., and Smith, E. J. 1990, The nonlinear response of AE to the IMF Bs, *Geophysical Research Letters* 17, 1990.

Tu, Z. L., Yang, M., and Zhang, Z. J. 2020, Superflares on solar-type stars from the first year observation of TESS, *ApJ* 890, 46.

Turcotte, D. L. 1999, Self-organized criticality, *Reports on Progress in Physics* 62, 1377–1429.

Turolla, R., Zane, S., and Watts, A. L. 2015, Magnetars: The physics behind observations. A review, *RPPh* 78(11), id. 116901.

Ulyanov, A. S., Bogachev, S. A., Reva, A. A., Kirichenko, A. S., and Loboda, L. P. 2019, The energy distribution of nanoflares at the minimum and rising phase of solar cycle 24, *Astronomy Letters*, 45(4), 248–257.

Ukhorskiy, A. Y., Sitnov, M. I., Sharma, A. S., and Papadopoulos, K., 2002, Global and multiscale aspects of magnetospheric dynamics in local-linear filters, *Journal of Geophysical Research* 107(A11), 1369.

Ukhorskiy, A. Y., Sitnov, M. I., Sharma, A. S., and Papadopoulos, K., 2003, Combining global and multi-scale features in a description of the solar wind-magnetosphere coupling, *Annales Geophysicae* 21(9), 1913–1929.

Ukhorskiy, A. Y., Sitnov, M. I., Sharma, A. S., and Papadopoulos, K., 2004, Global and multiscale features of solar wind-magnetosphere coupling: From modeling to forecasting, *GRL* 31(8), Lo 8802.

Uritsky, V. M. and Pudovkin, M. I. 1998, Low frequency 1/f-like fluctuations of the AE-index as a possible manifestation of self-organized criticality in the magnetosphere, *Annales Geoophysicae* 16(12), 1580–1588.

Uritsky, V. M., Klimas, A. J., and Vassiliadis, D. 2001, Comparative study of dynamical critical scaling in the auroral electrojet index versus solar wine fluctuations, *GRL* 28(19), 3809–3812.

Uritsky, V. M., Klimas, A. J., Vassiliadis, D., Chua, D., Parks G., 2002, Scale-free statistics of spatiotemporal emissions as depicted by POLAR UVI images: Dynamic magnetosphere is an avalanching system, *JGR (Space Physics)* 107(A12), CiteID 1426.

Uritsky, V. M., Klimas, A. J., and Vassiliadis, D. 2003, Evaluation of spreading critical exponents from the spatiotemporal evolution of emission regions in the nighttime aurora, *GRL* 30(15), CiteID 1813.

Uritsky, V. M., Klimas, A. J., and Vassiliadis, D. 2006, Analysis and prediction of high-latitude geomagnetic disturbances based on a self-organized criticality framework, *Advances in Space Research* 37, 539–546.

Uritsky, V. M., Klimas, A. J., and Vassiliadis, D. 2006a, Analysis and prediction of high-latitude geomagnetic disturbances based on a self-organized criticality framework, *Advances in Space Research* 37, 539–546.

Uritsky, V. M., Klimas, A.. Vassiliadis, D. 2006b, Critical finite-size scaling of energy and lifetime probability distributions of auroral emission, *GRL* 33, 8102.

Uritsky, V. M., Paczuski, M., Davila, J. M., and Jones, S. I. 2007, Coexistence of self-organized criticality and intermittent turbulence in the solar corona, *Physical Review Letters* 99(2), 25001–25004.

Uritsky, V. M., Donovan, E., Klimas, A. J., and Spanswick, E. 2008, Scale-free scale-dependent modes of energy release dynamics in the nighttime magnetosphere, *GRL* 35(21), CiteID L21101.

Uritsky, V. M., Liang, J., Donovan, E., Spanswick, E., Knudsen, D., Liu, W., Bonnell, J., Glassmeier, K. H. 2009, Longitudinally propagating arc wave in the pre-onset optical aurora, *GRL* 36(21), CiteID L21103.

Uritsky, V. M., Donovan, E., Trondsen, T., Pineau, D., and Kozelov, B. V. 2010, Data-derived spatiotemporal resolution constraints for global auroral imagers, *JGR* 115, A09205.

Uritsky, V. M., Pouquet, A., Rosenberg, D., Mininni, P. D. and Donovan E. 2010, Structures in magnetohydrodynamic turbulence: detection and scaling, *Physical Review E* 82(5), 056326.

Uritsky, V. M., Davila, J. M., Ofman, L., and Coyner, A. J. 2013, Stochastic coupling of solar photosphere and Corona, *ApJ* 769, 62.

Uritsky, V. M. and Davila, J. M. 2014, Spatiotemporal organization of energy release events in the quiet solar corona, *ApJ* 795, 15 (5pp).

Valdivia, J. A., Sharma, A. S., and Papadopoulos, K. 1996, Evaluation of spreading critical exponents from the spatiotemporal evolution of emission regions in the nighttime aurora, *GRL* 30(15), 1813, SSC 7–1.

Validivia, J. A., Vassiliadis, D., Klimas, A., and Sharma, A. S. 1999a, Modeling the spatial structure of the high latitude magnetic disturbances and the related current systems, *Physics of Plasmas* 6(11), 4185–4194.

Validivia, J. A., Vassiliadis, D., Klimas, A., and Sharma, A. S. 1999b, Spatio-temporal activity of magnetic storms, *JGR* 104(A6), 12239–12250.

VallieRes-Nollet, M. A., Charbonneau, P., Uritsky, V., Donovan, E., and Liu, W. 2010, Dual scaling for self-organized critical models of the magnetosphere, *JGR (Space Physics)* 115, A12217.

Van Doorsselaere, T., Hoda, S., Debosscher, J. 2017, Stellar flares observed in long-cadence data from the Kepler mission, *ApJSS* 232(2), article ID 26, 13pp.

Van Hollebeke, M. A. I., Ma Sung L. S., and McDonald F. B. 1975, The variation of solar proton energy spectra and size distribution with heliolongitude, *SoPh* 41, 189–223.

Van Houten, C. J., van Houten-Groeneveld, I., Herget, P., and Gehrels, T. 1970, The Palomar-Leiden survey of faint minor planets, *AASS* 2(5), 339–448.

Vassiliadis, D., Sharma, A. S., Eastman, T. E., and Papadopoulos, K. 1990, Low-dimensional chaos in magnetospheric activity from AE index time series, *Geophysical Research Letters* 17(11), 1841–1844.

Vassiliadis, D., Sharma, A. S., and Papadopoulos, K. 1991, Lyapunov exponent of magnetospheric activity from AL time series, *Geophysical Research Letters* 18(8), 1731–1734.

Vassiliadis, D., Sharma, A. S., and Papadopoulos, K. 1993, An empirical model relating the auroral geomagnetic activity to the interplanetary magnetic field, *Geophysical Research Letters* 20(16), 1643–1646.

Vassiliadis, D., Klimas, A. J., Baker, D. N., and Roberts, A. 1995, A description of solar wind-magnetosphere coupling based on nonlinear filters, *JGR* 100, 3495.

Vassiliadis, D., Anastasiadis, A., Georgoulis, M., and Vlahos, L. 1998, Derivation of solar flare cellular automata models from a subset of the magnetohydrodynamic equations, *ApJ* 509, L53–L56.

Vassiliadis, D., 2006, Systems theory for geospace plasma dynamics, *Reviews of Geophysics* 44, RG2002.

Veltri, P., Malara, F., and Primavera, L. 1999, *Nonlinear Alfven Wave Interaction with Large-Scale Heliospheric Current Sheet*, Lecture Notes in Physics, Springer Verlag, Berlin, Vol. 536, 222.

Verbeeck, C., Kraaikamp, E., Ryan, D. F., and Podladchikova, O. 2019, Solar flare distributions: Lognormal instead of power law? *ApJ* 884 50.

Verma, A. K, Pandit, R., and Brachet, M. E. 2020, *Rotating self-gravitating Bose-Einstein condensates with a crust: a minimal model for pulsar glitches*, arXiv:2005.13310.

Veronig, A., Temmer, M., Hanslmeier, A., Otruba, W., and Messerotti, M. 2002a, Temporal aspects and frequency distributions of solar soft X-ray flares, *AA* 382, 1070–1080.

Veronig, A., Vrsnak, B., Temmer, M., and Hanslmeier, A. 2002b, Investigation of the Neupert effect in solar flares. I. Statistical properties and the evaporation model, *AA* 392, 699.

Veronig, A., Vrsnak, B., Temmer, M., Hanslmeier, A. 2002c, Relative timing of solar flares observed at different wavelengths, *SoPh* 208, 297.

Veronig, A. M., Brown, J. C., Dennis, B. R., Schwartz, R. A., Sui, L., and Tolbert, A. K. 2005, Physics of the Neupert Effect: Estimates of the effects of source energy and mass transport, and geometry, using RHESSI and GOES data, *ApJ* 621, 482–497.

Vespignani, A., Zapperi, S., and Pietronero, L. 1995, Renormalization approach to the self-organized critical behavior of sandpile models, *Physical Review E* 51(3), 1711–1724.

Vlahos, L., Georgoulis, M., Kluiving, R., and Paschos, P. 1995, The statistical flare, *AA* 299, 897–911.

Vlahos, L. and Georgoulis, M. K. 2004, On the self-similarity of unstable magnetic discontinuities in solar active regions, *ApJL* 603, L61–L64.

Vlahos, L. and Isliker, H. 2016, Complexity methods applied to turbulence in plasma astrophysics, *EPJST* 225, 977.

Walkowicz, L. M., et al. 2011, White-light flares on cool stars in the KEPLER quarter 1 data, *ApJ* 141, 50, 9p.

Wang, J. S., Wang, F. Y., and Dai, Z. G. 2017, Self-organized criticality in type I X-ray bursts, *MNRAS* 471(3), 2517–2522.

Wang, F. Y. and Dai, Z. G. 2013, Solar flare-like origin of X-ray flares in gamma-ray burst afterglows, *Nature Physics* 9(8), 465–467.

Wang, F. Y. and Dai, Z. G. 2017, A universal scaling law of black hole activity including gamma-ray bursts, *MNRAS* 470(1), 1101–1106.

Wang, F. Y., Dai, Z. G., Yi, S. X., and Xi, S. Q. 2015, Universal behavior of X-ray flares from black hole systems, *ApJSS* 216(1), id. 8, 8p.

Wang, F. Y. and Zhang, G. Q. 2019, A universal energy distribution for FRB 121102, *ApJ* 882(2), id. 108, 10p.

Wang, F. Y., Zhang, G. Q., and Dai, Z. G. 2021, Galactic and cosmological fast ratio bursts as scaled-up solar radio bursts, *MNRAS* 501(3), 3155–3161.

Wang, W. B., Li, C., Tu, Z. L., Guo, J. H., Chen, P. F., and Wang, F. Y. 2022, Avalanches of magnetic flux rope in the state of self-organized criticality, *MNRAS* 512(2), 1567–1573.

Wang, T. J., Innes, D. E., and Solanki, S. K. 2006, Fe XIX observations of active region brightenings in the corona, *AA* 455, 1105–1113.

Wanliss, J. A. and Weygand, J. M. 2007, Power law burst lifetime distribution of the SYM-H index, *GRL* 34(4), CiteID L04107.

Warszawski, L. and Melatos, A. 2008, A cellular automaton model of pulsar glitches, *MNRAS* 390(1), 175–191.

Warszawski, L. and Melatos, A. 2012, Knock-on processes in superfluid vortex avalanches and pulsar glitch statistics, *MNRAS* 428, 1911–1926.

Watkins, N. W., Chapman, S. C., Dendy, R. O., and Rowlands, G. 1999, Robustness of collective behaviour in strongly driven avalanche models: magnetospheric implications, *GRL* 26(16), 2617.

Watkins, N. W., Oughton, S. and Freeman, M. P. 2001, What can we infer about the underlying physics from distributions observed in an RMHD simulation? *Planetary and Space Science* 49(12).

Watkins, N. W. and Freeman, M. P. 2008, Natural complexity, *Science* 320, 323–324.

Watkins, N. W., Chapman, S. C., and Rosenberg, S. J. 2009, Comment on "coexistence of self-organized criticality and intermittent turbulence in the solar corona, *Physical Review Letters* 103(3), id. 039501.

Watkins, N. W., Pruessner, G., Chapman, S. C., Crosby, N. B., and Jensen, H. J. 2016, 25 Years of self-organized criticality: Concepts and controversies, *SSRv* 198, 3–44.

Wei, J. J., Wu, X. F., Dai, Z. G., Wang, F. Y., Wang, P., Li, D., and Zhang, B. 2021, Similar scale-invariant behaviors between soft gamma-ray repeaters and an extreme epoch from FRB 121102, *ApJ* 920(2), id. 153, 7pp.

Wheatland, M. S., Sturrock, P. A., and McTiernan, J. M. 1998, The waiting-time distribution of solar flare hard X-rays, *ApJ* 509, 448–455.

Wheatland, M. S. and Eddey, S. D. 1998, Models for flare statistics and the waiting-time distribution of solar flare hard X-ray bursts, in *Proceedings of Nobeyama Symposium, Solar Physics with Radio Observations* (eds. Bastian, T., Gopalswamy, N., and Shibasaki, K.), NRO Report 479, pp. 357–360.

Wheatland, M. S. 2000a, The origin of the solar flare waiting-time distribution, *ApJ* 536, L109–L112.

Wheatland, M. S. 2000b, Do solar flares exhibit an interval-size relationship? *SoPh* 191, 381–389.

Wheatland, M. S. 2000c, Flare frequency-size distributions for individual active regions *ApJ* 532, 1209–1214.

Wheatland, M. S. 2001, Rates of flaring in individual active regions, *SoPh* 203, 87–106.

Wheatland, M. S. and Litvinenko, Y. E. 2002, Understanding solar flare waiting-time distributions, *SoPh* 211, 255–274.

Wheatland, M. S. 2003, The coronal mass ejection waiting-time distribution, *SoPh* 214, 361–373.

Wheatland, M. S. 2010, Evidence for departure from a power-law size distribution for a small solar active region, *ApJ* 710: 1324.

White, E. P., Enquist, B. J., and Green, J. L. 2008, On estimating the exponent of power-law frequency distributions, *Ecology* 89, 905.

Wiesenfeld, K., Theiler, J., and McNamara, B. 1990, Self-organized criticality in a deterministic automaton, *Physical Review Letters* 65(8), 949–952.

Willis, J. C. and Yule, G. U. 1922, Some statistics of evolution and geographical distribution in plants and animals, and their significance, *Nature* 109(2728), pp. 177–179.

Wilkins H. P. 1946, *One Hundred Inch Reproduction of the Three Hundred Inch Map of the Moon*, Stanley.

Wilmot-Smith, A. L., Hornig, G., and Pontin, D. I. 2009a, Magnetic braiding and parallel electric fields, *ApJ* 696, 1339–1347.

Wilmot-Smith, A. L., Hornig, G., and Pontin, D. I. 2009b, Magnetic braiding and quasi-separatrix layers, *ApJ* 704, 1288.

Wolfram, S. 2002, *A New Kind of Science*, Wolfram Media.

Wu, C. J., Ip, W. H., and Huang, L. C. 2015, A study in the frequency distributions of the superflares of G-type stars observed by the Kepler mission, *ApJ* 798(3), article ID 92, 13pp.

Xapsos, M. A., Stauffer, C., Barth, J. L., and Burke, E. A. 2006, Solar particle events and self-organized criticality: Are deterministic predictions of events possible? *IEEE Transactions on Nuclear Science* 53(4), p. 1839–1843.

Xiao, S., et al. 2023a, Discovery of the linear energy dependence of the spectral lag of X-ray bursts from SGR J1935+2154. *MNRAS* 521(4), 5308–5333.

Xiao, S., et al. 2023b, The minimum variation time scale of X-ray bursts from SGR J1935+2154, *ApJSS* 268, 5 (9pp).

Xiong, Y., Witta, P. J., and Bao, G. 2000, Models for accretion-disk fluctuations through self-organized criticality including relativistic effects, *PASJ* 52, L1097–L1107.

Yan, D., Yang, S., Zhang, P., Dai, B., Wang, J., and Zhang, L. 2018, Statistical analysis of XMM-Newton X-ray flares of Mrk 421: Distributions of peak flux and flaring time duration, *ApJ* 864, 164 (16pp).

Yang, S., Yan, D., Dai, B., Zhang, P., and Zhu, Q. 2019, Statistical analysis of X-ray flares from the nucleus and HST-1 knot in the M87 jet, *MNRAS* 489(2), 2685–2693.

Yang, H. and Liu, J. 2019, The flare catalog and the flare activity in the Kepler mission, *ApJSS* 241 29.

Yashiro, S., Akiyama, S., Gopalswamy, N., and Howard, R. A. 2006, Different power law indices in the frequency distributions of flares with and without coronal mass ejections, *ApJ* 650, L143–L146,

Yeh, C. T., Ding, M. D., and Chen, P. F. 2005, Waiting time distribution of coronal mass ejections, *Chinese Journal of Astronomy and Astrophysics* 5, 193–197.

Yermolaev, Y. I., Lodkina, I. G., Nikolaeva, N. S., and Yermoaev, M. Y. 2013, Occurrence rate of extreme magnetic storms, *JGR (Space Physics)* 118(8), 4760–4765.

Yonehara, A., Mineshige, S., and Welsh, W. F. 1997, Cellular-automaton model for flickering of cataclysmic variables, *ApJ* 486, 388–396.

Young, M. D. T. and Kenny, B. G. 1996, Are giant pulses evidence of self-organized criticality? ASP Conference Series 105: IAU Colloq. 160: Pulsars: Problems and Progress, 179.

Yoshida, F., Nakamura, T., Watanab, J., Kinoshita, D., and Yamamoto, N., 2003, Size and spatial distributions of sub-km main-belt asteroids, *PASJ* 55, 701–715.

Yoshida, F. and Nakamura T. 2007, Subaru main belt asteroid survey (SMBAS) – size and color distributions of small main-belt asteroids, *Planetary and Space Science* 55, 113–1125.

Yuan, Q., et al. 2018, A systematic Chandra study of Sgr A, II. X-ray flare statistics, *MNRAS* 473, 306.

Zaslavsky, G. M., Edelman, M. N. Guzdar, P. N., Sitnov, M. I., and Sharma, A. S. 2007, Self-similarity and fractional kinetics of solar wind magnetosphere coupling, *Physica A* 321, 11–20.

Zaslavsky, G. M., Edelman, M. N. Guzdar, P. N., Sitnov, M. I., and Sharma, A. S. 2008, Multiscale behavior and fractional kinetics from the data of solar wind magnetosphere coupling, *Communications in Nonlinear Science and Numerical Simulation* 13, 314–330.

Zebker, E. A., Marouf, G. L., and Tyler, G. L. 1985, Saturn's rings particle size distributions for a thin layer model, *Ikarus* 64, 531–548.

Zeldovich, Ya. B. 1970, Gravitational instability: An approximate theory for large density perturbations, *A&A* 5, 85–89.

Zelenyi, L. M. and Milovanov, A. V. 2004, Reviews of topical problems: Fractal topology and strange kinetics: from percolation theory to problems in cosmic electrodynamics, *Physics Uspekhi* 47(8), R01.

Zhang, Y. C. 1989, Scaling theory of self-organized criticality, *Physical Review Letters* 63(5), 470–474.

Zhang, G. Q., Wang, P., Wu, Q., Wang, F. Y., Li, D., Dai, Z. B., and Zhang, B. 2021, Energy and waiting time distributions of FRB 121102 observed by FAST, *ApJ* 920L, 23.

Zhang, H. M., Zhang, J., Lu, R. J., Yi, T. F. Huang, X. L. and Liang E. W. 2018a, Flux and spectral variation characteristics of 3C 454.3 at the GeV band, *RAA* 18(4), article ID 040.

Zhang, H. M., Zhang, J., Lu, R. J., Yi, T. F., Huang, L., Liang, E. W. 2018b, Fast radio burst energetics and detectability from high redshifts, *ApJL* 867(2), idl. L21, 7pp.

Zhang, W. L., Yi, S. X., Yang, U. P., and Qin, Y. 2022, Statistical properties of X-Ray flares from the supergiant fast X-Ray transients, *Research in Astronomy and Astrophysics* 22(6), id. 065012, 8pp.

Zhang, et al. 2018c, Fast radio burst 121102 Pulse detection and periodicity: A machine learning approach, *ApJ* 866(2), id. 149, 18pp.

Zirker, J. B. 1993, Coronal heating (Invited review paper), *SoPh* 148, 43–60.

Zirker, J. B. and Cleveland, F. M. 1993a, Nanoflare mechanisms: Twisting and braiding, *SoPh* 144, 341–347.

Zirker, J. B. and Cleveland, F. M. 1993b, Avalanche models of active region heating and flaring, *SoPh* 145, 119–128.

Zou, Y., Heitig, J., and Kurth, J. 2013, Power laws of recurrence networks (Chapter 6), in *Self-Organized Criticality Systems* (ed. Aschwanden, M. J.), Open Academic Press GmbH Co., www.openacademicpress.de

Index

www.ingramcontent.com/pod-product-compliance
Lightning Source LLC
LaVergne TN
LVHW062201200926
848484LV00006B/1292

* 9 7 8 1 0 0 9 5 6 2 9 3 5 *